T0381065

Cambridge atmospheric and space science series

Physics and chemistry of the upper atmosphere

Physics and chemistry of the upper atmosphere

M. H. Rees

University of Alaska, Fairbanks

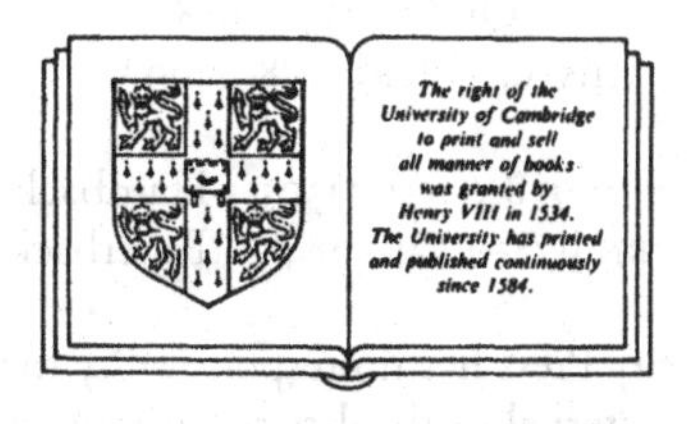

CAMBRIDGE UNIVERSITY PRESS
Cambridge
New York Port Chester Melbourne Sydney

CAMBRIDGE UNIVERSITY PRESS
Cambridge, New York, Melbourne, Madrid, Cape Town, Singapore,
São Paulo, Delhi, Dubai, Tokyo, Mexico City

Cambridge University Press
The Edinburgh Building, Cambridge CB2 8RU, UK

Published in the United States of America by
Cambridge University Press, New York

www.cambridge.org
Information on this title: www.cambridge.org/9780521368483

First published 1989

A catalogue record for this publication is available from the British Library

Library of Congress Cataloguing in Publication Data

Rees, M.H.
Physics and chemistry of the upper atmosphere/M.H. Rees.
P. cm.—(Cambridge atmospheric and space science series)
Includes bibliographies and index.
ISBN 0-521-32305-3
1. Atmosphere, Upper. 2. Atmospheric physics. 3. Atmospheric
chemistry. I. Title. II. Series.
QC879.R44 1988
551.5′14—dc19 88-9499 CIP

ISBN 978-0-521-32305-5 Hardback
ISBN 978-0-521-36848-3 Paperback

Contents

Preface

This text focuses on the physics and chemistry of the Earth's upper atmosphere, bounded at the bottom by the pressure level at which most (though not all) of the incoming ionizing radiation has been absorbed, and bounded at the top by the level at which escape of the neutral gas becomes important. The region defined by these boundaries contains a partially ionized gas. The principal ionization sources are solar ultraviolet radiation and auroral energetic particles. Thus, an ionosphere is embedded in the upper atmosphere.

The plan is to identify the multitude of processes that operate in the upper atmosphere and to relate observed input and output parameters by detailed physical and mathematical descriptions of the governing processes. The properties and behaviour of the atmosphere are a consequence of the interaction of processes that span a wide range of commonly identified disciplines; radiation physics and chemistry, transport phenomena, gas phase chemistry, fluid dynamics, optics and spectroscopy, and others. Basic disciplines are drawn upon in attempting to understand the upper atmosphere. It is hoped that this book will bridge the gap between those texts read by students taking courses in the standard disciplines of physics and the research literature in upper atmosphere physics and chemistry. Research papers all too often assume that the reader already has the background required to appreciate the new development reported in the article.

Each basic discipline has its convention on identifying parameters by certain symbols, and the same letter is frequently used to denote completely different quantities. For example, k identifies Boltzmann's constant, the wave number, the reaction rate coefficient, as well as being used as an index; α stands for the thermal diffusion coefficient, the atomic or molecular polarizability, and the recombination rate coefficient. The appropriate meaning should emerge in context.

The choice of units invariably represents a compromise between uniformity, convention and ease of usage, made especially difficult by the range of disciplines represented by phenomena of the upper atmosphere. For example, photon or electron intensities are specified per $cm^2\,s\,sr\,\Delta\lambda$ or eV rather than in SI units. The literature is divided between nm and Å as units of wavelength. Cross sections may be given in units of the area of the first Bohr orbit, in $Å^2$, m^2 or cm^2; the latter unit has been chosen.

Parameters in several equations in Chapter 4 are specified in Hartree atomic units. The guideline adopted in this book is to follow the convention of each discipline in the current research literature and in the standard reference works. Where the guidelines are not obvious I have adopted my own (arbitrary) choice. Units are always given when equations include numerical values; they are also included if helpful in identifying the physical parameters described by an equation. Various units used throughout the book are listed in Appendix 7.

Tables of numerical data are given in appendices. They include results of measurements acquired in the laboratory, geophysical measurements, and empirical representations of observations. While the concepts and equations presented in the body of the text should remain valid over time, some of the numerical results presented in appendices may change as better results become available.

A description of the coupled nature of the problem is presented in the first chapter. Subsequent chapters provide the physical and mathematical details. Our understanding is both based on and tested by observations, and progress often hinges on a crucial measurement. The complementary aspects of measurements and analyses is our theme, but neither instrumentation and measurement techniques nor purely theoretical concepts are discussed. The broad scope of the subject precludes a detailed treatment of each facet. The bibliographies provided with each chapter both complement and supplement the material presented. In addition, many figures throughout the book have been selected to guide the reader to the original research literature. Additional sources are required for solving some of the fifty or so problems, most of which were tackled by students in a course given during the 1986/7 academic year. I am grateful to T. Alcock, C. Grimm, D. Rice and Yong Shi for bringing to my attention discrepancies and suggesting clarifications in the presentation of the problems and the text.

The reason for writing this book is simple: I wanted to learn in some detail aspects of the physical and chemical processes in the upper atmosphere that are outside my narrow research area. A book allows me to share my enthusiasm with a wider audience than the few students who attend my course.

Fairbanks, 1987 M. H. R.

Acknowledgements

Several colleagues read selected chapters of the manuscript, made corrections, and suggested improvements in the presentation. I wish to thank Alex Dalgarno, Paul B. Hays, Dirk Lummerzheim, Marjorie Rees, Henry Rishbeth and William E. Sharp for their help. I am especially indebted to Raymond G. Roble for carefully and critically reading the entire manuscript.

Dr. Michael Rycroft, editor of the Series, provided encouragement and helpful suggestions from the beginning to the completion of this work.

Many colleagues have generously sent me original drawings and high quality prints of figures reproduced in the text. I hereby thank them collectively.

The manuscript was expertly typed by Sheila Finch and the original artwork was done by Deborah Coccia. Maureen Storey, copy editor at CUP, provided the final polish to the manuscript, as well as assuring consistency, continuity, correct spelling and grammar.

This undertaking began during a sabbatical leave spent as Visiting Fellow at Southampton University (UK) and I am most grateful to members of the Physics Department for their generous hospitality. Dr. Henry Rishbeth, leader of the Upper Atmosphere group was my host.

1

An Overview

1.1 Introduction

The definition of the term 'upper atmosphere' is not agreed upon by all scientists. To meteorologists it usually refers to the entire region above the troposphere, where the daily weather evolves. With this interpretation the upper atmosphere would include the stratosphere, mesosphere, and thermosphere, regions identified by their temperature structure, density, composition and the degree of ionization. Figure 1.1.1 illustrates schematically the altitude variation of these atmospheric parameters. Logarithmic scales are used for the abscissae and the ordinate to accommodate the large range of densities and heights.

The same laws of physics apply throughout the atmosphere, of course, but the relative importance of various processes varies widely between regions, accounting for the different behaviour of the characteristic parameters as a function of altitude. Thus, the temperature structure is governed by the absorption of solar radiation, and different wavelength bands are absorbed by various constituents in the region. For example, absorption by O_3 specifically accounts for heating in the stratosphere. Radiation transfer and terrestrial albedo also contribute to the temperature structure. Each region identified in Figure 1.1.1 warrants a book length treatment. In this work we study that portion of the upper atmosphere labelled the thermosphere and ionosphere.

For centuries, man's perception of the thermosphere was limited to the splendour of the Aurorae Borealis and Australis. It was the need for the propagation of radio waves over long distances that provided a practical stimulus to investigations of the thermosphere and the physical processes that cause and control the ionosphere. Compared with the density of the atmosphere at the Earth's surface, the thermosphere is a very high vacuum indeed, but not so perfect as to preclude drag and perturbations on satellites orbiting within the region. The need to predict thermospheric and ionospheric variability has arrived with the space age.

The region where

(1) energy input is dominated by solar ultraviolet (UV) photons and auroral energetic particles, electric fields and electric currents;

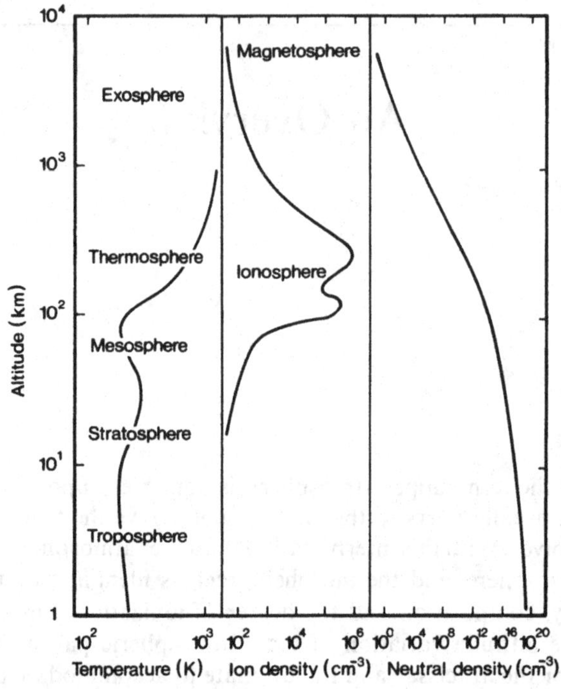

Fig. 1.1.1 Schematic representation of the thermal structure, the ion density and the neutral density of the Earth's atmosphere. The approximate altitude regimes of various named regions are indicated.

(2) the major neutral constituents are O, N_2 and O_2, and the minor constituents are NO, N, H, He, Ar and CO_2; and
(3) neutral atmosphere dynamics are strongly influenced by plasma motions

is identified as the thermosphere. It is, however, unrealistic to define precisely the boundaries of this region since the three characteristics are not sharply confined in altitude. For example, a small fraction of the UV radiation as well as high energy charged particles penetrate to lower levels in the atmosphere.

While it is possible to identify a region for the purpose of detailed study, nature does not allow us to decouple the region from its neighbours. Thus, the thermosphere is coupled energetically, dynamically, and chemically to the mesosphere at its lower bound, and to the exosphere and magnetosphere at its upper bound. Escape of gas becomes a dominant process in the exosphere, while the degree of ionization becomes high in the magnetosphere and the motion of ions is no longer dominated by collisions with the neutral gas. The complexity of neutral and ion chemistry increases substantially in the mesosphere due to a host of minor constituents, both neutrals and ions, the formation of large molecules, and influences propagating up from the troposphere.

There is one attribute that distinguishes the thermosphere from its neighbours, above and below. It is the region where most radiations from the atmospheric atoms and molecules in the visible spectrum originate, i.e., the nightglow, the dayglow and the aurora. All three phenomena are visible from a space platform. The nightglow is generally too faint and structureless to be seen by the naked eye from the ground while the dayglow is overwhelmed by scattered sunlight in the lower atmosphere. The aurora is the visual manifestation of processes that we shall study in detail in subsequent chapters.

Instruments have been developed that are capable of measuring such radiation from the upper atmosphere. Analysis of various emission features has yielded insight into chemical processes, energy balance and the dynamic behaviour of the upper atmosphere. This has been accomplished by passive remote sensing, i.e., without perturbing the medium or process under study. Radar techniques have been used extensively to investigate the properties and behaviour of the ionized component of the thermosphere and to derive several important dynamical and compositional properties of the region. Direct, *in situ* measurements by rocket and satellite borne instruments have provided a wealth of data on the composition, structure, and dynamics of the upper atmosphere over a wide range of altitude and large geographic extent. Many instruments and widely different experimental methods that are complementary have been adopted to make observations of and collect measurements about the thermosphere, thus providing inputs to models which, ultimately, will test our understanding of one part of our environment.

1.2 Physical properties of the upper atmosphere

An appreciation of the complexity of the physics and chemistry of the upper atmosphere may be acquired by listing the various processes that occur simultaneously and the resulting observable quantities. Adopting the organization that will be followed in succeeding chapters we identify processes with the basic conservation laws of physics, as shown in Figure 1.2.1, noting that the various processes and their mathematical descriptions are coupled.

Atmospheric composition: The neutral composition of the thermosphere is governed to a large extent by the evolution of the planet, a topic that is outside the scope of this text. Thus, processes that establish the abundance of N_2, O_2, Ar, He, H and CO_2 are not considered. We are interested in O because it is produced by photodissociation and energetic particle impact dissociation. We are also interested in the minor species, N and NO, produced by molecular dissociation and various chemical reactions.

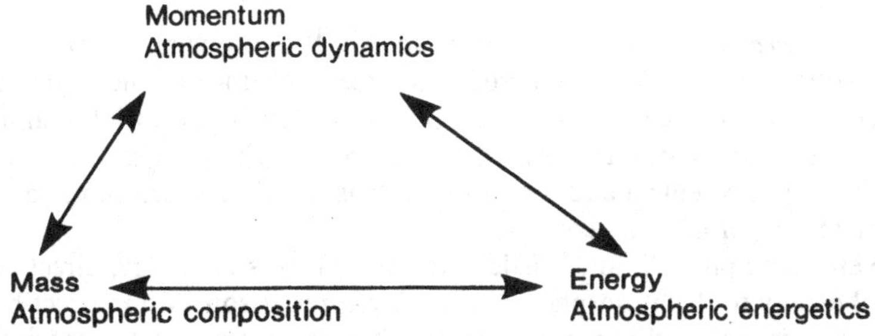

Fig. 1.2.1 Coupling between atmospheric composition, energetics and dynamics by the conservation laws of physics.

Of special interest are atoms and molecules in excited states producing the optical emissions that are the signature of important processes in the upper atmosphere. (The term 'optical emissions' includes radiation in the visible as well as the UV and infrared (IR) regions of the spectrum.) Excited state atoms and molecules are produced by UV photon impact as well as charged particle impact collisions and by subsequent chemical reactions. The minor neutral composition and the excited state production are closely related to the ion production and composition. UV and particle impact ionization of the major thermospheric species, N_2, O_2, and O, initiates a chain of ion chemistry reactions that determine the time-varying ion composition and abundance. Thus, while NO molecules are only a minor neutral species in the thermosphere, as a result of the ion chemistry NO^+ ions are a major constituent in the E- and lower F-regions of the ionosphere. The typical ion density–height profile in the centre panel of Figure 1.1.1 shows that the ion density occurs in layers.

Atmospheric dynamics: The thermosphere and ionosphere form a dynamic system that responds to a variety of forces or momentum sources. Diurnal and latitudinal variations of neutral gas heating cause pressure gradients which, together with the Coriolis effect, force zonal and meridional winds. Ions are driven by electric fields causing them to convect from one region of the ionosphere to another, constrained by the Earth's magnetic field. Collisions between ions and neutrals couple their motion so that ion drag participates in establishing the pattern of the winds. The resulting velocity vectors are asymmetric over the globe because the electric field configuration is governed by the geomagnetic field, whose poles are displaced from the geographic poles, while the Coriolis effect depends on the rotational symmetry of the Earth.

Vertical motion is governed largely by gravitational acceleration and pressure gradients. Molecular diffusion causes the density of individual neutral constituents to vary independently with altitude. The effect becomes important above about 120 km, resulting in a rapid decrease in the densities of heavier molecular species with altitude compared with the lighter atomic species. Another process that occurs in the atmosphere is turbulence which causes small scale eddy motions that are embedded in the large scale mean motion characterized by winds. Turbulence can be produced by thermal gradients and by wind shear, both effects existing in the lower thermosphere. Transport of the small eddies results in turbulent diffusion which tends to smooth out differences between air parcels and maintain a mixed gas. The effect dominates below 100 km. In addition to the effects of an externally applied electric field, ions are subject to dynamo action when driven by the neutral wind in the geomagnetic field.

Atmospheric energetics: The interaction of solar UV photons and energetic particles with the thermosphere and ionosphere is a source of heat for the neutral and ionized constituents. Ionospheric currents driven by electric fields cause frictional heating of the neutrals and ions that is a major energy source at high latitudes. Thus, differential motion between the neutral and ionized species is both a momentum and an energy source for the upper atmosphere.

While energetic photons and particles are the carriers of energy, direct collisional transfer of energy to the atmosphere is small by comparison with indirect heating by exothermic chemical and ionic reactions. Photoionization and dissociation, and electron impact ionization and dissociation, produce ions and atoms that may be converted to different species and that eventually recombine. The most likely reactions are exothermic, resulting in the conversion of photon and electron energy into kinetic energy of the neutral and ionized gas. These are the processes that yield the observed temperature variation with altitude. A small fraction of the energy absorbed in the thermosphere is radiated to space or to lower levels of the atmosphere and to the ground. The radiation is in the form of spectral features, i.e., atomic lines, molecular bands and recombination continua that originate from numerous excited states of several species.

Interdependence between composition, dynamics, and energetics emerges from the descriptive presentation given in the preceding paragraphs. The coupled nature of the problem is also demonstrated by its mathematical formulation.

Unlike lower levels of the atmosphere, the constituents of the thermosphere are not fully mixed. It becomes necessary, therefore, to describe the behaviour of each species by its own continuity, momentum and energy equations. Each species has its own production and loss rates, transport properties and thermal characteristics, and is subject to different forces. However, there is also interaction between the constituents through collisional processes involving momentum and energy transfer as well as chemical reactions, and through electromagnetic and gravitational forces.

The concentration of the jth constituent, n_j, is given by the continuity equation,

$$\partial n_j/\partial t + \nabla\cdot(n_j\mathbf{v}_j) = P_j - L_j \tag{1.2.1}$$

where P_j and L_j are the production and loss rates and $\mathbf{v}_j$ is the velocity vector. Each term in the continuity equation represents a rate of change of the concentration of species j per unit volume with time, t. The second term on the left side of Equation (1.2.1) represents the flux divergence.

Motion of the jth constituent is given by the momentum equation,

$$\rho_j \mathrm{D}_j\mathbf{v}_j/\mathrm{D}t + \nabla\cdot\bar{\mathbf{P}}_j = \sum_j n_j\mathbf{F}_j \tag{1.2.2}$$

The inertial term and the generalized pressure tensor term on the left side of the equation are balanced by the external forces, $\mathbf{F}_j$, on the right side. ρ_j is the mass density, $\bar{\mathbf{P}}_j$ is the pressure tensor and the external forces are gravitational and electromagnetic. (The convective derivative is $\mathrm{D}_j/\mathrm{D}t = \partial/\partial t + \mathbf{v}_j\cdot\nabla$.) Each term in the momentum equation represents a force per unit volume.

The thermal properties are specified by the energy equation,

$$n_j k \mathrm{D}_j T_j/\mathrm{D}t = Q_j - L_j^{\mathrm{T}} - n_j k T_j \nabla\cdot\mathbf{v}_j - \nabla\cdot\mathbf{q}_j + \sum_j n_j \mathbf{F}_j\cdot\mathbf{v}_j \tag{1.2.3}$$

Each term represents a time rate of change of energy per unit volume. The net rate of change (left side of the equation) is equal to the sum of the local heating rate, Q_j, heat loss rate, L_j^{T}, compressional heating (or cooling), divergence of the heat flow, $\mathbf{q}_j$, and the work done on species j.

The equation of state

$$p_j = n_j k T_j \tag{1.2.4}$$

relates parameters that appear in the three conservation equations and defines the partial pressure of the jth species; k is Boltzmann's constant. At the temperatures that prevail in the upper atmosphere the macroscopic properties of the gas obey Boltzmann statistics and the energy of the system can be described by classical mechanics. The ideal gas law (Equation (1.2.4)) is therefore an appropriate equation of state. In the upper atmosphere the transfer of molecular properties can be analysed by a set of transport equations. Particle motion is specified in terms of a distribution function in phase space and a flux vector is associated with each macroscopic property. The distribution function incorporates streaming and external forces. Including interactions amongst the particles of the system, collisions, leads to the Boltzmann equation for the distribution. Moments of the distribution function yield the transport properties; the first three correspond to the transport of mass, momentum, and kinetic energy, the conservation equations (Equations (1.2.1), (1.2.2) and (1.2.3)).

Application of the conservation laws to the various neutral and ion species yields the height profiles of the respective concentrations and temperatures, as well as the wind fields. However, the development of three-dimensional time-dependent models of the thermosphere and ionosphere has only recently begun. Even as these words are written, work on this problem is progressing at a rapid pace and the status reported in this book will have been superseded at the time of its publication. We endeavour, therefore, to emphasize the physics and chemistry of the basic processes that occur in the upper atmosphere.

Topics in the succeeding chapters are arranged to thread the following argument. The production and loss mechanisms in the continuity and energy equations, Q_j and L_j^T, are the microphysics that underlie the transport phenomena which govern the thermodynamics of the upper atmosphere. We therefore begin with the influence of solar UV radiation and energetic particle bombardment on the composition of the thermosphere and the ionosphere, including subsequent chemical and ionic reactions. This leads us into investigating various energy dissipation mechanisms for the incoming radiations. These include optical emissions that make up the airglow and aurora which are a conspicuous energy sink but only account for a very small fraction of the total energy. By far the largest fraction is dissipated locally, driving the wind system and establishing the temperature profile, our next topic. Ions are driven by electric fields and by the neutral wind. In the upper atmosphere ion and neutral gas motions are coupled, so that the ions become both a momentum and a heat source for the neutrals. Electric fields influence both the dynamics and energetics of the atmosphere. Electric fields drive current systems that provide the link between the ionosphere and the magnetosphere.

The properties of the upper atmosphere that are measured *in situ* or remotely comprise the body of facts from which models are constructed. Even in the 25 years in which the thermosphere has been studied in some detail we have come to realize that the structure and composition of this region is highly variable. This is a consequence of the dynamic behaviour of the atmosphere so that the state at any given time is a function of the preceding state over wide ranging time scales. The atmosphere has a memory and is continually evolving. Evolution of the atmosphere on geologic time scales is a book length topic by itself and will not be dealt with here (see references at the end of the chapter). The longest time scale that we will consider is related to the systematic variability of solar UV photon and energetic particle emission over an 11 year cycle of solar activity, commonly measured by the sunspot number. This variability is sufficient to produce significant changes in thermospheric parameters and thereby furnish a starting point in delineating the processes that determine its properties. Thus, the temperature and density profiles shown in Figure 1.1.1 are illustrative and typical but do not refer to a specific situation. There are significant systematic diurnal, seasonal, and positional (both geographic and geomagnetic) variations in density and temperature. By contrast, variations associated with the fluctuating solar wind are large and more difficult to predict. A quantitative description of thermospheric and ionospheric structure, dynamics and energetics therefore requires time-dependent solutions of the governing equations.

We shall not stray far from observable quantities throughout this work. Indeed, the analyses will be guided by them. The relative importance of various processes, however, changes markedly for different observables. For example, understanding the origin of certain auroral optical radiations requires only solution of a one-dimensional transport equation for electrons, with a measured initial phase space distribution function, and application of the excited state chemistry of some ionic and neutral species. Of much greater complexity is the modelling of the global neutral wind system which involves a coupled three-dimensional solution of all the conservation equations.

Bibliography

The physical and chemical processes to which the Earth's atmosphere has been subjected in the geologic past, leading to its present composition, are described in

J.C.G. Walker, *Evolution of the Atmosphere*, Collier Macmillan Publ. Co., New York, 1977.

A. Henderson-Sellers, *The Origin and Evolution of Planetary Atmospheres*, Adam Hilger Ltd., Bristol, 1983.

2

The interaction of energetic solar photons with the upper atmosphere

The major neutral molecular constituents in the thermosphere, N_2 and O_2, are a legacy of the evolution of the Earth's atmosphere, and the present day abundance is controlled largely by geochemical and biological processes, as is the concentration of the minor molecular species, CO_2. The major atomic constituent, O, is produced from dissociation of O_2 by solar UV photons and by energetic particle impact. These processes also account for the presence of N atoms. The minor species NO, O_3 and OH result from chemical reactions, while H_2O, He, H_2, Ar, CH_4 and CO_2 are transported into the thermosphere from the mesosphere and lower levels of the atmosphere.

All the charged species that make up the ionosphere are produced either directly by photoionization and impact ionization of neutral atoms and molecules, or indirectly by subsequent ionic–chemical reactions. Photodetachment rapidly destroys any negative ions that may be formed in the thermosphere and the ionic population consists of several species of positive ions and an electron density equal to the sum of the positive ion densities.

In this chapter we compute the direct production rates of O and N atoms, and of the ions N_2^+, O_2^+, N^+ and O^+ by energetic photons. This requires knowledge of the energy distribution of the solar UV intensity, various cross sections (defined in Chapter 4), and densities of the major molecular species. Electronically excited states of atoms, molecules and ions will be treated because excited states play a major role in chemical reactions, spectroscopic emissions, and the energy balance in the thermosphere.

2.1 The solar irradiance

The transfer of solar radiation in the atmosphere does not involve emission sources or scattering into the original beam (with some important exceptions that are taken up in Chapter 7), and the process is well described by the Lambert–Beer exponential absorption law,

$$I(\lambda) = I_\infty(\lambda) \exp[-\tau(\lambda)] \tag{2.1.1}$$

At any point in the atmosphere the intensity at wavelength λ is $I(\lambda)$ and $\tau(\lambda)$ is the optical depth. In this section we focus on the irradiance at 1 AU 'at the top of the atmosphere', i.e., $I_\infty(\lambda)$. The optical depth will be taken up in Section 2.2.

The wavelength region with which we concern ourselves is the UV portion of the solar spectrum which is absorbed in the atmosphere above about 80 km. Space borne instruments are therefore required to measure the radiation and this has only become possible since the advent of rocket and satellite technology. Solar UV irradiance measurements extend over solar cycle 20 and, to date, about half of cycle 21. Measurements made with instruments carried on board satellites yield data over extended periods of time but the inability to perform in-flight calibration and the inevitable degradation of detectors lead to uncertainties in both absolute and relative values of the irradiance at different wavelengths. Rocket borne instruments yield snapshots of the solar irradiance, but less than two dozen measurements have been made since 1960, most experiments covering only a limited wavelength interval. Even the limited data sets that have been obtained clearly show considerable variability in the UV spectral intensities, with the magnitude of the variability related to other variable solar parameters such as the decametric radiation, cyclical variations, and flare activity.

The solar UV spectrum between 14 Å and 2000 Å obtained on 23 April 1974 is shown in Figure 2.1.1. The wavelength resolution is 1 Å and the flux magnitude, usually specified in units of photons $cm^{-2}\,s^{-1}\,Å^{-1}$, is just photons $cm^{-2}\,s^{-1}$ in this case. This

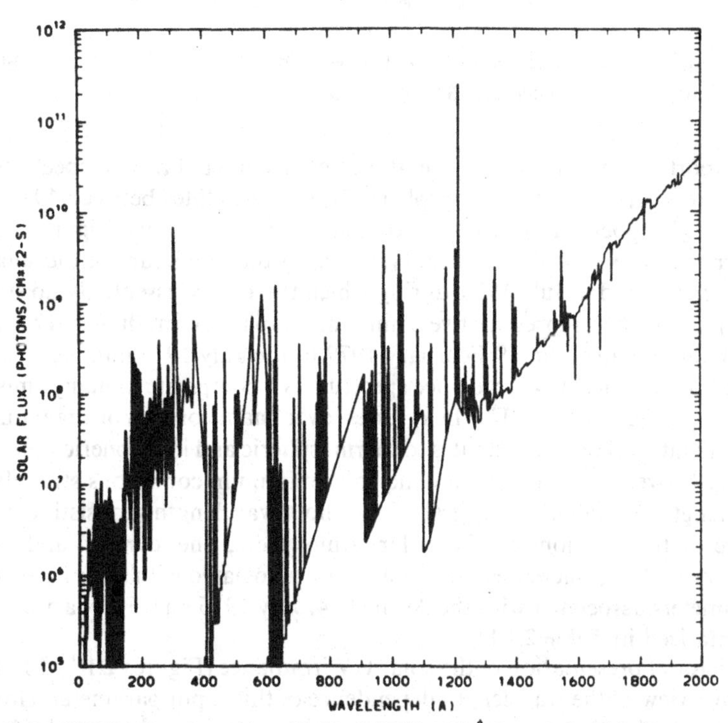

Fig. 2.1.1 Solar UV flux measured on 23 April 1974 at 1 Å intervals. This is the aeronomical reference spectrum F74113 for the Atmosphere Explorer satellite solar flux measurements. (R.G. Roble and B.A. Emery, *Planet. Space Sci.*, **31**, 597, 1983.)

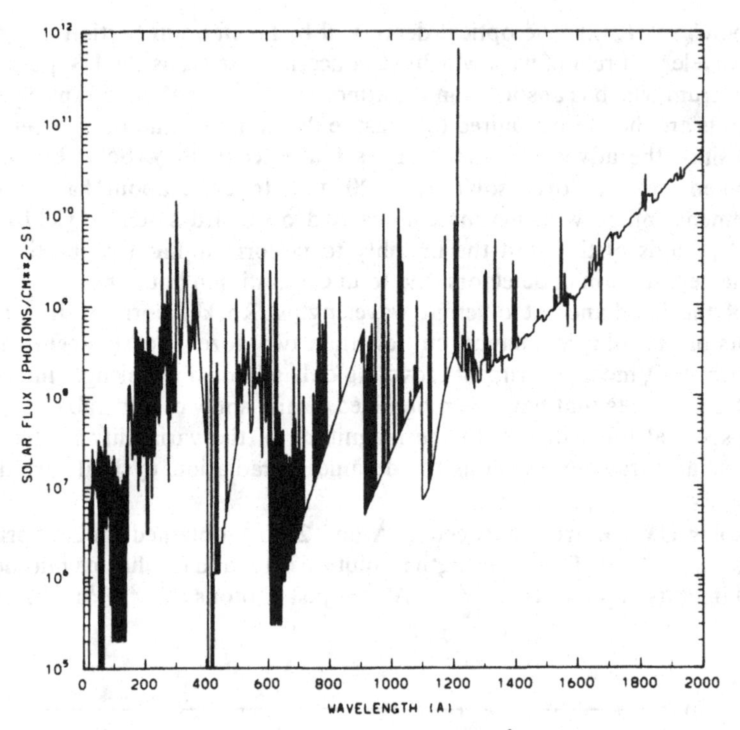

Fig. 2.1.2 Solar UV flux measured on 19 February 1979 at 1 Å intervals. (R.G. Roble and B.A. Emery, *Planet. Space Sci.*, **31**, 597, 1983.)

is an important spectrum because it served as a calibration spectrum for the spectrometers carried by the Atmosphere Explorer satellites between 1974 and 1981. The April 1974 epoch is representative of solar minimum activity. Figure 2.1.2 shows a similar spectrum obtained five years later during the maximum of the solar activity cycle. A brief period in July 1976, during which there were essentially no sunspots on the disc, is adopted as representative of minimum solar UV irradiance. The ratio of the solar UV flux obtained on 19 February 1979 to the July 1976 minimum is shown in Figure 2.1.3. The solar UV irradiance spectrum is a composite of many emission lines and continua (Problem 1). The flux varies over many orders of magnitude in the wavelength interval that contributes to thermospheric and ionospheric processes. Flux variability between solar maximum and solar minimum conditions strongly depends on wavelength. Variability is largest at the short wavelengths consisting of emission from the hottest region of the solar atmosphere, the corona and the upper chromosphere. Large increases of irradiance are associated with solar flares. Geophysical parameters associated with the April 1974, July 1976 and February 1979 periods are summarized in Table 2.1.1.

The pictorial presentation of the solar UV irradiance (Figures 2.1.1–2.1.3) provides a good overview of the wavelength dependence of this input parameter. However, for computations it is more convenient to present the measurements in tabular form and to combine photon intensities into larger wavelength intervals. Intensities of the brighter

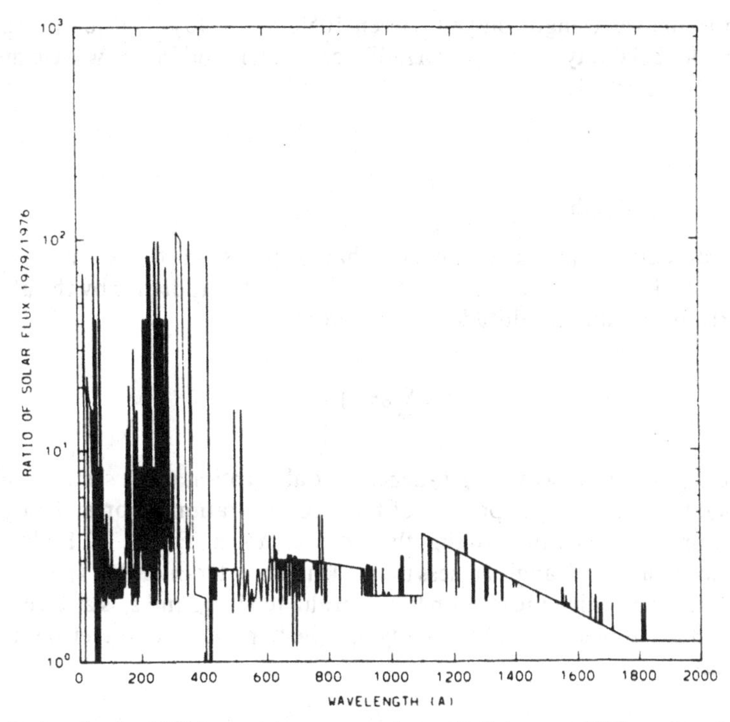

Fig. 2.1.3 Ratio of solar UV intensities measured on 19 February 1979 to the observations obtained during July 1976 on days of spotless solar conditions. (R.G. Roble and B.A. Emery, *Planet. Space Sci.*, **31**, 597, 1983.)

Table 2.1.1 *Geophysical parameters for various periods*

Day	F 10.7	$\overline{\text{F 10.7}}$	E_f	A_p	K_p
23 April 1974	74.5	86.1	1.99	29	5^- 4^- 4^- 5 4^- 4^- 4^-
13–28 July 1976	66.4–69.1	71.8	2.07	3–19	1^- 1^- 0 1^- 2^- 1^- 0 (21st)
(average = 67.6)			(average = 7 (15th))		2^- 2^- 2 2 2^- 4^+ 6^- 4^-
19 February 1979	243.3	187.7	7.19	15	4^+ 3 3^+ 2^+ 4 2^+ 3^+ 2

$\underline{\text{F 10.7}}$ – the 10.7 cm radiowave flux from the sun in units of ($\times 10^{-22}$ W m^{-2} Hz^{-1}) for the previous day.
F 10.7 – the 3 month solar rotation average of the F 10.7 solar flux centred on the day indicated.
E_f = solar UV energy flux in erg cm^{-2} s^{-1} integrated in wavelength from 14.25 to 1017 Å.
A_p = daily planetary geomagnetic index.
K_p = three hourly values of the planetary magnetic index on a quasi-logarithmic scale ranging from 0 to 9^+.

emission lines are still listed separately. One such tabulation is shown in Appendix 2 for five representative days in solar cycle 20. The wavelength range between 50 and 1050 Å includes the ionization energy of all major atomic and molecular constituents of the thermosphere (Section 2.3). Longer wavelength photons are capable of dissociating O_2 molecules, as well as other species; this will be discussed in Section 2.4. The reference spectrum in the wavelength range between 1050 and 1940 Å is listed in Appendix 2. Considerable variability of the solar irradiance is still found in this wavelength region, as shown in Figure 2.1.3.

2.2 The optical depth

The optical depth, $\tau(\lambda)$, a parameter that appears in the Lambert–Beer law (Equation (2.1.1)), specifies the attenuation of the solar irradiance by the atmosphere. For a vertical sun, and at altitude z_0 it is, simply,

$$\tau_{z_0}^{v}(\lambda) = \sum_j \sigma_j^{a}(\lambda) \int_{z_0}^{\infty} n_j(z)\,dz \tag{2.2.1}$$

where the $\sigma_j^a(\lambda)$s are the wavelength-dependent absorption cross sections of species j and the $n_j(z)$s are the height profiles of their concentrations. For a slant path on a spherical earth the equation is slightly more complex. Figure 2.2.1 illustrates the geometry for solar zenith angles χ_0 less than 90° and greater than 90°. The earth's radius is R, the altitude at which the photon flux is evaluated is z_0, and z_s is a screening height below which the atmosphere is essentially opaque to photons of wavelength λ. For the $\chi_0 < 90°$ case,

$$\frac{\sin\chi}{R+z_0} = \frac{\sin\chi_0}{R+z} \tag{2.2.2}$$

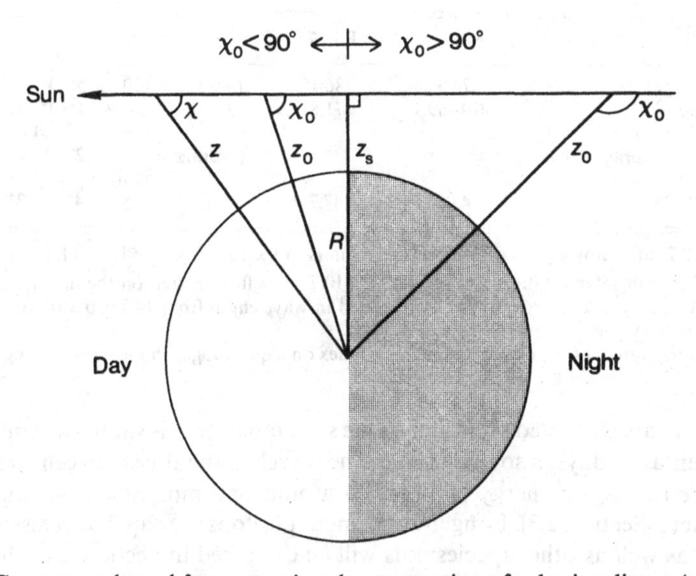

Fig. 2.2.1 Geometry adopted for computing the attenuation of solar irradiance (not to scale).

from which

$$\sec\chi = \left[1 - \left(\frac{R+z_0}{R+z}\right)^2 \sin^2\chi_0\right]^{-1/2} \tag{2.2.3}$$

and the optical depth is

$$\tau(\lambda, \chi_0, z_0) = \sum_j \sigma_j^a(\lambda) \int_{z_0}^{\infty} n_j(z)\left[1 - \left(\frac{R+z_0}{R+z}\right)^2 \sin^2\chi_0\right]^{-1/2} dz \tag{2.2.4}$$

Substantial effort has been devoted in the past toward expressing the integral by some analytic function, generically known as the Chapman function, or in tabular form. This hardly seems necessary any longer considering the ease of obtaining a solution numerically using a computer. For a solar zenith angle χ_0 less than about 75° it is found that $\sec\chi \approx \sec\chi_0$ for values of z_0 corresponding to ionospheric levels. The integral must still be retained if the altitude profiles of the concentrations of various species are arbitrary and not expressed in some analytic form.

For a solar zenith angle $\chi_0 > 90°$ the optical depth is

$$\tau(\lambda, z_0, \chi_0) = \sum_j \sigma_j^a(\lambda)\left\{2\int_{z_s}^{\infty} n_j(z)\left[1 - \left(\frac{R+z_s}{R+z}\right)^2\right]^{-1/2} dz - \int_{z_0}^{\infty} n_j(z)\left[1 - \left(\frac{R+z_0}{R+z}\right)^2 \sin^2\chi_0\right]^{-1/2} dz\right\} \tag{2.2.5}$$

The derivation is left as a problem (Problem 2). When $\chi_0 = 90°$ Equation (2.2.5) reduces to Equation (2.2.4) since $z_0 = z_s$, and for $\chi_0 = 0°$, an overhead sun, both Equation (2.2.4) and Equation (2.2.5) reduce to Equation (2.2.1).

The altitude dependence of the concentrations of the major thermospheric neutral species (together with the temperature profile) is known as a neutral atmosphere model. Several empirical models have been developed based on measurements obtained over long periods of time, a range of altitudes, latitudes, longitudes and levels of 'solar activity'. One goal of upper atmosphere research has been, and still is, to test our ability to model the response of the thermosphere and ionosphere to various inputs, such as a variable solar irradiance, i.e., to predict the changes in density, composition, temperature, etc. that are observed. Here we are faced with a dilemma. We must assume a model atmosphere to compute the optical depth which then enables us to derive the solar flux as a function of altitude and the multitude of effects that it has on the atmosphere. Among these effects will be changes in the model parameters that were assumed to begin the process. Moreover, rocket and satellite borne instruments do not measure the solar irradiance at the 'top of the atmosphere'; instead, $I(\lambda)$ is measured over a range of altitudes. To infer $I_\infty(\lambda)$ from the measurements a model atmosphere must be assumed. This interdependence of physical parameters is characteristic of the system being studied, the upper atmosphere, and we will be faced with similar situations throughout this book. Appendix 1 is devoted to empirical neutral atmosphere models.

The remaining parameter in the expression for the optical depth is the absorption cross section. The photoabsorption cross sections for the major thermospheric species are shown in Figures 2.2.2–2.2.4. The wavelength region relevant to our purpose, computing ionization and dissociation rates, is determined by the following related

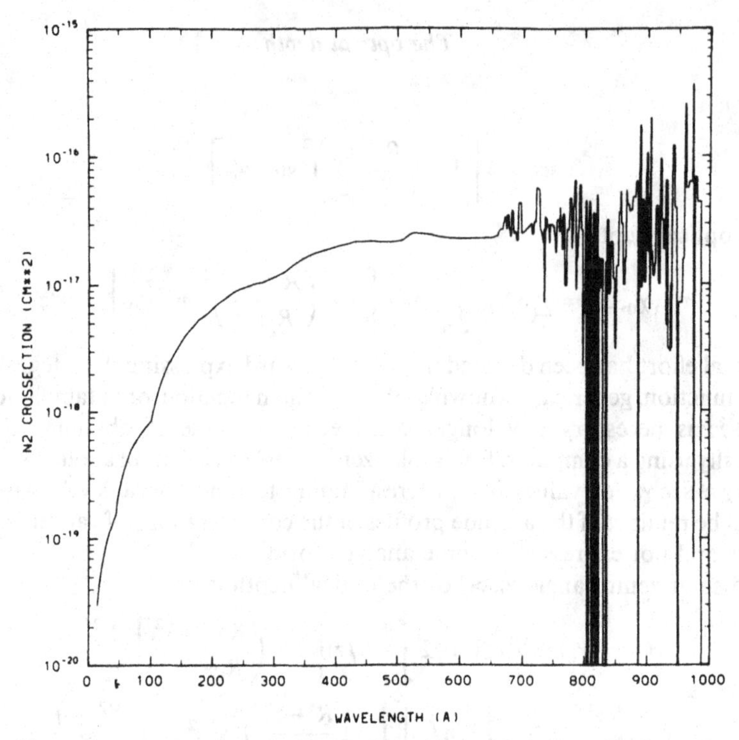

Fig. 2.2.2 Absorption cross sections for N_2. (R.G. Roble and B.A. Emery, *Planet. Space Sci.*, **31**, 597, 1983.)

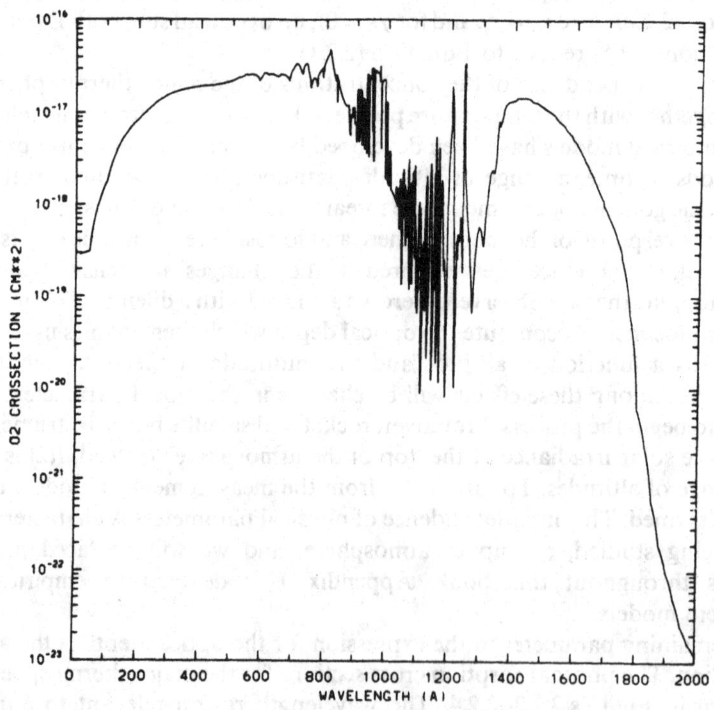

Fig. 2.2.3 Absorption cross sections for O_2. (R.G. Roble and B.A. Emery, *Planet. Space Sci.*, **31**, 597, 1983.)

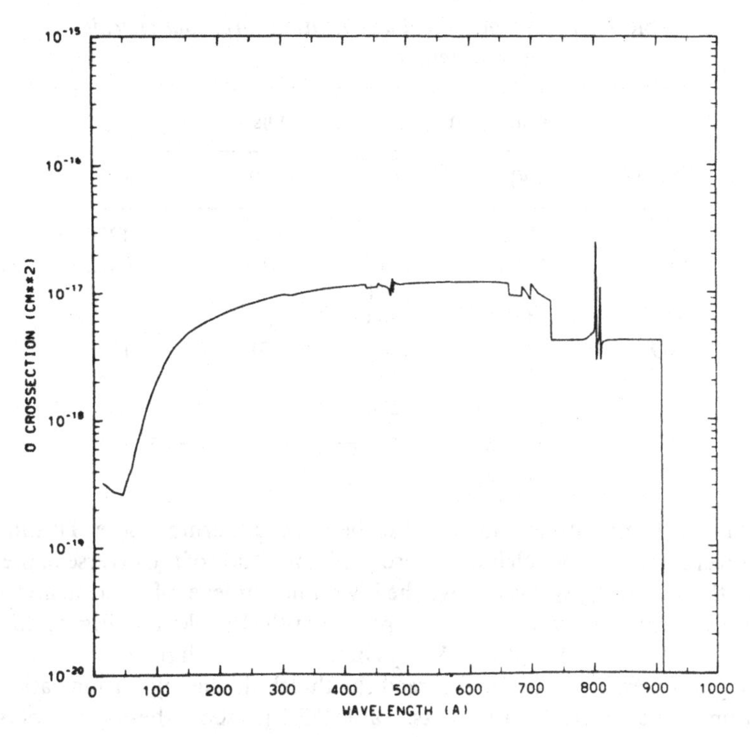

Fig. 2.2.4 Absorption cross sections for O. (R.G. Roble and B.A. Emery, *Planet. Space Sci.*, **31**, 597, 1983.)

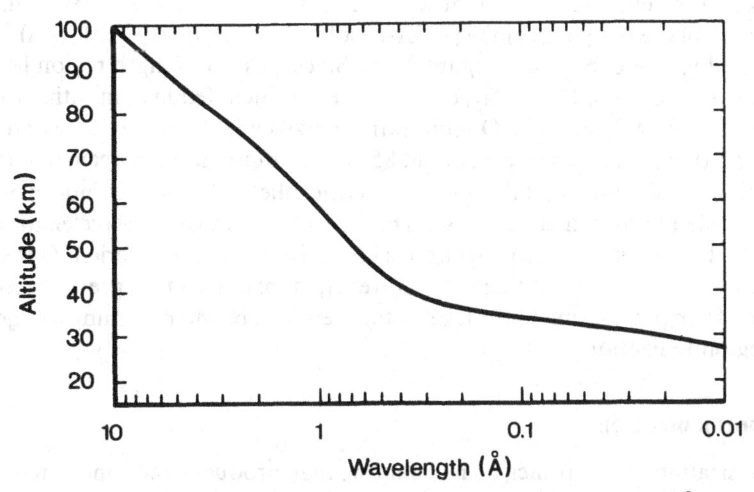

Fig. 2.2.5 Altitudes of the unit optical depth for $0.01 \leqslant \lambda \leqslant 10$ Å.

Table 2.2.1 *Ionization and dissociation threshold energies and wavelengths*

	Ionization		Dissociation	
Species	(eV)	λ(Å)	(eV)	λ(Å)
N_2	15.58	796	9.76	1270
O_2	12.08	1026	5.12	2422
O	13.61	911		
N	14.54	853		
NO	9.25	1340	6.51	1905
H	13.59	912		
He	24.58	504		

considerations. The photons must be absorbed in the thermosphere. This implies an absorption cross section which is of the order of magnitude of the inverse of the column density of the absorbing species above the lowest height level of interest, about 90 km, resulting in approximately unit vertical optical depth (Problem 3). Figure 2.2.5 shows the altitude of unit optical depth for X-rays in the wavelength region between 0.01 and 10 Å. The photon energy must also exceed the threshold energy for ionization and/or dissociation of the neutral constituents. Table 2.2.1 presents these thresholds for the thermospheric and ionospheric species. Absorption cross sections are given in tabular form in Appendix 2. Examination of the curves in Figures 2.2.2–2.2.4 shows that in the wavelength region of the continua the absorption cross section varies relatively smoothly with wavelength but that molecular band absorption is highly structured. In the region between the Schumann–Runge continuum and the photoionization limit ($\sim$1350–1026 Å) in O_2 especially large changes in the magnitude of the cross section occur, varying by three powers of ten within a range of a few ångstroms. Fortuitously, there is a deep minimum in the absorption cross section ($\sim 1 \times 10^{-20}\,\text{cm}^2$) at the wavelength of the H Lyman alpha (H-Ly-α) solar emission line ($\sim$1215.7 Å) which has a large and variable intensity (Figure 2.2.6). Since this wavelength region lies beyond the absorption edge of O and N_2, only O_2 need be included in computing the optical depth at 1216 Å. Adopting the O_2 concentration given in Appendix 1 we find that unit optical depth for O_2 is at a level of about 85 km. It would therefore seem that the H-Ly-α radiation should pass right through the thermosphere. This would indeed occur were it not for the minor constituent NO. The ionization threshold wavelength of NO is 1340 Å and the ionization cross section is large. Thus, thermospheric NO is efficiently ionized by H-Ly-α. The NO density, however, is insufficient to prevent most of the radiation from penetrating to the level of the mesosphere and becoming a major source of D-region ionization.

2.3 Photoionization

Photoionization is the principal mechanism that produces the ionosphere. For the three major thermospheric species we have

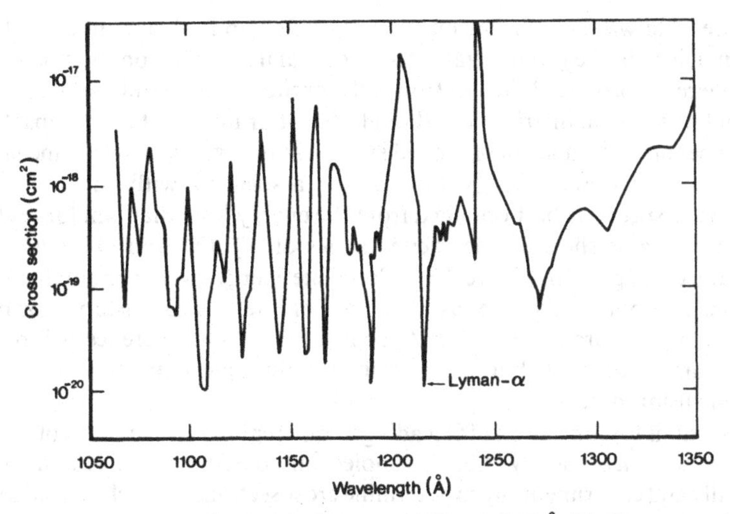

Fig. 2.2.6 Absorption cross sections of O_2 between 1050 and 1350 Å. (K. Watanabe, *Advances in Geophysics*, Eds. H.E. Landsberg and J. Van Mieghen, **5**, 153, 1958.)

$$N_2 + h\nu(<796\,\text{Å}) \rightarrow N_2^+ + e \tag{2.3.1}$$

$$O_2 + h\nu(<1026\,\text{Å}) \rightarrow O_2^+ + e \tag{2.3.2}$$

and

$$O + h\nu(<911\,\text{Å}) \rightarrow O^+ + e \tag{2.3.3}$$

The wavelengths specified in Reactions (2.3.1)–(2.3.3) correspond to the ionization thresholds, λ_{th}, listed in Table 2.2.1 for the production of ions in their ground electronic state. The photoionization rate of species j at altitude z is

$$P_z^i(j,l) = n_z(j)\int_{\lambda_{th}}^{0} I_z(\lambda)\sigma_j^i(\lambda)p_j^i(\lambda,l)\,d\lambda \qquad (\text{ions cm}^{-3}\,\text{s}^{-1}) \tag{2.3.4}$$

Quantities not previously defined are the ionization cross section, $\sigma_j^i(\lambda)$, and the branching ratios $p_j^i(\lambda, l)$ that identify production of ions in various electronically excited states, l, or in the ground state. For each constituent,

$$\sum_j p_j^i(\lambda, l) = 1 \tag{2.3.5}$$

The integration in Equation (2.3.4) is carried from the ionization threshold wavelength, λ_{th}, to the shortest wavelength (not zero, as has been shown) at which the photon flux is sufficient to contribute to ion production. X-ray photons of wavelength shorter than about 10 Å are absorbed largely below 100 km (Figure 2.2.5). The photon flux is small but highly variable.

Dissociative ionization is an additional source of atomic ions,

$$O_2 + h\nu(<662\,\text{Å}) \rightarrow O^+ + O + e \tag{2.3.6}$$

and

$$N_2 + h\nu(<510\,\text{Å}) \rightarrow N^+ + N + e \tag{2.3.7}$$

Photons with sufficient energy to both ionize and dissociate the molecule are required.

The threshold wavelengths for reactions specified up to this point are valid for the formation of ions in the ground state. A substantial fraction of ions and neutrals in the thermosphere is produced in electronically excited states and with an enhanced vibrational population distribution, though the latter not by photon impact. Various consequences of the formation of excited states will be discussed in this and succeeding chapters. An overview of the various states associated with each of the major thermospheric species is best obtained from the energy level diagram for each species. Simplified diagrams showing the excited states of N_2^+, O_2^+ and O^+ relevant to the present subject are given in Figure 2.3.1. Complete energy level diagrams for atoms and atomic ions and potential curves as a function of internuclear distance for molecules and molecular ions are presented in Appendix 3; frequent reference will be made to these diagrams. The contribution of double or multiple ionization is too small to warrant inclusion here.

Cross sections for ionization of O leading to different electronic states of the ion have been computed theoretically. For the molecular ions, however, it is difficult both theoretically and experimentally to determine cross sections for each excited ionic state. It has been common practice to specify the total photoionization cross section, $\sigma_j^i(\lambda)$, and the branching ratios, $p_j^i(\lambda, l)$, using compatible wavelength intervals. The different states of the ions are shown in Figure 2.3.1, including the respective threshold energies. The total photoionization cross sections and branching ratios are given in

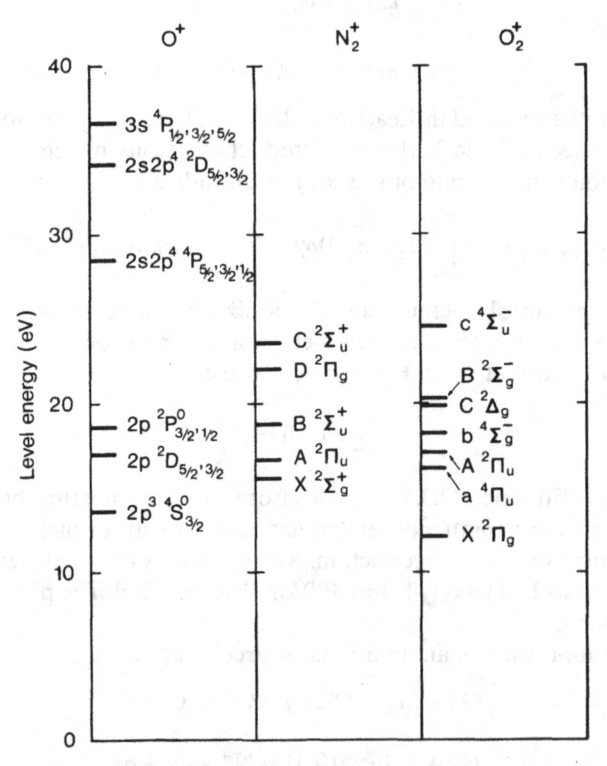

Fig. 2.3.1 Partial energy level diagrams of the major ionizable thermospheric species. The lowest level corresponds to the ionization energy of the neutral species.

Appendix 2, as is the fraction of ionization events going into dissociative ionization. For computational convenience the same wavelength intervals are adopted that have been used for the solar irradiance spectrum and for the absorption cross sections. We will learn in Chapter 7 how several excited states produced by photoionization lead to optical emissions that are part of the dayglow spectrum (Problem 4).

2.4 Photodissociation

Photodissociation of O_2 accounts for O being the major neutral species in the thermosphere above about 200 km. Two-body recombination is very slow so that O atoms must diffuse downward into the denser regions of the atmosphere where three-body recombination is the destruction mechanism, i.e., below approximately 100 km. The time constant for downward diffusion of O atoms is long compared with the diurnal cycle of solar illumination, preventing any significant night time decrease in the thermospheric abundance of O atoms.

The dissociation threshold energy of O_2 is only 5.12 eV (Table 2.2.1) so that photons at wavelengths less than 2422 Å are effective in the process

$$O_2 + h\nu(<2422\,\text{Å}) \rightarrow O + O \tag{2.4.1}$$

However, solar irradiance beyond 1800 Å is absorbed below the thermosphere and most of the thermospheric dissociation of O_2 is caused by photons in the wavelength region of the Schumann Runge continuum. Taking account of the entire spectral range, the rate of photodissociation at altitude z is

$$P_z^d(O_2) = n_z(O_2)\int_{2422\,\text{Å}}^{650\,\text{Å}} I_z(\lambda)\sigma_{O_2}^d(\lambda)\,d\lambda \quad (\text{cm}^{-3}\,\text{s}^{-1}) \tag{2.4.2}$$

where $\sigma_{O_2}^d(\lambda)$ is the photodissociation cross section. The short wavelength limit of integration, 650 Å, corresponds to the photoionization limit, i.e., photon absorption leads almost entirely to ionization (Appendix 2).

In Section 2.3 we noted that photoionization leads to several electronically excited states of the ions. Photodissociation likewise can lead to excitation of the products.

$$O_2 + h\nu(<1749\,\text{Å}) \rightarrow O(^1D) + O(^3P) \tag{2.4.3}$$

The excitation energy of the $O(^1D)$ state is 1.97 eV (see Appendix 3). We mentioned earlier that it is the solar irradiance at wavelengths less than of about 1800 Å that is absorbed by O_2 in the thermosphere. Photodissociation of O_2 in the thermosphere is therefore a copious source of excited O atoms. This has important consequences which will be explored in succeeding chapters.

N_2 molecules also photodissociate. However, the process is quite different from dissociation of O_2 which occurs primarily by absorption of continuum radiation. N_2 molecules are first excited into several high lying valence and Rydberg states, called predissociation states which decay largely by dissociation of the molecule instead of by radiation. The absorption bands responsible for this process lie in the wavelength region between about 800 and 1000 Å (Fig. 2.2.2), i.e., just beyond the ionization threshold. Although not the major source of N in the thermosphere, photodissociation of N_2 via excitation of predissociation levels does contribute to the production of this minor constituent. It is generally assumed that the difference between the total absorption cross section of N_2 and the ionization cross section (Appendix 2) is the effective dissociation cross section that may be used to compute the N atom production rate by an equation similar to Equation (2.4.2) (Problem 5). The predissociation mechanism is described in greater detail in Chapter 4 in connection with electron impact dissociation of N_2.

2.5 Photoelectrons

In our discussion of photoionization (Section 2.3) we indicated by Reactions (2.3.1)–(2.3.3) that the process leads to the production of ion–electron pairs, and this also applies to dissociative ionization, Reactions (2.3.6) and (2.3.7). The threshold wavelengths associated with each reaction specify the minimum photon energy that is required for the reaction to proceed; however, the photoionization cross section is larger at wavelengths shorter than the threshold wavelength (Appendix 2). The reactions, therefore, proceed at a higher rate in association with excess energy. As already mentioned, there is the probability (cross section) that the excess energy leads to internal excitation of the product ions (e.g., $N_2^+\ A^2\Pi_u$, $O_2^+\ a\ ^4\Pi_u$, $O^+\ ^2D$) but much of the excess energy is channelled into kinetic (translational) energy of the products. Momentum conservation shows that most of the energy is imparted to the lighter electrons. Photoionization, therefore, is a source of energetic photoelectrons. The energy spectrum of photoelectron production is governed by the UV photon flux, a multitude of photoionization cross sections and the neutral gas composition of the atmosphere. The production rate of photoelectrons with energy E at altitude z is given by

$$P_z^e(E) = \sum_j \sum_l \sum_\lambda n_z(j)\sigma_j^i(\lambda)p_j^i(l, \lambda_{th}, \lambda)I_z(\lambda) \qquad (cm^{-3}\,s^{-1}) \qquad (2.5.1)$$

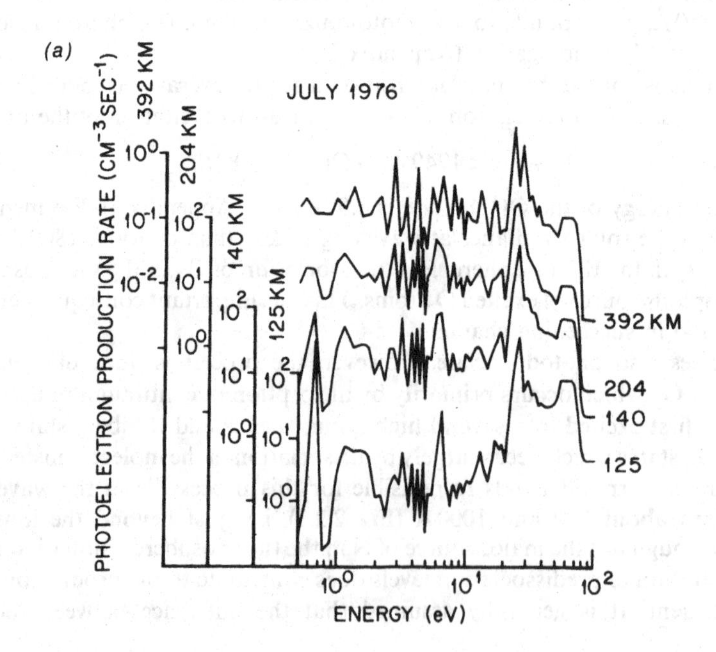

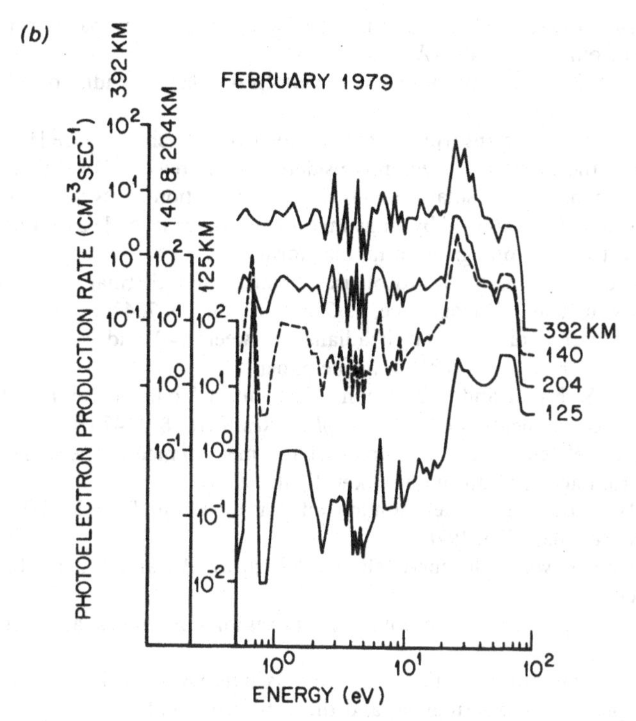

Fig. 2.5.1 The production rate of photoelectrons for typical solar minimum (*a*) and solar maximum (*b*) conditions as a function of electron energy at selected altitudes.

The electron energy will henceforth be specified in energy units, with

$$E = hc(1/\lambda_{\text{th}}(j, l) - 1/\lambda) \text{ (eV)} \tag{2.5.2}$$

Planck's constant is h and c is the speed of light. The threshold wavelength (or energy) is a function of the electronic state, l, of species, j. As noted in Section 2.1, the intensity of solar irradiance, $I_z(\lambda)$, may be specified per angstrom (Figures 2.1.1 and 2.1.2) or in finite wavelength intervals (e.g. 50 Å) plus discrete emission lines, as in Appendix 2. The summation over λ in Equation (2.5.1) must be treated accordingly. The branching ratios and excitation thresholds for the electronic states of the three major species in the thermosphere are listed in Appendix 2, corresponding to the states shown in Figure 2.3.1. Variability in the solar irradiance, $I_\infty(\lambda)$, with solar activity results in a corresponding variation of the photoelectron production rate, $P_z^e(E)$ (Problem 6). The production rates of photoelectrons as a function of energy for solar minimum (July 1976) and solar maximum (February 1979) conditions are shown in Figure 2.5.1 at selected altitudes. The structure in these curves is a consequence of structure in the solar irradiance as a function of wavelength.

Bibliography

The solar UV irradiance is one of the fundamental parameters in an investigation of the upper atmosphere and considerable effort has been expended in acquiring measurements of the absolute fluxes and the variability with solar activity. A recent review paper evaluates published irradiance data from 1350 to 4000 Å:

P.C. Simon and G. Brasseur, Photodissociation effects of solar UV radiation, *Planet. Space Sci.*, **31**, pp. 987–99, 1983.

This paper also discusses the absorption cross section of O_2 which is required to interpret the flux measurements in the wavelength region considered in the review. Over eighty references are given to publications on the solar irradiance and absorption cross sections. The results of irradiance measurements obtained by many investigators over several years are compared, using as reference spectrum the one given in tabular form by

G. Brasseur and P.C. Simon, Stratospheric chemical and thermal response to long-term variability in solar UV irradiance, *J. Geophys. Res.*, **86**, 7343–62, 1981.

A critical evaluation of the solar irradiance between 140 and 1850 Å acquired by the Atmosphere Explorer satellites is given in the paper:

H.E. Hinteregger, K. Fukui, and B.R. Gilson, Observational, reference and model data on solar euv, from measurements on AE-E, *Geophys. Res. Lett.*, **8**, 1147–50, 1981.

This paper gives references to the sources of the measurements, specifically, the so-called reference spectrum acquired during a rocket flight:

L. Herox and H.E. Hinteregger, Aeronomical reference spectrum for solar UV below 2000 Å, *J. Geophys. Res.*, **83**, 5305, 1978.

The measurements on which the curves shown in Figures 2.1.2 and 2.1.3 are based are given in the paper

H.E. Hinteregger, Representations of solar EUV fluxes for aeronomical applications, *Adv. Space Res.*, **1**, 39, 1981.

Photoionization and photoabsorption cross sections in the wavelength region 34 Å to 1027 Å are given in a compilation of experimental and theoretical results by

K. Kirby–Docken, E.R. Constantinides, S. Babeu, M. Oppenheimer and G.A. Victor, Photoionization and photoabsorption cross sections of He, O, N_2 and O_2 for aeronomic calculations, *At. Data Nuclear, Data Tables*, **23**, 63–82, 1979.

The above data compilation also gives the branching ratios into excited states of O, N_2 and O_2 and the dissociative ionization yield for the molecular species. In the longer wavelength region absorption cross sections are summarized in the review article

K. Watanabe, Ultraviolet absorption processes in the upper atmosphere, *Adv. in Geophys.*, **5**, 153, 1958.

The density and composition of the neutral atmosphere may be specified by empirical models, based on very large data sets acquired by satellite borne instruments and ground based measurements. One such model includes data compiled over many years' observations, allowing for different levels of solar activity, magnetic activity, geographic and diurnal variations:

A.E Hedin, MSIS-86 thermospheric model, *J. Geophys. Res.*, **92**, 4649, 1986.

The above cited paper gives references to preceeding publications on empirical model atmospheres.

Problems 1–6

Problem 1 Identify some prominent line emissions in the solar spectra shown in Figures 2.1.1 and 2.1.2. Use the tables in Appendix 2 as well as the references. Which lines show large differences in intensity between solar minimum and maximum conditions?

Problem 2 Derive Equation (2.2.5) for the optical depth for solar zenith angles greater than 90°.

Problem 3 Compute the altitude at which $I(\lambda)$ is reduced to 1/e of the irradiance at the top of the atmosphere, $I_\infty(\lambda)$, as a function of wavelength for a solar zenith angle of 40°. Use the absorption cross sections in Appendix 2.

Problem 4 Compute the photoionization rate for solar maximum conditions when the solar zenith angle is 40°. Compute the altitude profiles of the production rate of various ions of the major thermospheric species. Assume that ions produced by dissociative ionization are formed in the ground state.

Problem 5 Compute the N atom production rate as a function of altitude by photodissociation of N_2 for solar maximum conditions and a solar zenith angle of 40°.

Problem 6 How much energy does a photoelectron acquire when produced by HeI photons ionizing N_2? Compute the photoelectron production rate due to the HeI and HeII solar emission lines at 170 km altitude for day 113 in 1974 and day 50 in 1979 for a solar zenith angle of 40°.

3

The interaction of energetic electrons and ions with the upper atmosphere

Solar UV radiation is the principal cause for the existence of the ionosphere and the physical processes that underlie the production of photoelectrons are presented in the preceding chapter. Since photoelectrons are produced within the atmosphere the term commonly applied to this population is an embedded source of ionization. We have not avoided the observational aspect of photoelectrons at this point, rather, a production rate is not a measurable quantity. Detectors measure the intensity or the flux, the number of particles or photons passing through or absorbed by unit area per unit time and solid angle. Photoelectron intensities will be derived in this chapter but, first, an additional source of ionization is discussed, a source that is due to energetic charged particle precipitation of solar and magnetospheric origin, associated principally with the aurora. Unlike photoelectrons, primary auroral electrons are a source external to the atmosphere and their intensity is a measurable quantity, analogous to the solar photon flux at the top of the atmosphere. Carrying the analogy one step further, the primary auroral electrons ionize the atmospheric gases producing secondary electrons that are the equivalent of the photoelectrons produced by photoionization.

In this chapter we first survey briefly the characteristics of energetic electron fluxes that have been measured by rocket and satellite borne detectors. We next analyse the processes that operate when a stream of energetic electrons penetrates into the atmosphere from the tenuous regions of the magnetosphere. These processes involve angular scattering and energy degradation of the electrons and the term commonly used to describe them is electron transport. Auroral electrons (primaries, secondaries, tertiaries, etc.) undergo numerous collisions in their passage through the atmosphere and we end the chapter by focusing on the ionization that they produce.

3.1 Characteristics of primary auroral electrons

We have already noted the large variability in the solar UV irradiance that, as will emerge in succeeding chapters, is responsible for variability in several parameters that characterize the thermosphere and ionosphere; examples are the density and composition at a given pressure level, the neutral and plasma temperature and the ion density and composition. There are patterns associated with the solar irradiance such as the diurnal, seasonal and geographic variations as well as the long term (~ 11 yr) solar activity cycle. There is also a component of the energetic charged particle flux that penetrates into the atmosphere which has some regularity. This is the cosmic ray precipitation, consisting primarily of protons with energy in the megaelectron volt to hundreds of megaelectron volts range. These high energy particles penetrate deeply into the atmosphere, reaching the mesosphere and stratosphere. Cosmic rays are the dominant source of ionization in the stratosphere and lower mesosphere.

Compared with the ionization sources described in the preceding paragraph, aurorally associated charged particle precipitation is characterized by its storm-like behaviour: long-term unpredictability, highly variable strength and spatial inhomogeneity. These properties make up what is generally called auroral morphology. An adequate description of auroral morphology would require at least an entire chapter but will not be given in this book; instead, the reader is referred to the bibliography. We only note here that aurorae are, in general, high latitude phenomena, and there is a circumpolar band of maximum occurrence. Since charged particle motion in the collisionless magnetosphere is controlled by the geomagnetic field the auroral precipitation pattern is aligned by the eccentric dipole field. A discussion of thermospheric neutral and ion dynamics, presented in a subsequent chapter, does require some appreciation of auroral morphology, however, our present focus is on the physical and chemical processes that describe the interaction of energetic particles with the upper atmosphere. The auroral characteristics that are needed to study the interaction processes are the identity of the charged particles, electrons, protons, or heavier ions, the energy spectrum and the angular distribution in the stream.

The phase space distribution of auroral electrons is not unlike the human population: many individual events exist but no two are exactly the same. To be sure, there are similarities and types but observations show that the many thousands of auroral spectra that have been recorded by numerous satellite and rocket borne detectors are all different. To familiarize the reader with various display modes of the experimental data some examples are given in Figures 3.1.1–3.1.5. The variables are the electron intensity, the energy and the angle with respect to some reference axis, generally the direction of the geomagnetic field. Azimuthal symmetry about this direction is usually assumed because an electron gyrates many times about the magnetic field before undergoing a collision that may alter its direction. The angle between the electron's velocity vector and the geomagnetic field direction is called the pitch angle. The energy and angular variables may be replaced by the velocity components parallel and perpendicular to the magnetic field direction (Figures 3.1.3–3.1.5). Several detectors sampling different directions are necessary to obtain a complete phase space distribution of electron intensity. Detectors on a spinning spacecraft accomplish the same result with a loss of temporal and spatial resolution. The energy distribution is obtained by scanning a detector through the desired range of energy or by acquiring continuous measurements of intensity at discrete energies with several detectors. Electron spectra are highly variable in time and space and a large array of detectors is required to minimize temporal and spatial smearing. The importance and difficulty of obtaining 'snapshots' of the phase space distribution may best be appreciated by noting that auroral structures have been measured to be less than 100 m in width and that satellites used to acquire electron data travel at several kilometres per second. Rockets travelling approximately parallel to the magnetic field

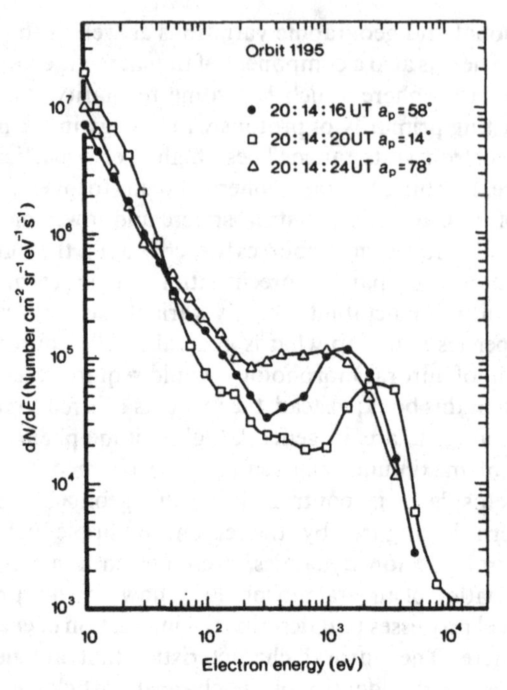

Fig. 3.1.1 Differential electron spectra observed in the northern polar cap on May 16, 1969. (J.D. Winningham and W.J. Heikkila, *J. Geophys. Res.*, **79**, 949, 1974.)

sample primarily the temporal and altitudinal variations of electron intensity while satellites provide the spatial variation along the orbital path.

The phase space distribution of auroral electrons depends on the altitude at which the measurements are made, on how much scattering the stream has undergone, on what fraction has been scattered in the conjugate hemisphere, and on the effect of a reflecting or accelerating electric field above the region where the measurement is made. To avoid adding unnecessary complications to the basic physical processes that govern electron transport in the collisionally dominated atmosphere we assume a primary auroral electron distribution at the 'top of the atmosphere' and investigate its subsequent evolution.

3.2 Electron transport in the atmosphere

Passage of solar irradiance, $I(\lambda)$, through the atmosphere was described in the preceding chapter, and it was stated that absorption of the photons is only a function of the optical depth. The change of intensity with optical depth, τ, is therefore

$$dI(\lambda)/d\tau = -I(\lambda) \qquad (3.2.1)$$

subject to the boundary condition $I(\lambda) = I_\infty(\lambda)$ at $\tau = 0$. This leads to the Lambert–Beer absorption law (Equation (2.1.1)), which implies that photons only suffer complete destruction, are not multiply scattered and do not reappear as photons at a different wavelength. Electron transport cannot be treated so simply.

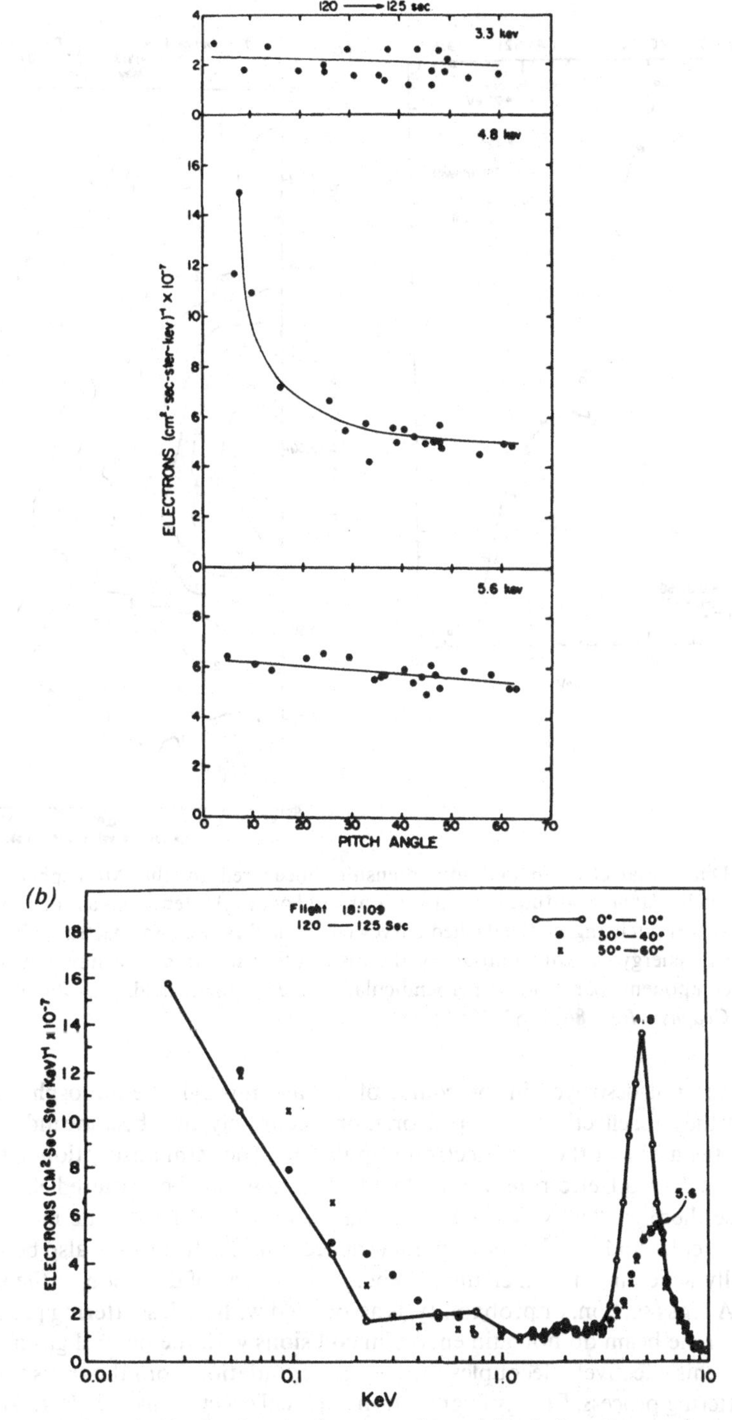

Fig. 3.1.2 (*a*) Pitch angle distributions at three energies and (*b*) pitch angle sorted spectra for 120 to 125 s flight time for flight 18:109. (R.L. Arnoldy *et al.*, *J. Geophys. Res.*, **79**, 4208, 1974.)

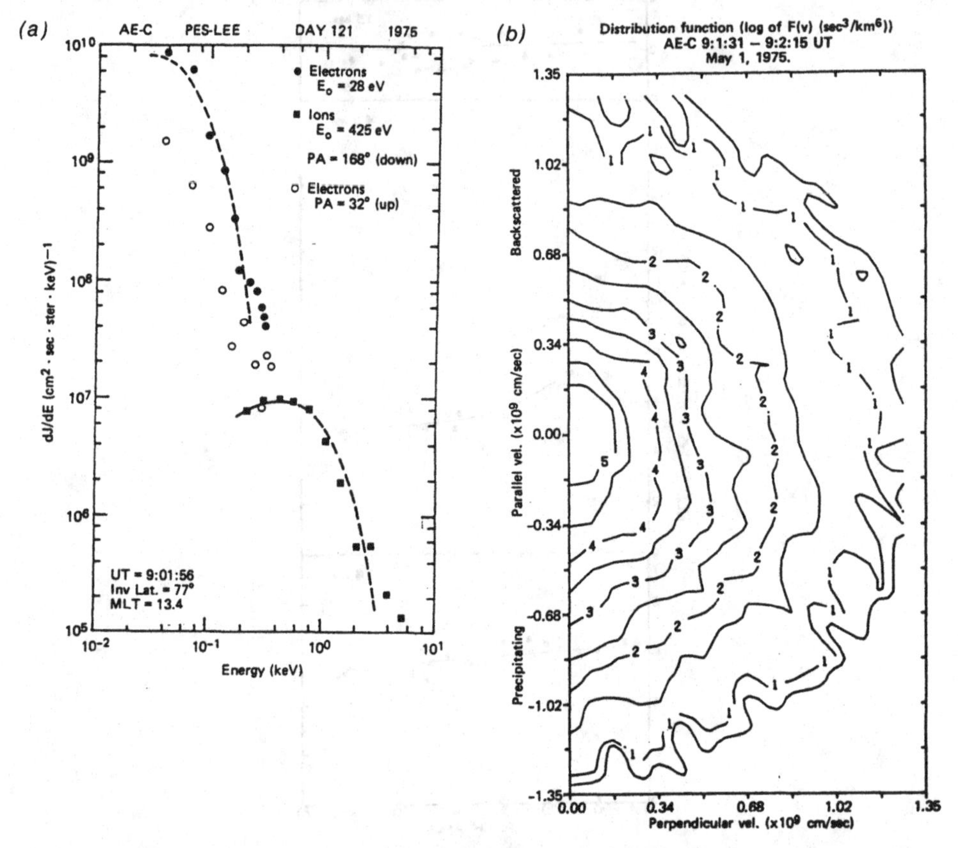

Fig. 3.1.3 Differential electron and ion intensities measured by the Atmosphere Explorer satellite at high latitude in the dayside auroral oval. (*a*) Intensity as a function of energy at two pitch angles; the dashed curves represent Maxwellian spectra with characteristic energy, E_o. (*b*) Contours of the distribution function as a function of velocity components parallel and perpendicular to the magnetic field. (L.J. Zanetti, *et al.*, *J. Geophys. Res.*, **86**, 8957, 1981.)

Electrons are not destroyed in the course of passage through the atmosphere, though eventually they are effectively 'stopped' or, more accurately, they become indistinguishable from the ambient thermal electron population. The term absorption is therefore inapplicable. Instead, electrons are scattered. Electrons may be scattered elastically, in which case there is effectively no energy change in colliding with the more massive atoms, molecules and ions but only angular deflection. Electrons may also be scattered inelastically which involves a change of energy as well as of direction of the scattered electron. A cross section, or probability, is associated with each scattering process. The electrons in the beam do not gain energy in collisions with the neutral gas atoms and molecules; this effectively decouples the energy degradation from the transport aspect of the scattering process. Energy degradation is a local effect. Thus, for elastic scattering the cross section is written in the form

$$\sigma_e(E', E; \mu', \mu) = \delta(E - E')P_e(E; \mu', \mu)\sigma_e(E) \quad (3.2.2)$$

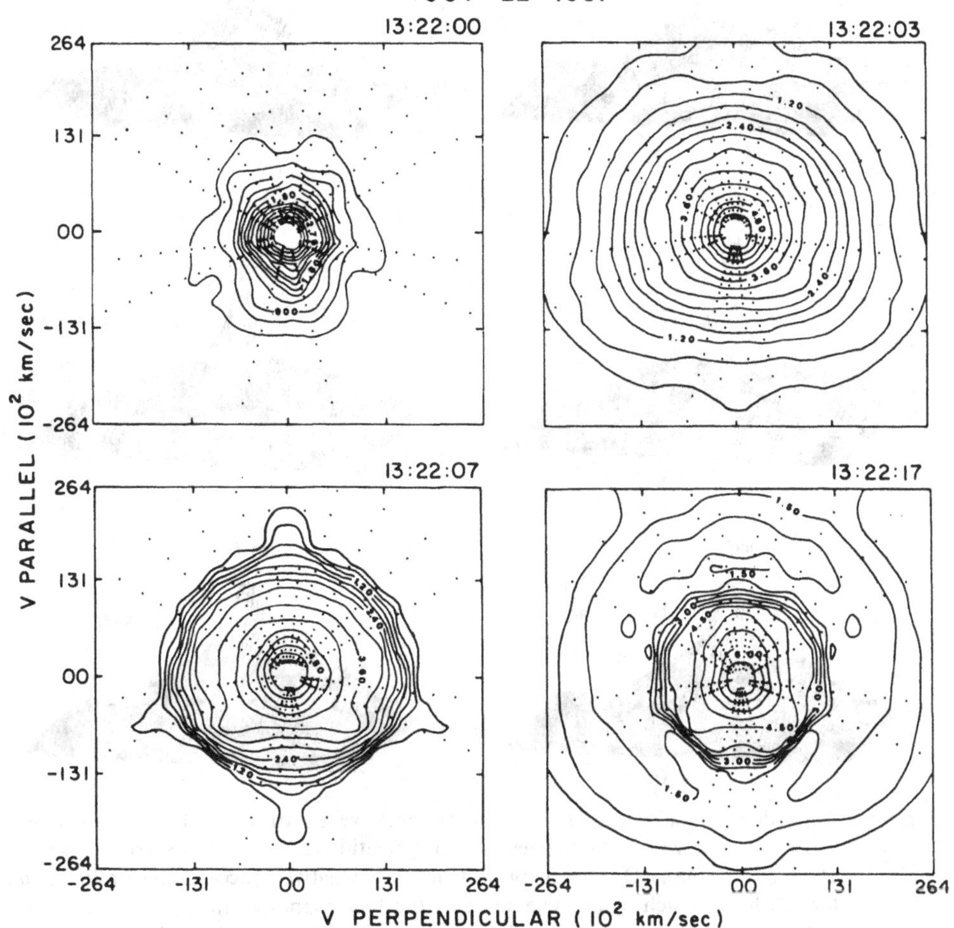

Fig. 3.1.4 Energetic electron distribution functions obtained for various times as a function of velocity components parallel and perpendicular to the magnetic field. (R.A. Heelis *et al.*, *J. Geophys. Res.*, **89**, 3893, 1984.)

to show explicitly that it does not involve a change in energy but only angular deflection. $P_e(E;\mu',\mu)$ is the phase function for angular deflection at energy E with μ' and μ the cosine of the pitch angles before and after the collision. E' and E refer to the electron energy before and after the collision and $\sigma_e(E)$ is the magnitude of the cross section at energy E. The Kröneker-δ has the usual meaning: it has the value zero except at $E = E'$ for which its value is one. Further discussion of cross sections and numerical values is presented in Chapter 4.

Inelastic scattering involves a change in the internal energy of the atmospheric collision partner. This could be excitation of electronic, vibrational and rotational states accompanied by a discrete energy loss and ionization and molecular

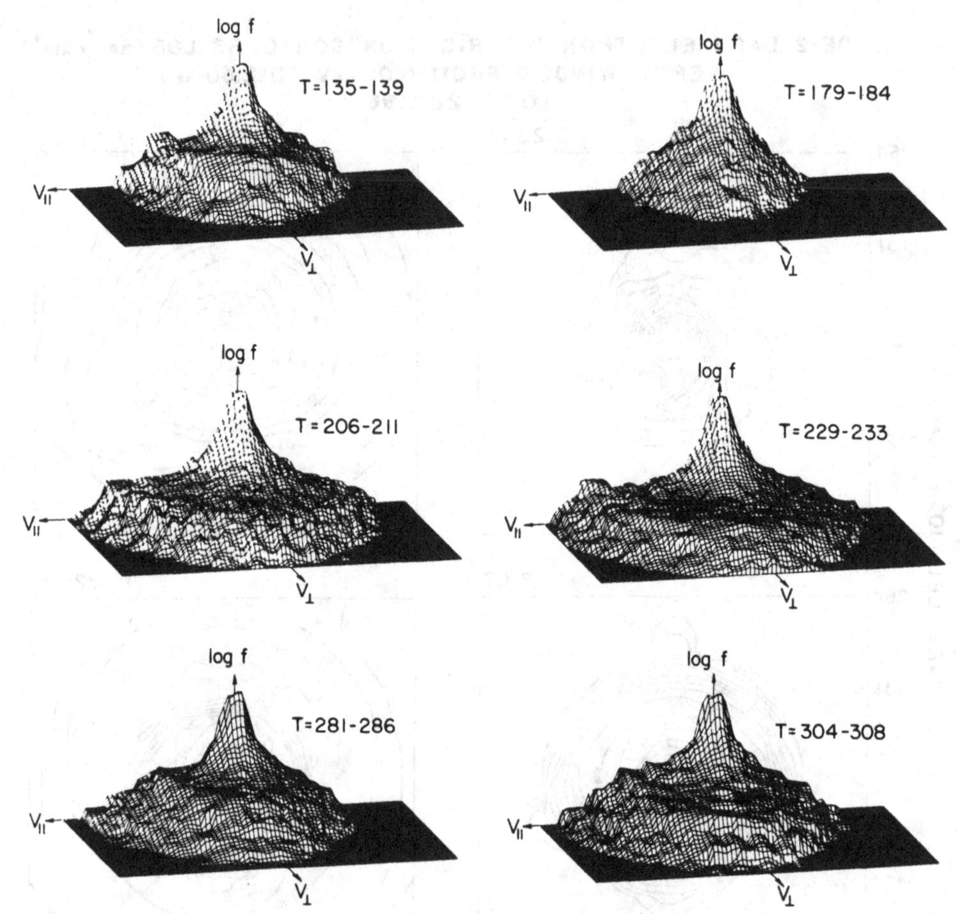

Fig. 3.1.5 Three-dimensional plots of several representative electron distribution functions. Each plot is shown as if it were viewed from a position 75° from the positive $v_{\parallel}$ axis and 15° above the plane. Downcoming electrons have positive $v_{\parallel}$, accounting for fluxes on the left half of each figure. The edges of the base planes are at $v_{\parallel} = 60 \times 10^8$ and -60×10^8 cm s^{-1} and $v_{\perp} = 0$ and 60×10^8 cm s^{-1}. The vertical axis is the common logarithm of $f(v)$, and the base plane was introduced at $\log f = -1.0$ to aid in visualization. (Kaufmann, *et al.*, *J. Geophys. Res.*, **83**, 586, 1978.)

dissociation accompanied by continuous energy loss above a threshold. Ionization is not only an energy loss process but also a source of secondary electrons that have their own phase space distribution. The auroral electron transport problem therefore involves both an external and an embedded source of ionization and excitation. Individual electrons are not labelled, of course, as being degraded primaries or fresh secondaries and measurements of electron intensities cannot distinguish between them. Indeed, the total number flux for a given event may be larger within the atmosphere than at the top but the energy spectra would be very different. The cross section for excitation of discrete states is

$$\sigma_{\text{in}}^{\text{d}}(E',E;\mu',\mu) = \sum_j P_j^{\text{d}}(E;\mu',\mu)\delta(E' - E - W_j)\sigma_j(E) \tag{3.2.3}$$

where W_j is the excitation threshold energy of state j and the meaning of the other symbols is analogous to those in Equation (3.2.2). Since the energy of the degraded primary electron, E_p, in an ionizing collision is generally not the same as the energy of the secondary electron, E_s, it is convenient to define separate cross sections for ionization and for production of secondary electrons, realizing that the two processes occur simultaneously. Thus, for ionization

$$\sigma^{\mathrm{I}}_{\mathrm{in}}(E'_p, E_p; \mu', \mu) = \sum_j P^{\mathrm{I}}_j(E_p; \mu', \mu) \frac{\mathrm{d}\sigma}{\mathrm{d}W}(E'_p, W = E'_p - E_p) \tag{3.2.4}$$

and for secondary electron production

$$\sigma^{\mathrm{s}}_{\mathrm{in}}(E'_p, E_s; \mu', \mu) = \sum_j P^{\mathrm{s}}_j(E_s; \mu', \mu) \frac{\mathrm{d}\sigma}{\mathrm{d}W}(E'_p, W = E_s + I) \tag{3.2.5}$$

The energy loss, W, is the sum of the ionization potential, I, and the energy of the secondary electron, E_s. Energy conservation requires that $E_p + E_s + I = E'_p$. The differential cross section, $\mathrm{d}\sigma/\mathrm{d}W$, is used for continuous energy loss processes. Not only is the energy not generally the same for the degraded primary and for the secondary electrons but the phase functions, P^{I} and P^{s}, that specify the angular scattering are also different, the primaries being scattered preferentially in the forward direction while the secondaries are ejected isotropically. The cross section for molecular dissociation, $\sigma^{\mathrm{D}}_{\mathrm{in}}$, has the same form as the ionization term, Equation (3.2.4), but the numerical values of the differential cross section, the energy loss and the threshold energies are not the same. This cross section only applies to molecular species and does not result in the production of secondary electrons. Another process, dissociative ionization of molecules, is a source of secondary electrons and is included in Equation (3.2.5). The total scattering cross section is then

$$\sigma_{\mathrm{tot}} = \sigma_{\mathrm{e}} + \sigma^{\mathrm{d}}_{\mathrm{in}} + \sigma^{\mathrm{I}}_{\mathrm{in}} + \sigma^{\mathrm{D}}_{\mathrm{in}} \tag{3.2.6}$$

We can now define a scattering depth, analogous to the optical depth,

$$\tau = \sum_j \sigma_{\mathrm{tot}} \int_z^{\infty} n_j(z)\mathrm{d}z \tag{3.2.7}$$

in terms of the altitude, z, and the densities of species $n_j(z)$. The geometry is shown in Figure 3.2.1.

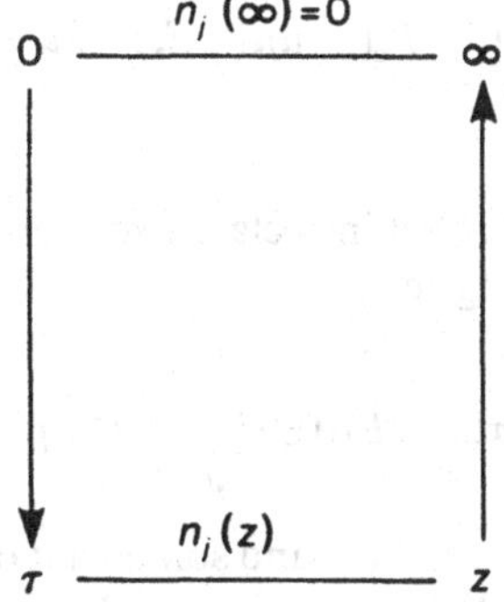

Fig. 3.2.1 The scattering depth τ is a function of the height varying number density of species $n_j(z)$ as defined by Equation (3.2.7).

In the upper atmosphere the electron density is always considerably less than the neutral gas density and the precipitating energetic electrons therefore collide primarily with the neutrals. As the primary and secondary beam electrons continue to lose energy, they are left in an energy regime just above the thermal population; the term suprathermal electrons is applied to them. The number of inelastic collision processes in the suprathermal regime is limited to excitation of low vibrational levels and rotational levels in molecules and fine structure levels in O. Consequently, the energy of suprathermal electrons is shared with the less abundant ambient thermal electrons by elastic collisions, resulting in an increase of the electron temperature. This topic will be discussed in Chapter 6; in the electron transport problem, electron–electron collisions represent a continuous energy loss process and do not involve an angular dependence. The expression for the rate of energy transfer from suprathermal to thermal electrons is

$$-n_e(\partial/\partial E)[L(E)I(\tau,E,\mu)] \tag{3.2.8}$$

where n_e is the ambient electron density and $L(E)$ is the stopping cross section (eV cm^2). The derivations of Expression (3.2.8) and of the stopping cross section are given in references listed at the end of the chapter.

The various scattering processes described above do not allow us to use the simple Lambert–Beer law for electron transport. $I(\tau,E,\mu)$ is the intensity of electrons at scattering depth τ, with energy ΔE about E, and polar angle with cosine $\Delta\mu$ about μ with respect to the local vertical. The change in intensity with scattering depth, $\mathrm{d}I/\mathrm{d}\tau$, involves elastic scattering of electrons with energy E, inelastic scattering of electrons with energy greater than E into the ΔE bin about E, production of electrons at energy E, elastic scattering with neutrals and ions both into and out of $\Delta\mu$ about μ, and electron–electron collisions. The transport equation therefore should have not only attenuation terms but also source terms. The source terms include primary beam electrons degraded from a higher energy and newly created secondary electrons at a specified energy. These processes may be expressed mathematically by the following one-dimensional equation,

$$\mu\frac{\mathrm{d}I}{\mathrm{d}\tau} = I(\tau,E,\mu) - \frac{n_e}{\sum_j \sigma_{\mathrm{tot}} n_j}\frac{\partial}{\partial E}[L(E)I(\tau,E,\mu)$$

$$-\frac{\sum_j \sigma_e n_j}{\sum_j \sigma_{\mathrm{tot}} n_j}\int_{-1}^{+1} P_e(\mu',\mu)I(\tau,E,\mu')\,\mathrm{d}\mu' - \frac{1}{\sum_j \sigma_{\mathrm{tot}} n_j}Q(\tau,E,\mu) \tag{3.2.9}$$

The source term has been separated into elastic scattering, the third term on the right side, and the inelastic scattering term,

$$Q(\tau,E,\mu) = \int \sigma_{\mathrm{in}}(E',E)\,\mathrm{d}E'\cdot\tfrac{1}{2}\int_{-1}^{+1} P_{\mathrm{in}}(\mu',\mu)I(\tau,E',\mu')\,\mathrm{d}\mu' \tag{3.2.10}$$

which actually represents three separate and somewhat different source terms, one each for excitation of discrete states, ionization and dissociation, and production of secondary electrons.

We have written the electron transport equation on the basis of physical arguments only, adhering to our aim in this book of not devoting too many pages to a primarily mathematical development. The reader is urged, however, to consult the references cited at the end of the chapter in which the electron transport equation is derived from the continuity equation for the electron distribution function. The relationship between the electron transport equation and the Boltzmann equation is also shown. Equation (3.2.9) rests on basic physical principles. The equation is applicable to the transport of both photoelectrons and auroral electrons. Photoelectrons are primarily an embedded source in the atmosphere but at times there is also a small photoelectron flux from the geomagnetically conjugate hemisphere which becomes an external intensity source. Auroral electron precipitation is primarily an external source with the secondary electrons becoming the embedded source. In Section 3.1 examples of observations are presented in two different units, the intensity $I(\mathrm{cm}^{-2}\,\mathrm{s}^{-1}\,\mathrm{sr}^{-1}\,\mathrm{erg}^{-1})$ or the distribution function $f(\mathrm{cm}^{-6}\,\mathrm{s}^3)$. These are related by

$$I(\mathbf{x},E,\mathbf{\Omega},t)=\frac{v^2}{m}f(\mathbf{x},\mathbf{v},t) \qquad (3.2.11)$$

where $\mathbf{x}$ is the generalized position vector, $\mathbf{\Omega}$ is a unit vector in the direction of the velocity $\mathbf{v}$, E is the energy and m is the electron mass.

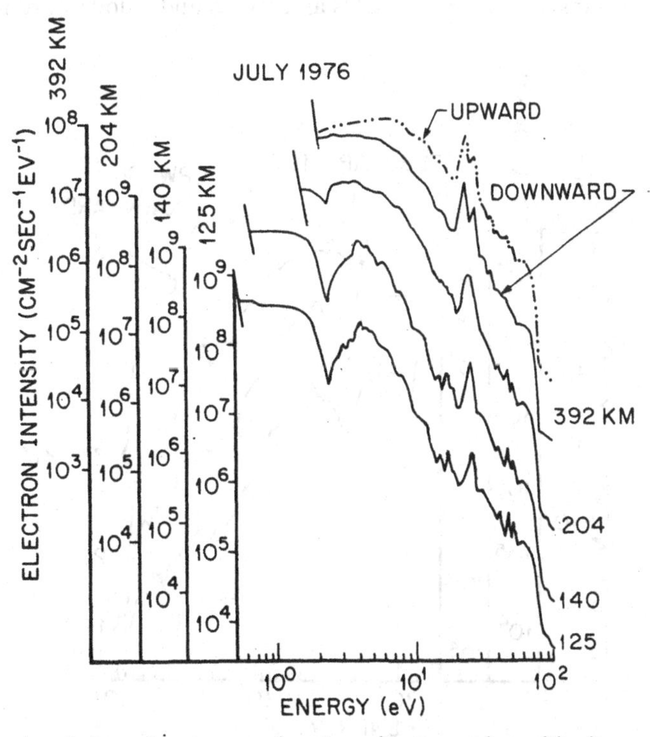

Fig. 3.2.2 Intensity of photoelectrons as a function of energy at four altitudes computed with a two-stream solution of the transport equation. The escape flux at 392 km, labelled upward, is shown by a dash–double dot line. Representative solar minimum UV fluxes yielding an imbedded source of photoelectrons were adopted in the computations.

Even a casual perusal of the papers on electron transport theory reveals that solution of the integro-differential Equation (3.2.9) in a multicomponent atmosphere is a non-trivial problem. Nevertheless, it is the intensity $I(\tau, E, \mu)$ that is basic for testing our understanding of the physical processes that govern photoelectron and auroral electron transport because it is the parameter which is measured by rocket and satellite borne detectors. The intensity is also the quantity that enables us to compute ionization rates (analogous to photoionization, Equation (2.5.1)), dissociation rates, excitation rates, and electron heating rates. Therefore, considerable effort has been and still is being devoted to obtaining adequate solutions to the electron transport equation. To date, all available solutions require some approximations and large numerical computational schemes. Monte Carlo type solutions have also been carried out.

We end this section by showing the results of computations using the electron transport equation. The production rates of photoelectrons for solar minimum (July 1976) and solar maximum (February 1979) conditions were derived in Section 2.5 and the results were shown at four altitudes in Figure 2.5.1.

The angular distribution of photoelectrons produced by solar UV photons is nearly isotropic so that, initially, there are as many photoelectrons ejected into the upward hemisphere as in the downward hemisphere. Elastic scattering causes the photoelectrons to undergo deflections in angle. If it is assumed that the electrons are scattered into either the upward or downward hemisphere with specified probabilities then the solution to the transport equations yields an upward and a downward intensity only.

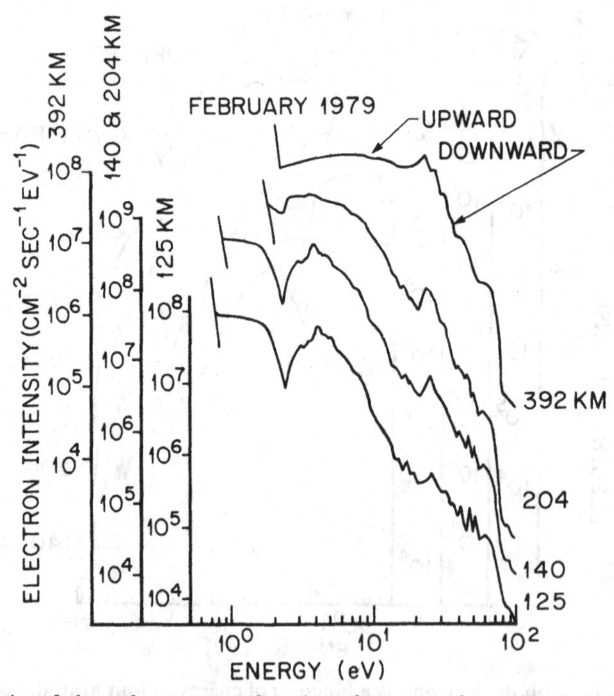

Fig. 3.2.3 Intensity of photoelectrons as a function of energy at four altitudes. The escape flux at 392 km is labelled upward. Solar maximum UV fluxes measured on 19 February 1979 by the Atmosphere Explorer satellite were adopted in the computations.

This has been called the two-stream solution to the electron transport equation. Production rates are not measurable quantities but are a source term in the transport equation. The computed photoelectron intensities are shown in Figures 3.2.2 and 3.2.3 at four altitudes as a function of energy. The downward and upward intensities are identified, but only at the highest altitude do they become distinguishable on the scale of the figure. Zero conjugate photoelectron flux was assumed in the computation with the upward directed hemispherical intensity therefore exceeding the downward intensity at this altitude. An example of auroral electron intensities is shown in Figure 3.2.4. The input spectrum, shown on the panel labelled 'downward flux', was measured by an electron spectrometer on board the Dynamics Explorer 2 satellite at 454 km. The large intensity at the low energy end of the distribution is due to copious production of secondary electrons. With decreasing energy the intensity eventually merges with the thermal electron distribution, i.e., the high energy tail of a Maxwellian distribution at a few thousand degrees Kelvin. This appears in the figure at the 454 km level of the upward flux. The structure that appears in the flux as a function of energy is due to a multitude of inelastic scattering processes. The upward directed flux at 454 km represents an escape flux to the conjugate hemisphere. Primary auroral electrons enter at the 'top of the atmosphere' and represent an external source, i.e., a boundary condition to the transport equation. Secondary, tertiary, etc. electrons are an imbedded source that appears in the quantity Q in Equation (3.2.10).

Theory and experiments show that electrons are scattered over a wide range of angles. Auroral measurements likewise show a diversity of anisotropic distributions. Indeed, the intensity $I(\tau, E, \mu)$ is an explicit function of the scattering depth, τ, (or the altitude) energy, E, and pitch angle, μ, (assuming azimuthal symmetry). In practice, the transport equation is solved for a preselected number of streams to represent scattering in a finite number of directions. The lowest approximation is that of assuming two streams only, a downward and an upward stream. Multistream solutions are necessary whenever more detailed angular information is required.

3.3 Electron energy deposition in the atmosphere

In the course of their passage through the atmosphere photoelectrons and auroral electrons lose energy by a variety of inelastic collisions identified by general type in the preceding section. These collision reactions are discussed in more detail in subsequent sections because they are responsible for the physical and chemical changes in the atmosphere that are a consequence of electron impact, including the production of optical emissions (UV, visible, IR, X-rays). In this section we are concerned with the overall energy loss suffered by the electrons, independent of details, which is called the rate of energy deposition at various height levels in the atmosphere. It is given by

$$\varepsilon(z) = \sum_j \sum_k n_j(z) \int_{W_{j,k}}^{\infty} W_{j,k} \sigma_j^k(E) I(E, z) \, \mathrm{d}E \tag{3.3.1}$$

The summation extends over all inelastic processes, k, of all species, j, for which the cross sections are $\sigma_j^k(E)$ and the energy loss is $W_{j,k}$. The electron intensity, $I(E,z)$, has been integrated over all directions and the species concentrations are denoted by $n_j(z)$. Integration over altitude gives the total energy deposited within the atmosphere by the incoming electron or photon flux less the outgoing (or backscattered) energy flux.

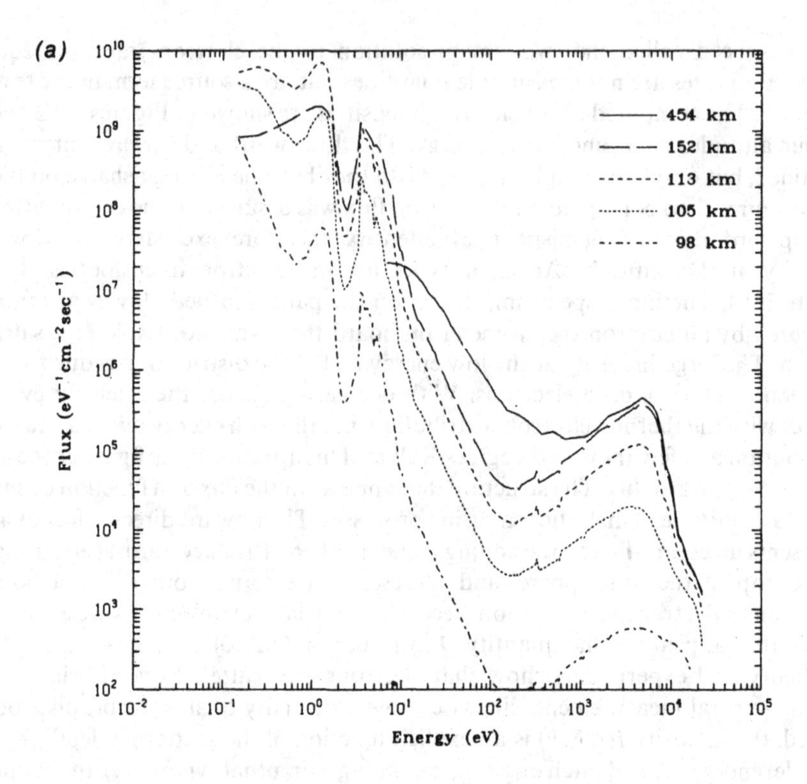
(a)
454 km
152 km
113 km
105 km
98 km
Flux (eV^{-1}cm^{-2}sec^{-1})
Energy (eV)

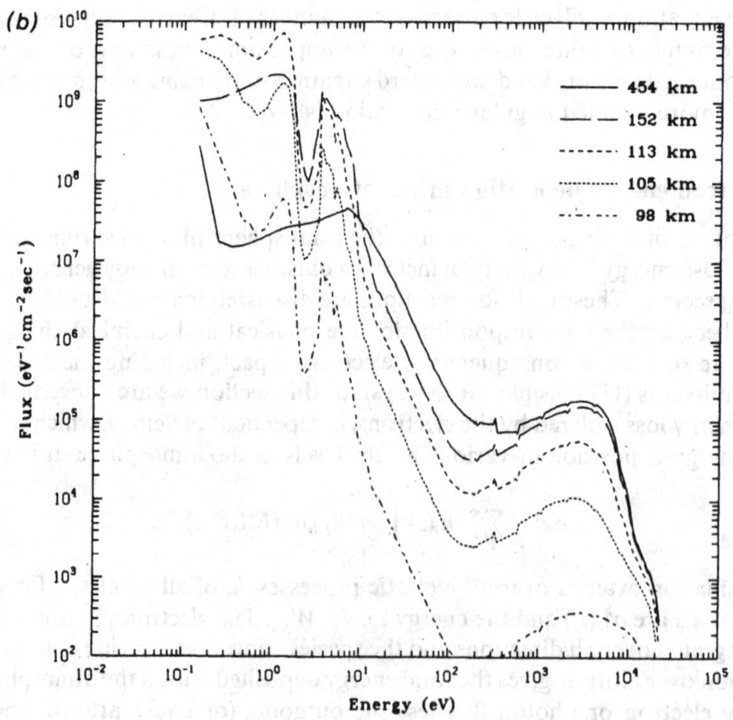
(b)
454 km
152 km
113 km
105 km
98 km
Flux (eV^{-1}cm^{-2}sec^{-1})
Energy (eV)

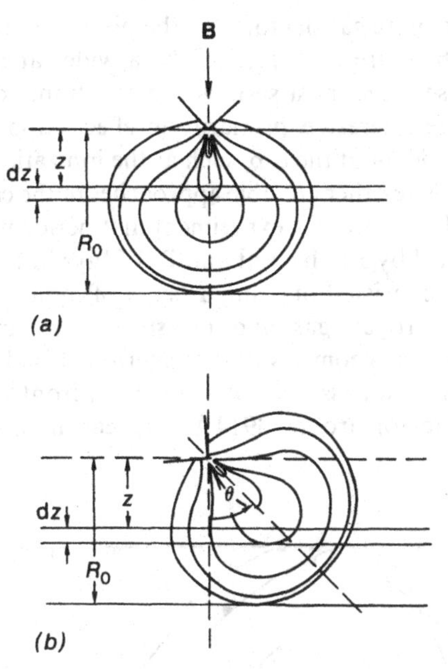

Fig. 3.3.1 (*a*) Laboratory measurement (Grün, 1957) of N_2^+ emission intensity contours produced by monoenergetic, unidirectional beams of electrons in air. In a magnetic field, *B*, the electron paths are confined to helices around the dashed line. (*b*) Same as (*a*) but with direction of incidence at angle θ to *B*. The beam would still be confined close to the dashed line if a field is present. (A. Vallance-Jones, *Aurora*, Fig. 4.16 p. 100, D. Reidel Publ., 1974; A.E. Grün, *Zeitschrift fur Naturforschung*, **12a**, 89, 1957.)

Energy conservation must prevail. We note that as far as local energy deposition (and all processes which are included therein) is concerned the angular distribution of intensity is not required. After identifying all the inelastic cross sections of all states for every species it would be a straightforward procedure to compute the energy deposition rate using Equation (3.3.1), if the electron intensity were known as a function of energy and altitude in the atmosphere. We concluded in the preceding section, however, that the intensity is not easily computed, although we should qualify this statement by noting that the approximations that are generally made to obtain a solution to the

---

Fig. 3.2.4 Auroral electron intensities as a function of energy at five altitudes computed with a two-stream solution of the transport equations. The spectrum measured by the Dynamics Explorer 2 satellite at 454 km was adopted as the external source (or boundary condition to the solution); it is shown as the downward flux (*a*) at 454 km assuming angular isotropy. The low energy cutoff of the electron spectrometer was about 7 eV. Many inelastic scattering processes contribute to the structure in the curves, the most prominent at about 2 eV being due to vibrational excitation of N_2, discussed in Chapters 4 and 6. The upward flux (*b*) at 454 km represents an escape flux at that level.

transport equation are quite satisfactory for the photoelectron case. It is the auroral electron intensities, which are characterized by a wide range of energy and angular distributions, that present the most serious computational difficulties.

Suppose that we are interested primarily in effects upon the atmosphere of the auroral electron flux incident at the top, such as the ionization that is produced or the optical emissions. Are there other possible approaches to the calculations? One method makes use of results of a laboratory experiment first performed by A.E. Grün in 1957 and extended and refined by J.L. Barrett and P.B. Hays in 1976 (see Figures 3.3.1 and 3.3.2). The experiment consisted of firing a beam of monoenergetic electrons into a chamber filled with nitrogen gas and measuring the luminosity in the region surrounding the orifice. The geometry of the experiment is illustrated schematically in Figure 3.3.1. A narrow bandpass filter was placed in front of the optical detector to isolate the wavelength region around 3914 Å. The reason for selecting this wavelength

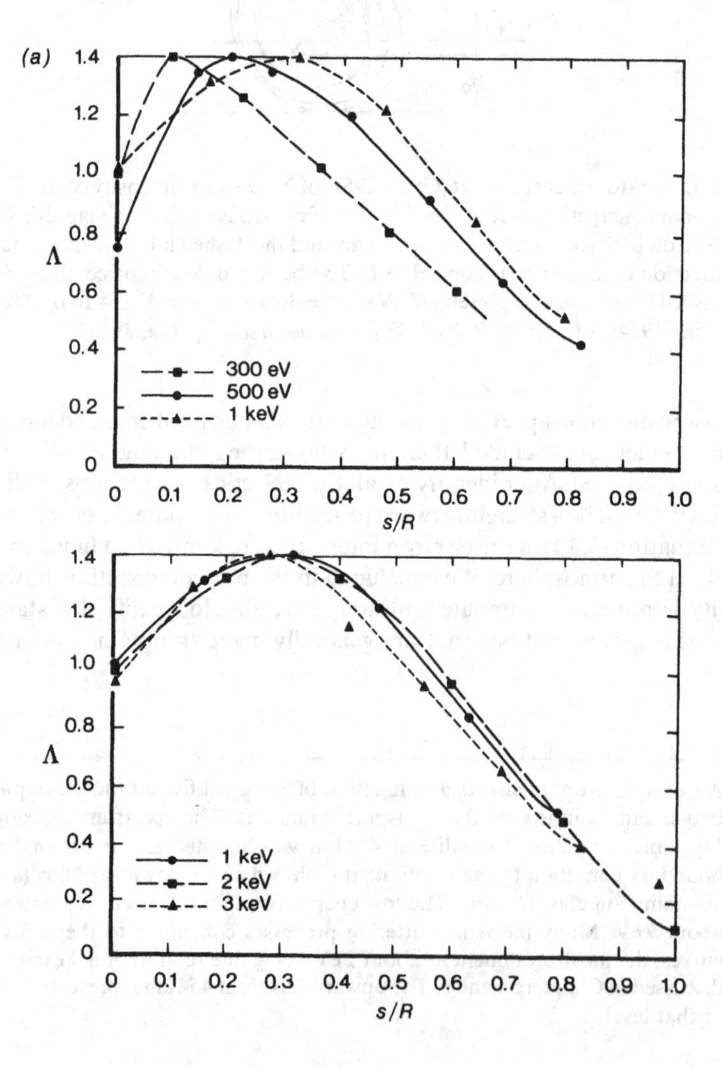

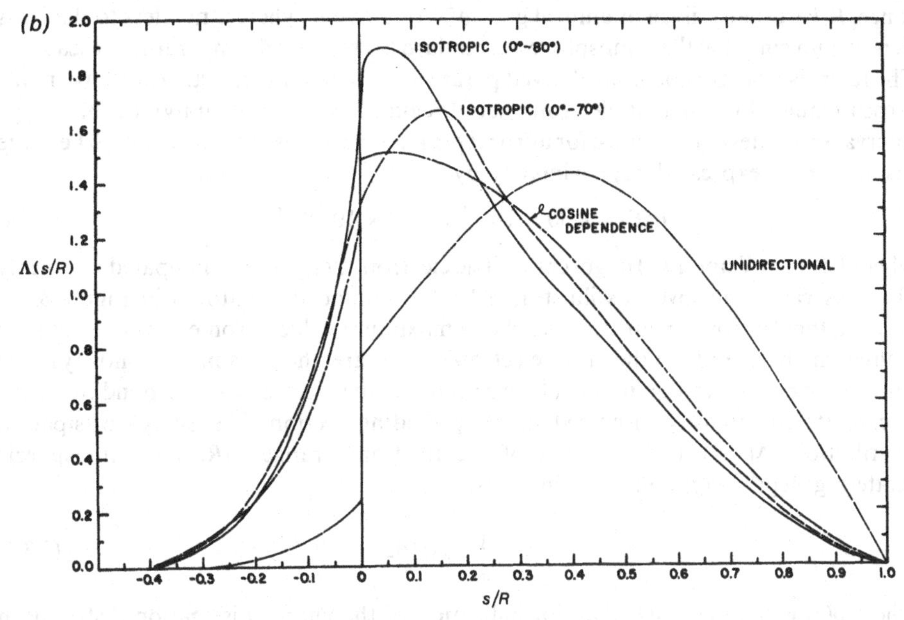

Fig. 3.3.2 (*a*) Normalized curves of energy deposition for different values of the electron beam energy. (J.L. Barrett and P.B. Hays, *J. Chem. Phys.*, **64**, 743, 1976.) (*b*) Energy dissipation distribution function for four angular dispersions of the incident electron stream normalized according to Equation (3.3.6). (M.H. Rees, *Planet. Space Sci.*, **11**, 1209, 1963.)

requires that we anticipate here some topics that will be treated at length in subsequent chapters.

A small fraction of molecular ions produced by electron impact ionization of N_2 will be in the excited electronic state $B^2\Sigma$. In the following chemical equation the primary electron is denoted by e* and the secondary electron by e,

$$e^* + N_2 \rightarrow e^* + N_2^+(B^2\Sigma) + e \qquad (3.3.2)$$

Decay to the ground state of the ion is by an allowed transition, and the resulting emission is in the First Negative (1 NG) system. The $v' = 0$ to $v'' = 0$ vibrational band emits at 3914 Å.

$$N_2^+(B\,^2\Sigma, v' = 0) \rightarrow N_2^+(X\,^2\Sigma, v'' = 0) + h\nu \qquad (3914\,\text{Å}) \qquad (3.3.3)$$

The emission cross section for the (0, 0) transition, the rate of formation of ion pairs per 3914 Å photon and, most importantly, the energy lost by the incident electron per ion pair formation are all well-established experimental parameters. Isophotes of the luminosity contours of the 3914 Å emission rate shown in Figure 3.3.1 are directly proportional to the ionization rate and the energy deposition rate. The variation of 3914 Å emission rate with distance from the orifice, the source, is a measure of the distribution of energy deposition along an effective path that includes implicitly all the elastic and inelastic scattering processes discussed in the preceding section that pertain to N_2 molecules. An electron with initial energy E at the source describes a convoluted path until it is finally stopped at a distance from the source identified as the effective range, $R(E)$, usually given in units of gm cm^{-2} or atm cm, where atm refers to the mass density (gm cm^{-3}) of the atmosphere at standard pressure and temperature at sea level. The range is an experimentally derived parameter and several investigators have made measurements for different electron energy regimes (see bibliography). In the energy interval of greatest importance for auroral primary electrons, 200 eV $< E <$ 50 keV, the results can be expressed approximately by

$$R(E) = 4.30 \times 10^{-7} + 5.36 \times 10^{-6} E^{1.67} \tag{3.3.4}$$

where E is in keV and $R(E)$ in gm cm^{-2}. The electron energy is not dissipated uniformly along its range, however, as illustrated by the luminosity contours in Figure 3.3.1. Instead, the distribution function reaches a maximum value at some distance from the source which depends on the initial electron energy, and there is some luminosity in the direction opposite to the initial velocity vector of the electrons, corresponding to the backscattered intensity discussed in the preceding section. The energy dissipation distribution, $\Lambda(s/R)$, is a function of the fractional range, s/R. The atmospheric scattering depth, s (gm cm^{-2}), is given by

$$s = \int_z^\infty \rho(z)\mathrm{d}z \tag{3.3.5}$$

where $\rho(z)$ is the mass density. Normalization of the energy dissipation distribution function

$$\int_{-1}^{+1} \Lambda(s/R)\,\mathrm{d}(s/R) \equiv 1 \tag{3.3.6}$$

conserves energy. The curves in Figure 3.3.2 (*a*) apply to an electron beam directed perpendicular to the source plane, corresponding to a unidirectional intensity normal to the atmosphere. Since auroral electrons frequently have an isotropic angular distribution the $\Lambda(s/R)$ function is shown for this and other angular distributions in Figure 3.3.2(*b*). The energy deposition (eV cm^{-3} s^{-1}) for monoenergetic electrons at energy, E_p, can be expressed by

$$\varepsilon(z, E_p) = q(z)\Delta\varepsilon_{\text{ion}} = FE_p\Lambda\left(\frac{s}{R}\right)\frac{\rho(z)}{R(E_p)} \tag{3.3.7}$$

F is the electron flux (cm^{-2} s^{-1}), E(eV) is the electron energy, $\rho(z)$ is the mass density, $R(E_p)$ is the range, and $\Lambda(s/R)$ is the energy dissipation function. The energy deposition rate is the product of the ionization rate, $q(z)$, and the energy loss per ion pair formation, $\Delta\varepsilon_{\text{ion}}$. For species in the upper atmosphere, mainly composed of N_2, O_2 and O, it is found experimentally that $\Delta\varepsilon_{\text{ion}}$ has about the same value for the major species, 37 eV and 33 eV for N_2 and O_2 respectively. It has been common practice to adopt a value $\Delta\varepsilon_{\text{ion}} = 35$ eV in computing the energy deposition rate and ionization rate using the method described above. It is assumed that $\Delta\varepsilon_{\text{ion}}$ has about the same value for O as for the molecular species, but the appropriate measurement has not been performed. These values apply only to electrons with energy above about 70 eV; at lower energies the loss per ion pair increases sharply. The reason for this will emerge from a discussion of cross sections in Chapter 4. The energy dissipation function, $\Lambda(s/R)$, was obtained in air ($\frac{4}{5}N_2, \frac{1}{5}O_2$) by Grün and in N_2 by Barrett and Hays. There appears to be little difference between the two results and little error is introduced by adopting the same Λ curve throughout the atmosphere. Dependence on electron energy below about 1 keV is significant and cannot be ignored (Problem 7).

The laboratory experiments described above and the electron transport equation derived in the preceding section did not consider the possible effects of a magnetic field on the motion of the electrons, while auroral electrons are guided into the atmosphere by the earth's magnetic field. In the auroral zone where the lines of force are nearly perpendicular to the horizontally stratified atmosphere the magnetic field will not affect the distribution function appreciably. The velocity component parallel to the magnetic field remains unaffected while the orthogonal component of electron motion only suffers a change in direction within a given layer of the atmosphere and there is neither gain nor loss of energy involved since the magnetic field is nearly uniform in the height interval considered. This is equivalent to unwinding the spiral of the electrons' motion leaving the total range unaffected. By contrast, in the magnetosphere electromagnetic forces dominate the motion of charged particles.

For an extended source, Equation (3.3.7) therefore yields the energy deposition rate, $\varepsilon(z, E_p)$, as a function of altitude for monoenergetic electron precipitation, within the range of validity of $\Delta\varepsilon_{ion}$, and available Λ functions. The total ionization rate is then

$$q(z, E_p) = \varepsilon(z, E_p)/\Delta\varepsilon_{ion} \quad (\mathrm{cm}^{-3}\,\mathrm{s}^{-1}) \tag{3.3.8}$$

Altitude profiles of the ionization rate produced by monoenergetic, unidirectional (field-aligned) beams of electrons are shown in Figure 3.3.3. A dependence of $q(z)$ on the density profile of the atmosphere (Equation (3.3.7)) leads to some variability in the

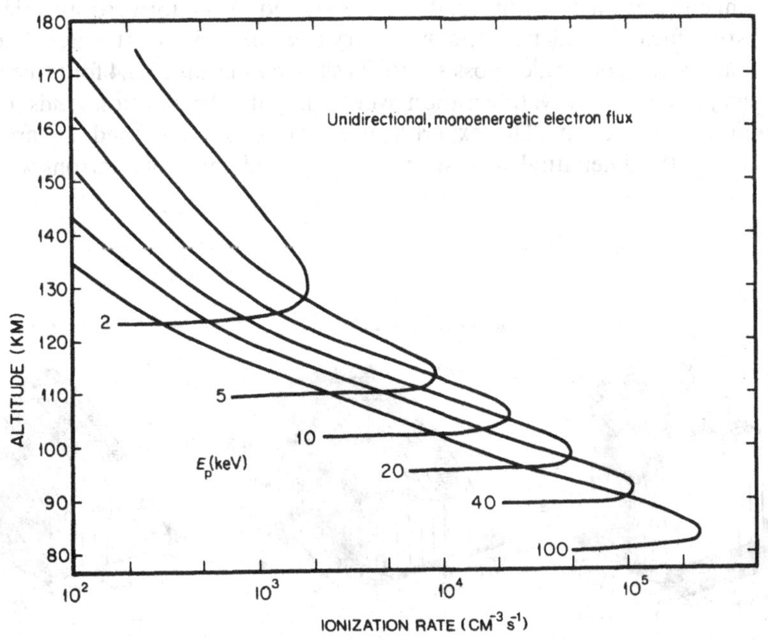

Fig. 3.3.3 Altitude profiles of the ionization rate due to a flux of 10^8 electrons $\mathrm{cm}^{-2}\,\mathrm{s}^{-1}$ at several initial values of energy, E_p (keV), precipitating along magnetic field lines into the Earth's atmosphere.

numerical results. The equations for monoenergetic electrons may be convolved with any arbitrary energy spectrum of electron precipitation to obtain energy deposition and ionization rates for observed or model spectra. A distribution of the form

$$N(E)\mathrm{d}E = E^{\gamma}\exp(-E/E_0)\mathrm{d}E \qquad (3.3.9)$$

has been found to represent closely many observed primary auroral spectra. An appropriate choice of the parameters γ and E_0 yields power laws, exponential variations or Maxwellian distributions. The ionization rate is

$$q(z) = \int_0^\infty q(z,E)N(E)\mathrm{d}E \qquad (\mathrm{cm}^{-3}\,\mathrm{s}^{-1}) \qquad (3.3.10)$$

Every ionization event gives rise to a positive ion and a secondary electron, but what is the energy of this newly produced electron? The question did not arise in the case of photoelectrons (Equation (2.5.1)) because the ionizing photon is 'destroyed' in the process leaving the electron to acquire all the excess kinetic energy. Secondary electrons are not produced at a single energy; instead, there is a distribution with energy given by a cross section, or a probability. We again make use of results of laboratory experiments, see references at the end of the chapter. The experiment consisted of firing a monoenergetic electron beam into a gas and measuring the differential cross section of the scattered and newly produced electrons. As mentioned previously, electrons are not labelled, and one can only distinguish between scattered beam electrons and newly produced secondary electrons by noting that the latter are most numerous at energies much less than the initial beam energy and that secondary electrons are emitted at all angles while the beam electrons tend to be scattered in the forward direction. These criteria are sufficient to identify the secondary electrons with confidence. The double differential (energy and angle) cross section is shown in Figure 3.3.4 for three values of the primary electron energy. Integration over the angular distribution yields the (single) differential cross section. The experimental results can be fitted by an analytic expression for the differential cross section for the production of electrons with energy

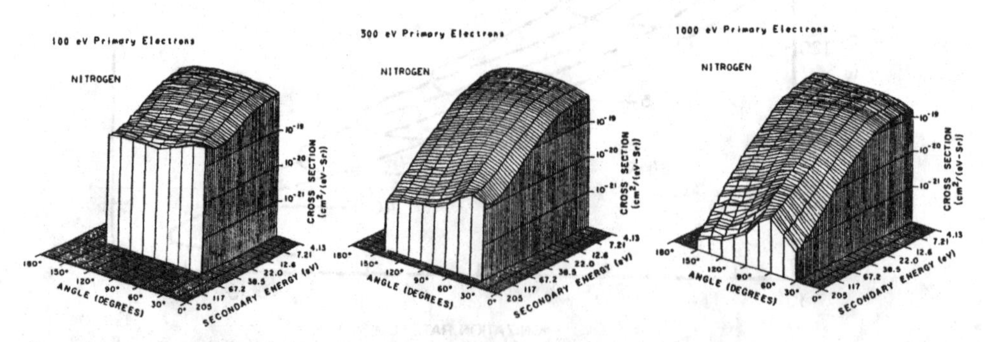

Fig. 3.3.4 Double differential cross section for the production of secondary electrons by impact ionization of primary electrons on N_2. Results for three values of primary electron energy are shown. (C.B. Opal, E. Beatty and W. Peterson, *Atomic Data*, **4**, 209, 1972.)

E_s by incident beam electrons with energy E'_p

$$\frac{d\sigma}{dW}(E'_p, E_s) = \frac{\sigma^I(E'_p)(1 + E_s/\bar{E})^{-2.1}}{\bar{E}\tanh[(E'_p - I)/2\bar{E}]} \qquad (\text{cm}^2\,\text{eV}^{-1}) \tag{3.3.11}$$

where $\bar{E}$ is a spectral shape parameter, I is the ionization potential, and $\sigma^I(E'_p)$ is the ionization cross section. The experiment was performed in several gases. The two relevant to the upper atmosphere are N_2 for which $\bar{E} = 13.0\,\text{eV}$, $I = 15.6\,\text{eV}$, and O_2 for which $\bar{E} = 17.4\,\text{eV}$, $I = 12.2\,\text{eV}$. The ionization cross section, $\sigma^I(E)$, has been measured experimentally by several investigators, and the relationship

$$\sigma^I(E'_p) = \int_0^{E'_p} (d\sigma/dW)(E'_p, E_s)\,dE_s \qquad (\text{cm}^2) \tag{3.3.12}$$

establishes the absolute magnitude of the differential cross section. Numerical values of ionization cross sections are listed in Appendix 4.

A secondary electron production experiment in O, the third major species in the upper atmosphere, has not yet been performed in the laboratory. Thus, we draw on theoretical work for the differential cross section of O. The general framework is the first approximation of the Born theory for inelastic collisions expressed in terms of a generalized differential oscillator strength and a theoretical relationship between this quantity and the photoionization and electron impact ionization cross sections. (This cryptic sentence is elaborated in Chapter 4 and in the references listed at the end of the chapter.) The differential cross section can be fitted by an empirical formula,

$$\frac{d\sigma}{dW}(E'_p, E_s) = \frac{\sigma^I(E'_p)}{WA(E'_p)}\exp\left[-\frac{W}{31.5} - 339\exp\left(-\frac{W}{2.49}\right)\right]$$
$$\times \ln\left[\frac{E'^{\frac{1}{2}}_p + (E'_p - W)^{\frac{1}{2}}}{E'^{\frac{1}{2}}_p - (E'_p - W)^{\frac{1}{2}}}\right] (\text{cm}^2\,\text{eV}^{-1}) \tag{3.3.13}$$

where $W = I + E_s$ and I is the ionization potential of O, 13.6 eV. The normalization factor, $A(E'_p)$, follows from the requirement that the integral over energy of the differential cross section equals the total ionization cross section,

$$\sigma^I(E'_p) = \int_I^{E'_p} (d\sigma/dW)(E'_p, W)\,dW \qquad (\text{cm}^2) \tag{3.3.14}$$

For $E'_p \gg W$, which is generally the case in aurorae, the normalization factor varies little with energy. The total ionization cross section is obtained from laboratory experiments so that the Bethe approximation in the limit of no momentum change of the colliding electron is used only to determine the shape of the differential cross section.

Since the differential cross sections for the production of secondary electrons are not the same for the three major species, N_2, O_2 and O, it becomes necessary to compute the ionization rate of each type from the total ionization rate $q(z)$ given by Equation (3.3.8). The partitioning depends on the relative abundance of the constituents, which is a function of altitude, and the relative magnitude of the effective ionization cross sections, averaged over energy. The latter parameters may be obtained from the cross sections given in Appendix 4 while the number densities are given by the adopted model atmosphere, Appendix 1. The expressions given below already incorporate factors that take account of the ionization cross sections and we have

$$q_z(\mathrm{N_2}) = q_z \frac{0.92\, n_z(\mathrm{N_2})}{0.92\, n_z(\mathrm{N_2}) + n_z(\mathrm{O_2}) + 0.56\, n_z(\mathrm{O})} \quad (\mathrm{cm^{-3}\,s^{-1}}) \tag{3.3.15}$$

with corresponding expressions for $q_z(\mathrm{O_2})$ and $q_z(\mathrm{O})$. Denoting the species by j, the altitude and energy dependent production rate of secondary electrons is given by

$$P_z(E_s, E_p) = \sum_j \frac{\mathrm{d}\sigma_j/\mathrm{d}W}{\sigma_j^{\mathrm{I}}} q_j(z, E_p) \quad (\mathrm{cm^{-3}\,s^{-1}\,eV^{-1}}) \tag{3.3.16}$$

This is analogous to the production rate of photoelectrons but results from rather different physical processes. Once produced, secondary electrons and photoelectrons become indistinguishable and auroral electron precipitation into a sunlit atmosphere contains both sources. The net production rate is simply the sum of the two which becomes part of the source term in the electron transport equation (Equation (3.2.10)) (Problem 8).

The intensity or flux, not the production rate, is the quantity measured by particle detectors; examples of such observations are shown in Figures 3.1.1–3.1.3. Rigorous computation of the intensity requires solution of the transport equation (Equation (3.2.9)) and we have attempted, in this section, to avoid this complexity. An approximate procedure is adopted, instead, which is valid for secondary electrons but which does not take account of the degraded primary electrons. The secondary electron production spectrum shows that most secondaries have much lower energy than the primaries, a characteristic that is illustrated in Figure 3.3.5 for an incident electron stream with energy $E'_p = 100\,\mathrm{eV}$. In order to calculate the curves shown in Figure 3.3.5, the differential cross section (Equation (3.3.9)) was used with parameters appropriate for $\mathrm{N_2}$. The two populations can therefore be distinguished by their energy, except in the region of overlap. The relative energy of secondary and primary electrons is of principal concern but the angular disposition is also different; primary electrons are

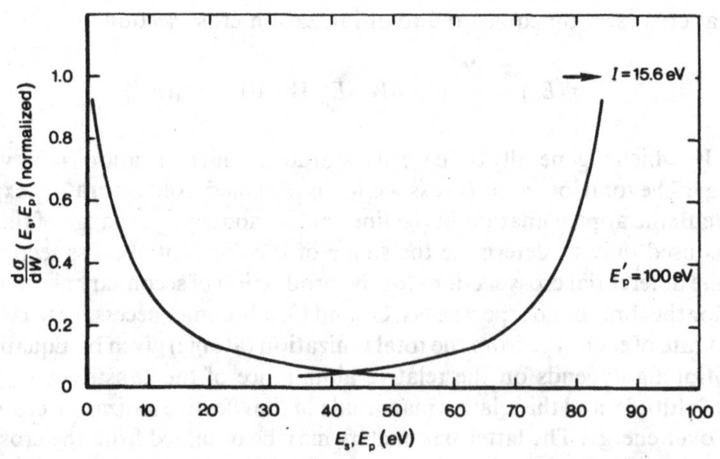

Fig. 3.3.5 Differential cross section for the production of ejected secondary and degraded primary electrons for a 100 eV primary electron flux.

scattered preferentially in the forward direction while secondary electrons are ejected nearly isotropically.

Electrons lose energy through several types of inelastic collisions (described in Section 3.2) with the ambient gas and the energy loss per unit path length is a function of the ionization, dissociation or excitation energy, W_{jk}, the cross section, $\sigma_j^k(E)$, and number density of all species, $n_j(z)$,

$$\mathscr{L}(E,z) = \sum_j \sum_k n_j(z) W_{j,k} \sigma_j^k(E) \qquad (\mathrm{eV\,cm^{-1}}) \tag{3.3.17}$$

Energy is also lost to the ambient electrons, as described in the preceding section. From Equation (3.2.8) the loss per unit path is $n_e(z)L(E,z)\,\mathrm{eV\,cm^{-1}}$. Assuming that the energy of secondary electrons is lost locally, the intensity at energy E_s is given by the production rate of secondaries at all higher energies divided by the loss rate

$$I(E_s, z) = \frac{\int_{E_s}^{\infty} P(E_s, z)\,\mathrm{d}E}{\mathscr{L}(E_s, z) + n_e(z) L(E_s, z)} \qquad (\mathrm{cm^{-2}\,s^{-1}}) \tag{3.3.18}$$

Equation (3.3.18) treats the energy degradation as a continuous process, known as the continuous slowing down approximation. Since inelastic scattering actually results in a finite amount of energy being lost by the electron a more accurate approximation is the so-called discrete energy loss method by which the electron intensity at energy E is computed stepwise from the intensity at energy $E + \Delta E$, adopting energy intervals ΔE that are small compared with differences in the various excitation levels of electronic and vibrational states,

$$I(E,z) = \frac{P(E,z)}{\sum_j \sum_k n_j(z) \sigma_j^k(E)} + \sum_j \sum_k \frac{\sigma_j^k(E + \Delta E)}{\sigma_j^k(E)} I(E + \Delta E, z) \tag{3.3.19}$$

Energy loss to the ambient electron gas is still a continuous process.

Numerical computations have shown that the approximate method given above for computing the secondary electron intensity agrees well with the rigorous solution for primary electrons with energies greater than a few hundred electron volts and secondaries with energies less than 100 eV. In general, the procedure is valid so long as the two populations illustrated in Figure 3.3.5 are well separated and local energy loss by secondaries exceeds transport effects.

To summarize, the procedure outlined in this section yields the intensity of secondary auroral electrons and photoelectrons as a function of energy and altitude but does not give any information about the degraded primary auroral spectrum. We will learn in succeeding chapters that excitation cross sections of many auroral spectral features peak at energies less than about 20 eV so that the secondaries account for much of the optical aurora and the dayglow.

3.4 Interaction of energetic ions and atoms with the atmosphere

3.4.1 Observational evidence

Observations carried out by instruments on board rockets and satellites have shown that energetic H^+ and O^+ ions precipitate into the atmosphere. The magnitude of the

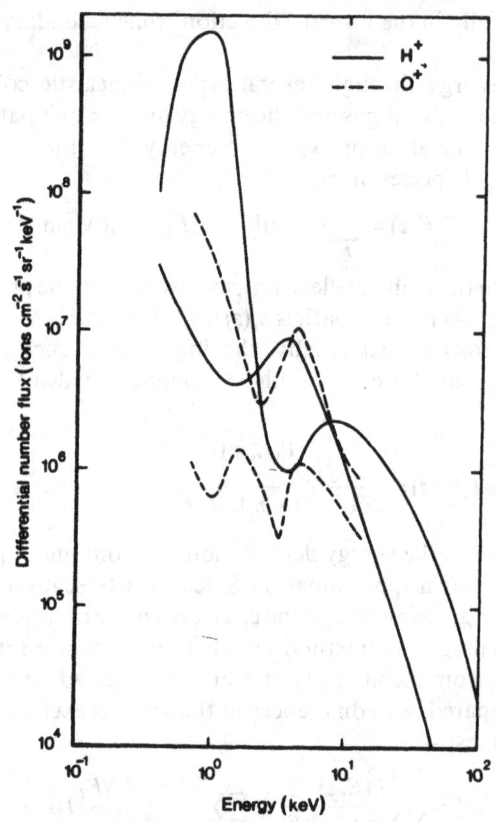

Fig. 3.4.1 Representative energy spectra of precipitating H^+ and O^+ auroral ion fluxes. (D.S. Evans, private communication, 1981; E.G. Shelley, R.G. Johnson and R.D. Sharp, *J. Geophys. Res.*, 77, 6104, 1972.)

fluxes is largest in the auroral region at high latitude but some precipitation has been observed at all latitudes. Typical auroral H^+ and O^+ ion precipitation events are shown in Figure 3.4.1. At low latitudes there is indirect evidence that precipitation is in the form of H and O atoms resulting from neutralization of the respective ions in the magnetosphere. Precipitation is characterized by large temporal and spatial variability. The origin of energetic H^+ and O^+ ions is a current research topic. Both species are transported into the magnetosphere by the polar wind and other transport processes (Chapter 8); they are therefore of ionospheric origin subject to acceleration in the magnetosphere. The solar wind is also a source of energetic protons. Spectroscopic evidence for proton streams precipitating into the atmosphere is discussed in Chapter 7 but optical evidence for energetic O^+ precipitation is neither expected nor observed. The difference emerges from a discussion of the collisional processes that are effective for each species.

There are reports in the research literature of the presence in auroral spectra of emission lines attributable to excited He atoms, presumably due to He^+ ion precipitation. The observations were carried out in the region of the polar cusp where the magnetic field topology favours direct access of the solar wind to the thermosphere.

Although these observations are of considerable interest to solar and magnetospheric studies the contribution of an external source of energetic He^+ ions to thermospheric and ionospheric structure, energetics and dynamics is negligible.

3.4.2 Proton transport in the atmosphere

Penetration of a stream of energetic protons into the atmosphere and interaction with atoms and molecules is described mathematically by a transport equation similar to that adopted for auroral electrons. There is an important difference between the two species, however, stemming from the propensity of protons to undergo charge changing collisions with the neutral gas. The processes are electron capture from an atmospheric species, M,

$$\mathrm{H^+ + M \rightarrow H + M^+} \tag{3.4.1}$$

and ionization-stripping

$$\mathrm{H + M \rightarrow H^+ + M + e} \tag{3.4.2}$$

Two coupled transport equations are therefore required, one for protons and the other for energetic H atoms. The proton intensity, $I_p(s, E, \mu)$, is a function of distance in the atmosphere, s, energy, E, and the cosine of the angle with respect to the magnetic field, $\mu = \cos\theta$, assuming azimuthal symmetry,

$$\begin{aligned} \mu \frac{\partial I_p(s, E, \mu)}{\partial s} &= -\sum_j n_j(s)\sigma^j_{p,\mathrm{tot}}(E) I_p(s, E, \mu) \\ &+ \sum_j n_j(s) \int\!\!\int \sum_k \sigma^j_{p,k}(E', \mu'; E, \mu) I_p(s, E', \mu')\,\mathrm{d}E'\mathrm{d}\mu' \\ &+ \sum_j n_j(s) \int\!\!\int \sigma^j_{01}(E', \mu'; E, \mu) I_H(s, E', \mu')\,\mathrm{d}E'\mathrm{d}\mu' \end{aligned} \tag{3.4.3}$$

The concentration of species j is $n_j(s)$ and only the major neutral constituents, N_2, O_2 and O, need to be included. The differential cross section for protons undergoing process k in species j, $\sigma^j_{p,k}$, is a function of energy and scattering angle. The total cross section, $\sigma^j_{p,\mathrm{tot}}$, is the sum of the cross sections for individual elastic and inelastic collisions at energy E. The ionization-stripping cross section for species j is σ^j_{01}.

The H atom intensity, $I_H(s, E, \mu)$, is

$$\begin{aligned} \mu\, \partial I_H(s, E, \mu)/\partial s &= -\sum_j n_j(s)\sigma^j_{H,\mathrm{tot}}(E) I_H(s, E, \mu) \\ &+ \sum_j n_j(s) \int\!\!\int \sum_k \sigma^j_{H,k}(E', \mu'; E, \mu) I_H(s, E', \mu')\,\mathrm{d}E'\mathrm{d}\mu' \\ &+ \sum_j n_j(s) \int\!\!\int \sigma^j_{10}(E', \mu'; E, \mu) I_p(s, E', \mu')\,\mathrm{d}E'\,\mathrm{d}\mu' \end{aligned} \tag{3.4.4}$$

where the cross sections for individual processes are labelled $\sigma^j_{H,k}$, the total cross section is $\sigma^j_{H,\mathrm{tot}}$, and the electron capture cross section is σ^j_{10}. The last term in each of the above transport equations contains the intensity of the other beam providing the coupling between the ion and neutral species, H^+ and H.

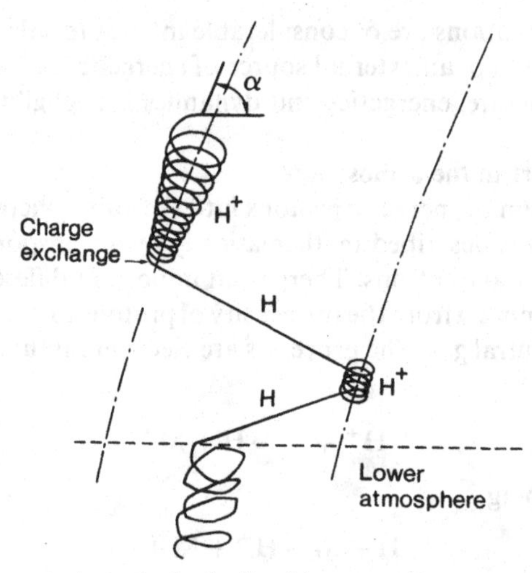

Fig. 3.4.2 H^+ and H atom trajectories in the Earth's atmosphere permeated by the geomagnetic field. (G.T. Davidson, *J. Geophys. Res.*, **70**, 1061, 1965.)

An added complication is due to the proton motion being confined to helical trajectories by the magnetic field while H atoms proceed in the direction acquired on collision, as illustrated in Figure 3.4.2. As a consequence, an initially narrow beam tends to spread out as it penetrates into the scattering atmosphere. The extent of the spreading is determined by the fraction of time that the incident proton spends as a charged particle, H^+, and as a neutral atom, H. An adequate approximation is given in terms of the cross section for charge changing collisions. Therefore, the equilibrium flux fraction of H^+ for species j is given by

$$F_1^j(E) = \sum_j \frac{\sigma_{10}^j(E)}{\sigma_{10}^j(E) + \sigma_{01}^j(E)} \tag{3.4.5}$$

and the fraction of hydrogen atoms is

$$F_0^j(E) = \sum_j \frac{\sigma_{01}^j(E)}{\sigma_{10}^j(E) + \sigma_{01}^j(E)} \tag{3.4.6}$$

where $\sigma_{10}^j(E)$ and $\sigma_{01}^j(E)$ are the energy-dependent electron capture and ionization-stripping cross sections for species j, respectively. The proton energy at the top of the atmosphere determines its depth of penetration and the number of charge changing collisions it undergoes until it is thermalized. The composition of the neutral atmosphere is a function of altitude and the effective flux fraction of H^+ is given by

$$F_1^{\text{eff}}(E) = f(N_2)F_1^{N_2}(E) + f(O_2)F_1^{O_2}(E) + f(O)F_1^{O}(E) \tag{3.4.7}$$

where the $f(M)$s give the fractional composition of the major species. An analogous expression gives the flux fraction of H atoms, $F_0^{\text{eff}}(E)$. The proton and H atoms flux fractions are shown in Figure 3.4.3.

The charge changing collisions, Equations (3.4.1) and (3.4.2), produce a charge

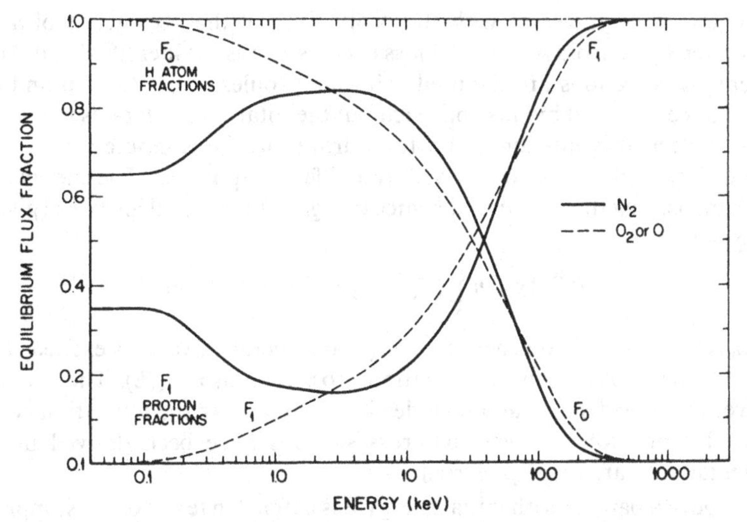

Fig. 3.4.3 Partition between protons and H atoms in a charge equilibrated beam in the Earth's atmosphere as a function of energy. (B. Van Zyl, *U. of Alaska report UAG R-265*, November, 1978.)

equilibrated H^+/H beam that can ionize atmospheric gases,

$$H^+/H + M \rightarrow H^+/H + M^{+*} + e_s \tag{3.4.8}$$

producing an ejected secondary electron, e_s, and leaving a fraction of the ions in excited states, denoted by a superscripted asterisk. Protons are unlikely to produce excitation without ionization but hydrogen atoms are effective projectiles

$$H + M \rightarrow H^* + M^* \tag{3.4.9}$$

The charge equilibrated beam also undergoes elastic, or momentum transfer collisions with the atmospheric atoms and molecules

$$H^+/H + M \rightarrow H'/H + M + \text{k.e.} \tag{3.4.10}$$

sharing kinetic energy and heating the neutral gas. Cross sections for the various reactions between H^+ and H and N_2 and O_2 have been measured in the laboratory over a large energy range while incomplete laboratory data are available for O. A list of cross sections is given in Appendix 4. The transport equation for H^+ and H should therefore be solvable, in principle. To date, however, severe simplifications have had to be adopted to achieve a solution, and different methods have been tried instead. The Monte Carlo method has been applied without simplifying physical assumptions but the many variables that need to be included in following the history of any one proton and H atom requires a very large particle sample to achieve a convergent solution. The Monte Carlo solution and the transport equations yield, in principle, a complete description of the phase space distribution of the H^+/H beam as a function of altitude, leading to energy deposition rates, ionization and excitation rates and heating rates.

If only parameters that are derivable from the energy deposition rate are required then simpler solutions may be adopted. One method uses a detailed accounting of all known energy loss processes with the probability of the occurrence of a specified reaction given by its cross section. All loss processes must be identified and the energy-dependent cross sections are required. Altitude profiles of production and emission rates cannot be obtained by this approach but the total emission rates in many atomic lines and molecular bands due to proton aurora may be computed.

Altitude information may be derived by a different approach. The energy loss by an H^+/H beam passing through unit distance of a gas of number density n is given by the stopping power

$$-\mathrm{d}E/\mathrm{d}x = n\sum_k (E_k - E_0)\sigma_k(E) \qquad (\mathrm{eV\,cm^{-1}}) \tag{3.4.11}$$

where E_0 is the ground state energy and E_k is the energy of various excited states. The energy-dependent excitation or ionization cross section is $\sigma_k(E)$. The summation is taken over all excited states and includes integration over the ionization continuum. Theoretical expressions for various cross sections have been derived but reliable experimental data are to be preferred.

The range of a particle with initial energy E is defined in terms of the stopping power

$$R(E) = \rho\int_0^E -(\mathrm{d}E/\mathrm{d}x)^{-1}\,\mathrm{d}E \qquad (\mathrm{gm\,cm^{-2}}) \tag{3.4.12}$$

where ρ is the mass density of the gas. We have already encountered the range concept in the preceding section in connection with the penetration of electrons into the atmosphere. Based on experimental data, the range of protons in air is given by

$$R(E) = 5.05 \times 10^{-6} E^{0.75} \qquad (\mathrm{gm\,cm^{-2}}),\ 1 < E < 100\,\mathrm{keV} \tag{3.4.13}$$

This empirical relationship is valid over the most important auroral proton energy range. Since energy loss occurs by a large number of discrete reactions, the range of an initially monoenergetic beam will be distributed about the mean range given by Equation (3.4.13), an effect known as straggling.

The mean energy, ε, expended in the production of an ion pair has, experimentally, been found to be nearly constant over a wide range of initial proton energies. The ion production per unit path length is therefore

$$P_\mathrm{i} = -\frac{1}{\varepsilon}\frac{\mathrm{d}E}{\mathrm{d}x} \qquad (\mathrm{cm^{-1}}) \tag{3.4.14}$$

which determines the relationship between the average ionization and the residual range of a large number of particles, known as the Bragg curve. Experiments have shown that ionization (and energy deposition) is largest near the end of the particle range, similar to the electron case (Figure 3.3.3). The normalized energy deposition function for an isotropic flux of protons has been derived from detailed Monte Carlo computations and is shown in Figure 3.4.4 as a function of the normalized scattering depth, s/R (see Equation (3.3.6)). Altitude profiles of the ionization rate by auroral protons as a function of their energy at the top of the atmosphere are shown in Figure 3.4.5.

Auroral proton precipitation is a source of ionization, neutral gas heating, and excitation of both H atoms and atmospheric species. The velocity of the H atoms is sufficiently large to produce a Doppler shift in the H emissions. Indeed, observations of

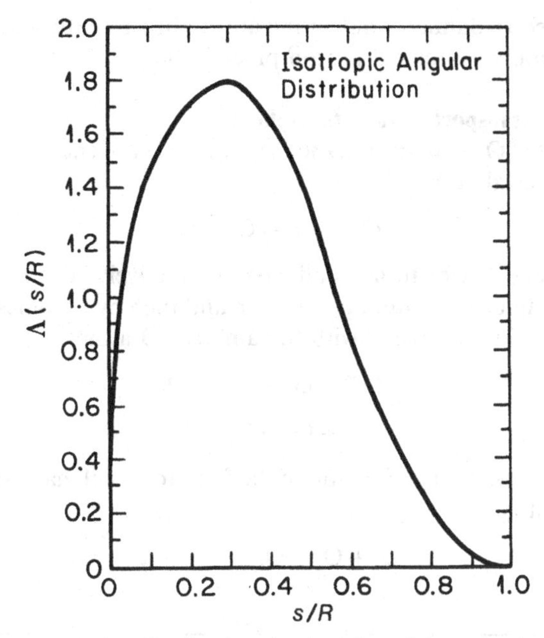

Fig. 3.4.4 The energy deposition distribution as a function of the scattering depth. The integral $\int_0^1 \Lambda\,(s/R)\,d(s/R) \equiv 1$. (M.H. Rees, *Planet. Space Sci.*, **30**, 463, 1982.)

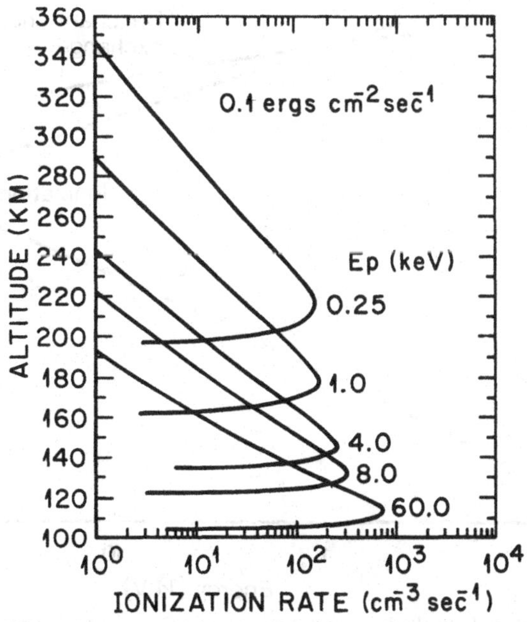

Fig. 3.4.5 Altitude profiles of the ionization rate produced by proton fluxes with initial energy, E_p, identified on each curve. An isotropic flux over the upper hemisphere is assumed and the energy flux is 0.1 erg cm^{-2} s^{-1} at every initial proton energy. (M.H. Rees, *Planet. Space Sci.*, **30**, 463, 1982.)

the Doppler shifted H-Balmer-β line in the aurora half a century ago provided direct evidence for aurorally associated proton precipitation.

3.4.3 Energetic O transport in the atmosphere

The unique aspect of O^+ ion precipitation is the rapid conversion by resonance charge transfer to fast neutral atoms,

$$O_f^+ + O_s \rightarrow O_f + O_s^+ \tag{3.4.15}$$

The subscripts f and s refer to fast and slow, respectively. The fast ion captures an electron from the ambient O atom population and thereby becomes a fast atom. Fast ions and atoms collide elastically with the ambient O atoms,

$$O_f^+ + O_s \rightarrow O_f^+ + O_f \tag{3.4.16}$$

$$O_f + O_s \rightarrow O_f + O_f \tag{3.4.17}$$

sharing kinetic energy. A small fraction of the fast atoms will have sufficient energy to ionize the ambient gas,

$$O_f + O_s \rightarrow O_f + O_s^+ + e_s \tag{3.4.18}$$

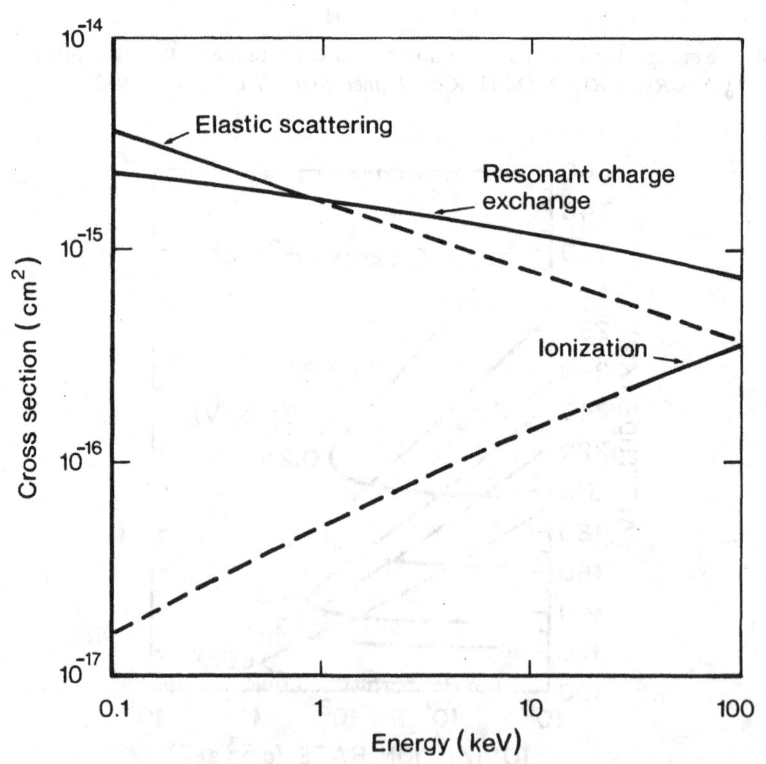

Fig. 3.4.6 Cross sections for collision processes between O^+ ions and atoms described in Section 3.4.3 as a function of the relative energy of the colliding particles. The dashed lines are extrapolations of available measurements. (M.H. Rees, *Physica Scripta*, **T18**, 249, 1987.)

and the process becomes a source of ionization. Another energy loss reaction is excitation by fast atom collisions,

$$O_f + O_s \rightarrow O_f + O_s^* \tag{3.4.19}$$

producing states that may decay by spectral emissions. The relative importance of the above processes is governed by their collision cross sections, shown in Figure 3.4.6. At primary energies below about 10 keV virtually all the energy goes into heating the atmosphere. Ionization becomes important only at very high O^+ energies, well above the value usually observed in space.

The collisional interactions described above leave neutral atoms as the dominant energetic O species. Fast O^+ ions are the primary source of fast atoms but it is sufficient to formulate the transport equation for fast O atoms. Elastic scattering (Equation (3.4.17)) is the important transport mechanism of fast atoms, and collisions occur primarily with ambient O atoms in the upper thermosphere. Denoting the intensity of fast atoms at scattering depth τ, energy E, and angle θ with respect to the vertical by $I(\tau, E, \theta)$ ($cm^{-2}\,s^{-1}\,sr^{-1}\,eV^{-1}$) the transport equation is

$$\mu \frac{dI(\tau, E, \theta)}{d\tau} = -I(\tau, E, \theta) + \frac{P(\tau, E, \theta)}{\sigma_e(E)n(O)} + \frac{Q(\tau, E, \theta)}{\sigma_e(E)n(O)} \tag{3.4.20}$$

where $\mu = \cos\theta$. Primary production of fast atoms by charge exchange (Equation (3.4.15)) and momentum transfer (Equation (3.4.16)) is denoted by $P(\tau, E, \theta)$ and the production of fast atoms by elastic collisions with faster atoms (Equation (3.4.17)) is given by $Q(\tau, E, \theta)$. The elastic scattering cross section is $\sigma_e(E)$ and $n(O)$ is the thermospheric O density. The O transport equation has been solved under simplifying assumptions. In the initial study, the intensity was obtained as a function of energy and scattering depth and direction of scattering but more recent work has adopted the two-stream approximation, i.e., scattering occurs either into the upward or the downward hemisphere.

Observed O^+ precipitation is at energies where the cross section for elastic scattering of O atoms is much larger than the ionization cross section, and most of the energy flux associated with O^+/O precipitation goes into heating the neutral atmosphere.

Bibliography

Section 3.1

A primary auroral electron spectrum includes a much broader energy range than one usually finds in any one research paper. This is due to the limited energy range that can be measured with any single instrument. Electron spectrometers yield differential intensity measurements and these instruments are used from the lowest energy (about 10 eV) to about 25 keV. Integral detectors such as geiger counters are employed in the high energy regime which extends to several hundred kiloelectron volts. The flux of auroral electrons above 100 keV is small but it is a part of the population that cannot be ignored in a comprehensive survey of auroral electrons. The high energy tail (> 100 keV) of an auroral electron spectrum deposits very little energy in the thermosphere but penetrates to the mesosphere, below our (arbitrary) lower boundary. The electron spectra shown in Figures 3.1.1–3.1.5 all contribute to thermospheric energy deposition. The original references are given in the figure captions.

A survey of the temporal and spatial characteristics of the total flux and the average energy of auroral electron precipitation measured by detectors on board the ISIS 2 satellite has been presented by

I.B. McDiarmid, J.R. Burrows, and E.E. Budzinski, Average characteristics of magnetospheric electrons (150 eV to 200 keV) at 1400 km, *J. Geophys. Res.*, **80**, 73–9, 1975.

Section 3.2

Derivation of the electron transport equation from the continuity equation for the electron distribution function may be found in the Appendix of

K. Stamnes and M.H. Rees, Inelastic scattering effects of photoelectron spectra and ionospheric electron temperature, *J. Geophys. Res.*, **88**, 6301–9, 1983.

Electron–electron interactions applied to photoelectrons and auroral secondary electrons are discussed by

K. Stamnes and M.H. Rees, Heating of thermal ionospheric electrons by suprathermal electrons, *Geophys. Res. Lett.*, **10**, 309–12, 1983.

A summary of the development of electron transport theory appears in

K. Stamnes, Analytic approach to auroral electron transport and energy degradation, *Planet. Space Sci.*, **28**, 427–41, 1980.

Important earlier papers on the subject are

P.M. Banks and A.F. Nagy, Concerning the influence of elastic scattering upon photoelectron transport and escape, *J. Geophys. Res.*, **75**, 1902–10, 1970.

A.F. Nagy and P.M. Banks, Photoelectron fluxes in the ionosphere, *J. Geophys. Res.*, **75**, 6260–70, 1970.

D.J. Strickland *et al.*, Transport equation techniques for the deposition of auroral electrons, *J. Geophys. Res.*, **81**, 2755–64, 1976.

M.J. Berger, Monte Carlo calculation of the penetration and diffusion of fast charged particles, in *Methods in Computational Physics*, Vol. 1, Academic Press, New York, 1963, pp. 135–215.

Extensions of the theory to include interhemispheric transport are discussed by

S.S. Prasad *et al.*, Auroral electron interaction with the atmosphere in the presence of conjugate field-aligned electrostatic potentials, *J. Geophys. Res.*, **88**, 4123–5130, 1983.

K. Stamnes, A unified theory of interhemispheric electron transport and energy degradation, *Geophys. Norveg.*, **33**, 41, 1985.

Section 3.3

The laboratory experiments that have provided measurements of the energy deposition rate were carried out by

A.E. Grün, Lumineszenz, photometrische Messungen der Energie Absorption in Strahlungsfeld von Electronenquellen eindemensionaler Fall in Luft, *Zeitschrift der Naturforschung*, **12a**, 89, 1957.

for electron energies between 5 and 54 keV. At lower electron energies, between 0.3 and 3 keV, the results that are adopted are by

J.L. Barrett and P.B. Hays, Spatial distribution of energy deposited in nitrogen by electrons, *J. Chem. Phys.* **64**, 743, 1976.

This paper also discusses the effective range of energetic electrons. The application of these laboratory measurements to auroral electrons was first presented by

M.H. Rees, Auroral ionization and excitation by incident energetic electrons, *Planet. Space Sci.*, **11**, 1209, 1963.

A more detailed exposition of this topic is presented by

A. Vallance-Jones, *Aurora* (Section 4.2), D. Reidel Publ. Co., Dortrecht, 1974.

The doubly differential cross section for the production of secondary electrons by electron impact has been measured in the laboratory by

C.S. Opal, W.K. Peterson and E.C. Beatty, Measurements of secondary electron spectra produced by electron impact ionization of a number of simple gases. *J. Chem. Phys.*, **55**, 4100, 1971; *Atomic Data*, **4**, 209, 1972.

Similar measurements have been made in N_2 by

T.W. Shyn, Doubly differential cross sections of secondary electrons ejected from gases by electron impact: 50–400 eV on N_2, *Phys. Rev.*, **A27**, 2388, 1983.

Electron impact ionization cross sections have been measured by several experimenters. The results have been collected by

H.S.W. Massey, E.H.S. Burchop and H.B. Gilbody, *Electronic and Ionic Impact Phenomena*, Vols. I and II, Clarendon Press, Oxford, 1969.

A discussion of the production of secondary electrons in aurora and the derivation of Equation (3.3.11) are given by

M.H. Rees, A.I. Stewart and J.C.G. Walker, Secondary electrons in aurora, *Planet. Space Sci.*, **17**, 1997, 1969.

Secondary electrons are also discussed in A. Vallance-Jones' monograph, *Aurora*, cited above.

Section 3.4

The classic review paper on auroral proton precipitation is

R.H. Eather, Auroral proton precipitation and hydrogen emissions, *Rev. Geophys.*, **5**, 207, 1967.

Formulation of the proton transport problem is given in

J.R. Jasperse and B. Basu, Transport theoretic solutions for auroral proton and H atom fluxes and related quantities, *J. Geophys. Res.*, **87**, 811, 1982.

and the Monte-Carlo approach was first described by

G.T. Davidson, Expected spatial distribution of low energy protons precipitated in the auroral zones, *J. Geophys. Res.*, **70**, 1061, 1965.

Detailed computations of photon emission rates based on comprehensive cross section data are given in

B. Van Zyl, M.W. Gealy and H. Neumann, Prediction of photon yields for proton aurorae in an N_2 atmosphere, *J. Geophys. Res.*, **89**, 1703, 1984.

A hybrid approach that focuses on atmospheric consequences of proton precipitation is given by

M.H. Rees, On the interaction of auroral protons with the Earth's atmosphere, *Planet. Space Sci.*, **30**, 463, 1982.

A discussion of the range and energy loss of energetic particles and a list of references is given in

A. Dalgarno, Range and energy loss, Chapter 15 of *Atomic and Molecular Processes*, Ed., D.R. Bates, Academic Press, New York, 1962.

Atmospheric effects of energetic O^+ ion precipitation on the atmosphere and the transport of fast O atoms were first analyzed by

M.R. Torr, J.C.G. Walker and D.G. Torr, Escape of fast oxygen from the atmosphere during geomagnetic storms, *J. Geophys. Res.*, **79**, 5267, 1974.

Additional work on O^+ precipitation has been reported in two articles

J.U. Kozyra, T.E. Cravens, and A.F. Nagy, Energetic O^+ precipitation, *J. Geophys. Res.*, **87**, 2481, 1982.

M. Ishimoto, M.R. Torr, P.G. Richards, and D.G. Torr, The role of energetic O^+ precipitation in a mid-latitude aurora, *J. Geophys. Res.*, **91**, 5793, 1986.

The contribution by proton and O^+ precipitation to ionization and heating of the atmosphere is discussed by

M.H. Rees, Modeling of the heating and ionizing of the polar thermosphere by magnetospheric electron and ion precipitation, *Physica Scripta*, **T18**, 249, 1987.

Problems 7–8

Problem 7: Compute the altitude profile of the energy deposition rate by downward isotropic precipitation of auroral electrons in a non-sunlit atmosphere for solar minimum conditions. Adopt a Maxwellian energy distribution, $N(E)\,dE = N_0 E \exp(-E/E_0)\,dE$, with total flux magnitude $\int_0^\infty N(E)\,dE$ of 1.56×10^8 electrons $cm^{-2}\,s^{-1}$ and an average energy of 4 keV, or $E_0 = 2$ keV.

Problem 8: Compute the secondary electron production rate spectrum at 120 km and at 170 km for solar minimum conditions for the aurora specified in Problem 7. Compare the result at 170 km with the result of Problem 6 which gives the photoelectron production rate for two lines only in the solar spectrum. Use a value of 2.71 for the normalization factor $A(E_p')$ in Equation (3.3.13).

4

Collisions and reactions

4.1 Introduction

This book deals with the outermost shell of the collisionally dominated gaseous envelope of the Earth, the thermosphere and ionosphere. In the preceding chapters collision cross sections have appeared prominently in the physical descriptions and in the equations that govern the interaction of energetic photons, ions and electrons with the atmosphere. We learned, for example, that the optical depth (Equation (2.2.1)) is proportional to the column density of absorbing molecules, with the proportionality factor identified as the absorption cross section. Photoionization (Section 2.3) involves ionization cross sections and electron transport (Section 3.2) depends on elastic and inelastic collision processes and the associated collision cross sections. It is therefore appropriate to assess our understanding of cross sections and expand upon the part which they play in physical and chemical processes in the atmosphere.

It has been found convenient to classify collisions into several types: elastic, inelastic and reactive. When two particles collide and only kinetic energy and linear momentum are exchanged (and the total of each is conserved) then the collision is elastic. If one or both of the collision partners undergoes a change in internal energy then the collision is inelastic. Reactive collisions are those that involve the production of new species; such collisions are also inelastic. Reactive collisions underlie the chemistry of the upper atmosphere. Elastic collisions govern the transport properties of the neutral and ionized components in the atmosphere, diffusion, viscosity and conduction. It has already been noted in Chapter 3 (without detail) that the angular distribution of auroral and photoelectron transport depends primarily on elastic scattering. Collisions between all the different constituents are of interest in atmospheric studies. These include collisions between various neutral atoms and molecules, between ions and neutrals, between electrons and neutrals, as well as ion–ion, electron–ion and electron–electron collisions. Two-body collisions predominate in the thermosphere; the three-body reaction involving the recombination of atomic O is the only such reaction that is important. Additional three-body interactions contribute to chemical reactions in the denser mesosphere.

Atoms and molecules in the atmosphere have random motion characterized by a temperature. Ions and electrons have, in addition, directed motion imparted by electric and magnetic fields. Atoms, molecules and ions are surrounded by a potential field characteristic of each species. It is intuitively evident that the probability of a collision should depend on the relative velocity between the particles as well as the geometry of the encounter: they must approach with sufficient speed to overcome any repulsive force but remain in close proximity to establish communication, or remain within each other's spheres, or interact.

The parameter defined to give a quantitative measure of the probability of some interaction upon the encounter of two 'particles' is the collision cross section. The differential cross section, $\sigma(E, \theta, \phi)$, is the probability that particles with relative kinetic energy, E, are scattered upon collision in the direction θ, ϕ, in centre of mass coordinates, into an element of solid angle $\mathrm{d}\Omega = \sin\theta \, \mathrm{d}\theta \, \mathrm{d}\phi$ (Figure 4.1.1). In the absence of external fields, all azimuthal angles are equivalent and $\mathrm{d}\Omega = 2\pi \sin\theta \, \mathrm{d}\theta$. The total cross section is then

$$\sigma(E) = 2\pi \int_0^\pi \sigma(E, \theta) \sin\theta \, \mathrm{d}\theta \tag{4.1.1}$$

Cross sections are the basic parameters for evaluating dynamical and chemical processes in the atmosphere. This motivates our interest in these quantities; especially because atmospherically important results of laboratory experiments and theoretical work are frequently given in terms of the cross section. Experimentally derived cross

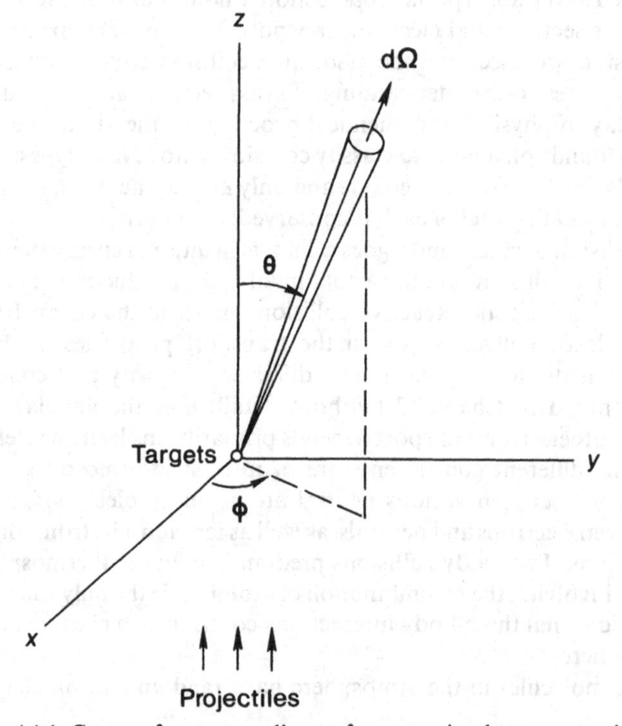

Fig. 4.1.1 Centre of mass coordinates for scattering between particles.

sections are measured in laboratory coordinates while theoretical derivations are formulated in centre of mass coordinates (Problem 9).

4.2 Elastic scattering of electrons

Theoretical analysis of electron scattering requires the quantum mechanical approach. Representing the projectile electron by a plane wave incident upon the atom or molecule the procedure involves the mathematical description of changes effected by the incident wave on the initial state of the target to produce a final state, taking account of the appropriate interaction potential. Elastic scattering is a special case that does not result in changes of the internal energy state of the target. Formulation of the problem and solution of the wave equation are given in several texts devoted to various aspects of collision phenomena and the reader is encouraged to consult these to gain a broader view of the field than will be given here. A few titles are given in the bibliography to this chapter. The differential scattering cross section is given by

$$\sigma(E,\theta) = |f(\theta)|^2 \tag{4.2.1}$$

where $f(\theta)$ is the scattered amplitude.

If the energy of the incident electron is large compared with the binding energy of the atomic or molecular electrons, then the solution known as the Born approximation gives results in good agreement with experiments. In this approximation

$$f(\theta) = -\frac{2M_R}{\hbar}\int_0^{\infty}\frac{\sin Kr}{Kr}V(r)r^2\,dr \tag{4.2.2}$$

In this equation M_R is the reduced mass,

$$M_R = \frac{mM}{m+M} \tag{4.2.3}$$

where m and M are the masses of the electron and the target, respectively. The wave number is

$$K = \frac{2M_R v}{\hbar}\sin\frac{\theta}{2} = \frac{2(2M_R E)^{1/2}}{\hbar}\sin\frac{\theta}{2} \tag{4.2.4}$$

where v and E are the electron velocity and energy respectively. The spherically symmetric scattering potential is $V(r)$ and, as stated previously in different words, it is an assumption in the Born approximation that $V(r)$ is small by comparison with the kinetic energy of the electron.

An approximation to elastic scattering of an electron in the field of an atom is the Coulomb potential which is given by

$$V(r) = -Ze^2/r \tag{4.2.5}$$

where Ze is the nuclear charge and r is the distance from the scattering centre. For Coulomb scattering the differential cross section is given by

$$\sigma(E,\theta) = Z^2e^4/4M_R^2v^4\sin^4(\theta/2) \tag{4.2.6}$$

in the centre of mass coordinate system. Only for Coulomb scattering is the differential cross section obtained by classical mechanics identical to the Born approximation quantal results. This type of scattering is also known as Rutherford scattering

(Problem 10). Equation (4.2.6) leads to an infinite total scattering cross section due to the infinite range of the Coulomb potential. A potential that represents quite well collisions between electrons (or ions) and neutral atoms is the screened Coulomb potential expressed mathematically by

$$V(r) = -(Ze^2/r)\exp(-\lambda_0 r) \tag{4.2.7}$$

where $1/\lambda_0$ is a positive parameter that has a magnitude which is of the order of the atomic dimension. Equation (4.2.7) represents the potential energy of an electron in the vicinity of a neutral atom. (The screened Coulomb potential will appear again in connection with collisions in a plasma i.e., electron–positive ion scattering.) With this modified potential, the differential scattering cross section becomes

$$\sigma(E, \theta) = \frac{Z^2 e^4}{M_R^2 v^4} \frac{1}{(1 + 2\eta - \cos\theta)^2} \tag{4.2.8}$$

where η is called the phase shift for reasons that will emerge in the succeeding paragraphs. The phase shift is related to the screening parameter by the approximate equation

$$2\eta = (\hbar\lambda_0/M_R v)^2 \tag{4.2.9}$$

where

$$\lambda_0 = Z^{1/3}/0.885a_0 \tag{4.2.10}$$

with a_0 the radius of the first Bohr orbit. More complex equations result for relativistic electron energies and for higher order Born approximations. Neither are required for the application to elastic collisions of fast electrons with the thermospheric gas (Problem 11).

If the projectile electron energy is comparable to the energy of the orbital electrons in the target atom the Born approximation to the scattering problem is no longer applicable and more complex methods of solving the wave equation of quantum mechanics must be used. The wave function of the incident electron generally represents a plane wave and the portion that is scattered by a radial potential is represented by a wave function of the form of outgoing spherical waves that decrease in amplitude with radial distance. Adopting the terminology already used in the preceding paragraphs, the scattering amplitude is $f(\theta)$ and the differential cross section is given by Equation (4.2.1). Several texts listed in the bibliography give detailed derivations of $f(\theta)$ by the so-called method of partial waves but we only give the result for the differential cross section for elastic potential scattering

$$\sigma(E, \theta) = \frac{\hbar^2}{2M_r E} \left| \sum_{l=0}^{\infty} (2l+1)\exp(i\eta_l)\sin\eta_l P_l(\cos\theta) \right|^2 \tag{4.2.11}$$

This cross section is valid for potentials that decrease more steeply than r^{-1} for large values of r. The total energy, E, is equivalent to the initial electron kinetic energy at large separation; $P_l(\cos\theta)$ is the Legendre polynomial of order l, θ is the scattering angle with $l = 0$ denoting s-wave scattering, $l = 1$ p-wave scattering, $l = 2$ d-wave scattering, etc. For scattering in the forward direction, $\theta = 0°$, the Legendre polynomials become unity for all values of l. The quantity η_l is the phase shift of the lth wave due to the scattering potential; its value is governed by E, $V(r)$ and M_R.

The total electron elastic scattering cross section is obtained by integrating

Equation (4.2.11) over a complete solid angle

$$\sigma_e(E) = \frac{4\pi\hbar^2}{2M_R E} \sum_{l=0}^{\infty} (2l+1)\sin^2\eta_l \tag{4.2.12}$$

The quantity $\hbar^2/2M_r E$ is the square of the wavelength of de Broglie waves. If the wavelength is of the order of a Bohr radius, a_0, then the cross section is $4\pi a_0^2$ suggesting that in the most simplistic model the surface area of a sphere with radius equal to the first Bohr orbit represents the atomic cross section. Observed structure in cross sections has been verified theoretically by the partial wave solution to the scattering problem. The total elastic scattering cross sections for electrons on N_2, O_2 and O are shown in Appendix 4, derived from laboratory measurements and theoretical computations. The mathematical approach for solving the wave equation for the phase shift depends on the interaction potential, $V(r)$. Solving the radial wave equation for a system with and without the interaction potential yields the phase shift for l waves. Approximate techniques have been developed for special cases, of which the Born approximation is an example. Several other theoretical methods for solving the wave equation of a collision process have been developed and these are described in several texts listed in the bibliography. Collision cross sections applied to atmospheric problems rely heavily on experimental measurements.

The scattering amplitudes and differential cross sections considered so far are applicable to atomic targets. Elastic scattering of electrons by molecules results in a perturbation of the scattered waves due to interference effects between the scattered waves of the individual atoms. For high energy electrons, as defined previously for the applicability of the Born approximation, it is sufficient to displace the phase reference of the atomic scattering centre while retaining the independent atomic scattering formulation. Ignoring molecular vibration, the differential cross section for homonuclear diatomic molecules (e.g., N_2, O_2) becomes

$$\sigma_M(E,\theta) = 2\sigma_A(E,\theta)\left(1 + \frac{\sin Kr_e}{Kr_e}\right) \tag{4.2.13}$$

where K is defined by Equation (4.2.4), r_e is the equilibrium nuclear separation, and σ_A is the elastic differential scattering cross section of each atom.

Chapter 3 is devoted in part to auroral electron and photoelectron transport and it was argued that under certain assumptions the differential cross section may be separated into a product of an energy degradation function and an angular scattering phase function (Equation (3.2.2)),

$$\sigma_e(E,\theta) = P(\cos\theta)\sigma_e(E) \tag{4.2.14}$$

where $\sigma_e(E)$ is the total elastic cross section at energy E and $P(\cos\theta)$ is the phase function, normalized to unity,

$$\int P(\cos\theta)\,d\Omega = 1 \tag{4.2.15}$$

The total cross section, $\sigma_e(E)$, (Equation (4.1.1)) may be evaluated numerically if an experimentally derived differential cross section is adopted (Figure 3.3.4) but if, for example, the screened Coulomb cross section (Equation (4.2.8)) is deemed adequate an analytic solution is obtained,

$$\sigma_e(E) = \frac{Z^2 e^4}{M_R^2 v^4} \frac{\pi}{\eta(1+\eta)} \tag{4.2.16}$$

Assuming azimuthal symmetry the phase function is obtained from Equations (4.2.8), (4.2.14) and (4.2.16)

$$P(\cos\theta) = \frac{2\eta(\eta+1)}{(1+2\eta-\cos\theta)^2} \tag{4.2.17}$$

(Problems 12, 13, 14).

4.3 Inelastic electron scattering

Collisions that result in a change of internal energy of the target atoms or molecules are inelastic. The process entails a change in one or more quantum numbers that identify the wave function of the target and is caused by the perturbation produced by the collision with the energetic electron. These may be transitions between discrete internal energy states, electronic, vibrational, and rotational, or the final state may be in the continuum as occurs for ionization and molecular dissociation. Extensive theoretical computations of inelastic cross sections have been made only for the simplest atomic structures involving H and He, but some work has also been done on more complex atoms such as N and O that are of atmospheric interest. The complexity of molecular structure, including N_2 and O_2, is considerable and virtually precludes obtaining excitation cross sections by *ab initio* quantum mechanical means. In practice, quantal, semiclassical, empirical, and experimental results have been combined to arrive at cross sections required for application to problems in the upper atmosphere.

In the Born approximation the differential cross section for transition from an initial state, i, to a final state, f, is

$$\sigma_{i,f}(E,\theta) = \frac{m^2 e^4}{4\pi^2 \hbar^4} \left(\frac{E_f}{E_i}\right)^{1/2} |M_{i,f}(\theta)|^2 \tag{4.3.1}$$

where E_i and E_f are the initial and final energies of the system and $M_{i,f}(\theta)$ is the transition matrix element which is a function of the scattering angle, θ, the initial and final wave functions and the interaction potential. The angular distribution of inelastically scattered electrons follows a trend similar to elastic scattering, peaking in the forward direction (Figure 4.3.1).

The total cross section is

$$\sigma_{i,f}(E) = 2\pi \int_0^\pi \sigma_{i,f}(E,\theta) \sin\theta \, d\theta \tag{4.3.2}$$

In the Born approximation and for optically allowed transitions, the cross section in the high energy limit varies as follows

$$\sigma_{i,f}(E) \propto E^{-1} \ln E \tag{4.3.3}$$

Ionization involves the ejection of an atomic electron; the differential scattering cross section is a function of the energy and angular distribution of the ejected as well as of the scattered electron. If the influence of the nucleus is neglected, then conservation of

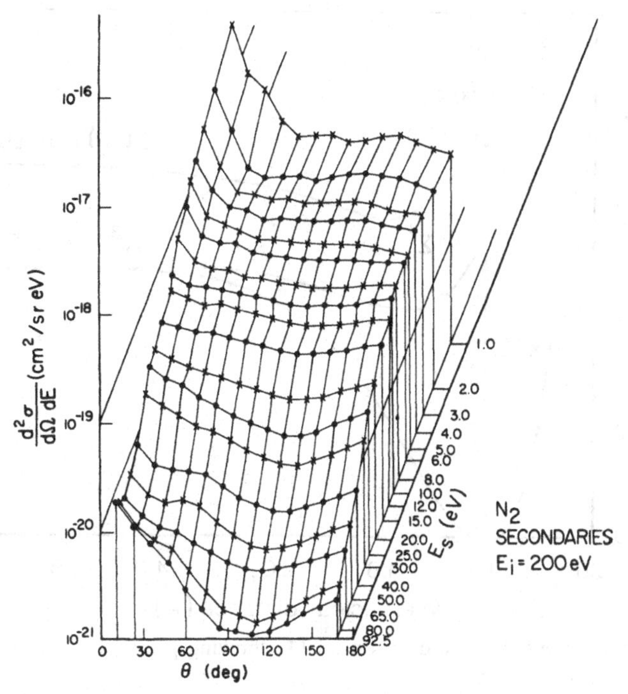

Fig. 4.3.1 Doubly differential cross section vs. energy and angle of the ejected electrons for 200 eV electrons on N_2. (T.W. Shyn, *Phys. Rev.*, **A27**, 2388, 1983.)

momentum between the incident and free atomic electrons gives an indication of the angular disposition of the ejected electrons. The velocity v'_p of the incident electron is related to the velocity v_p after scattering through angle θ_p by

$$v'_p \cos\theta_p = v_p$$

The velocity, v_s, of the ejected electron emitted at an angle θ_s with respect to the direction of the incident electron is related to v'_p by

$$v'_p \cos\theta_s = v_s$$

Most of the ejected electrons have much less energy than that of the incident electron, $v_s'^2 \ll v_p'^2$, so that the probability is large for θ_s to have a value near $\pi/2$. Similarly, the scattered electron energy will be close to the incident electron energy, $v_p'^2 \sim v_p^2$; this most probably gives a scattering angle, θ_p, near 0°, or in the forward direction. Perturbations due to the potential of the electronic structure of the atom account for the angular distributions observed in laboratory experiments.

Photoionization cross sections between atomic or molecular states are related to optical transition probabilities. Absorption of photons in some frequency range γ to $\gamma + d\gamma$ causes a transition from an initial state to some final state which may be in the ionization continuum, ejecting photoelectrons. The transition probability is a function of the photon energy, or wave frequency, the transition matrix which involves the wave functions of the initial and final states of the atom and the position vectors of the electrons relative to the nucleus. Extensive theoretical work has been done on the

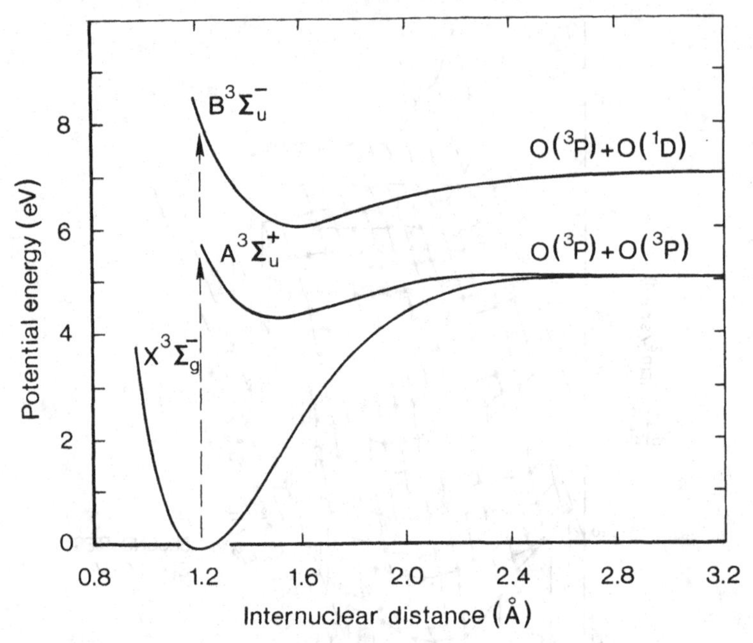

Fig. 4.3.2 Partial potential curve diagram for O_2 showing principal states that contribute to dissociation.

relation between optical transitions and collision theory. This is described in several books listed in the bibliography.

Two types of inelastic collisions have been described so far, excitation of discrete states and ionization. A third process is molecular dissociation which may be caused by photon impact or by electron impact. The mechanisms by which the major thermospheric molecular species, O_2 and N_2, are dissociated differ in detail. The potential energy curves for the states in O_2 that take part in dissociation are shown in Figure 4.3.2. All three are bound states but the potential minima occur at different internuclear distances. A transition from the ground state, X $^3\Sigma_g^-$, to the excited states requires only a very short time by comparison with vibration or relative motion of the nuclei that would change the internuclear distance. According to the Frank–Condon principle, based on quantum mechanical results, the transitions will occur at the initial nuclear separation, exciting vibrational levels in the upper electronic states at an energy that lies above the dissociation energy of the molecule. Dissociation via the A $^3\Sigma_u^+$ state yields two ground state O atoms while O_2 in the B $^3\Sigma_u^-$ state dissociates into an excited O(^{1}D) atom and a ground state O(^{3}P) atom. The X $^3\Sigma_g^-$ – B $^3\Sigma_u^-$ transition is associated with an electric dipole moment and is optically allowed. However, the X $^3\Sigma_g^-$–A $^3\Sigma_u^+$ transition violates one of the selection rules for dipole transitions, i.e., a $\Sigma^- \rightarrow \Sigma^+$ transition is not allowed. Dissociation through the A state is therefore much weaker than through the B state as already noted in Section 2.4, where it was found that absorption of solar UV radiation in the Schumann Runge continuum ($\lambda < 1749$ Å) is substantially stronger than that in the Herzberg continuum ($\lambda < 2422$ Å). A more detailed potential energy curve diagram for O_2 is given in Appendix 3 showing several combinations of dissociation products. The potential energy curves for O_2^+ show that the molecule is subject to dissociative ionization by electron impact, producing several combinations of atoms and atomic ions.

The photon absorption spectrum of N_2 (Figure 2.2.2) shows an ionization continuum but, unlike O_2, there is no absorption continuum associated with dissociation of the molecule. There are only bands in the 800–1000 Å region of the spectrum. Absence of a dissociation continuum indicates that there are no electronic states that have potential energy curves for which transitions from the ground state (that obey the Frank–Condon principle) can populate vibrational levels above the dissociation energy of the molecule. (Appendix 3 presents a detailed potential energy curve diagram of N_2.) There are, however, several high lying electronic states of N_2 in $^1\Pi_u$ and $^1\Sigma_u^+$ electronic configurations that are readily excited by electron impact and photon absorption through dipole allowed transitions. There is also a weak absorption into the $a^1\Pi_g$ state. These states are called predissociation states because the probability of a radiationless transition to states that lead to dissociation is much higher than that leading to radiation. The mechanism is illustrated in Figure 4.3.3 for one such predissociation state, the $b\,^1\Pi_u$. The vertical transition, in this illustrative example, is to the $v' = 4$ level, whence coupling to the shallow $C'\,^3\Pi_u$ state is followed by dissociation into an $N(^2D)$ excited atom and a ground state $N(^4S)$ atom. Dissociative ionization yielding an N^+ ion and an N atom occurs directly (as in O_2) without formation of intermediate states at an onset energy of 24.3 eV, which is the sum of the

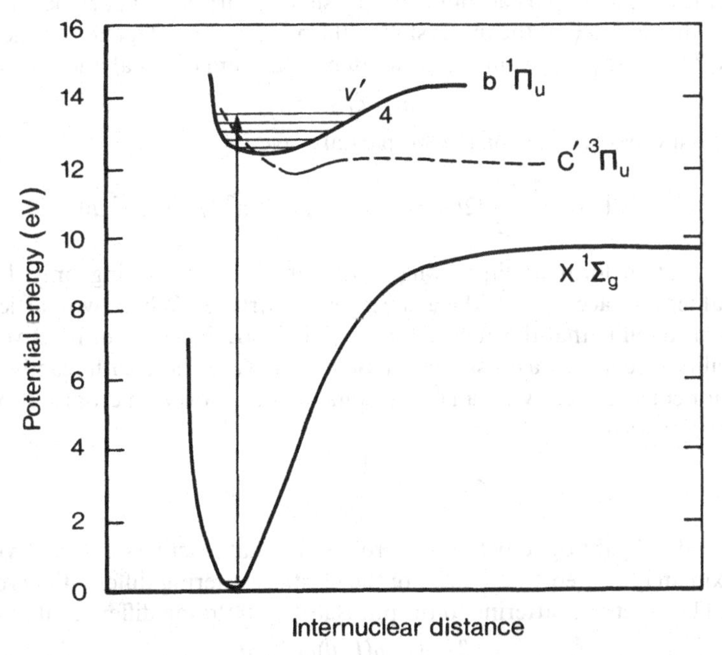

Fig. 4.3.3 Partial potential diagram for N_2 showing states involved in dissociation of N_2 by excitation of a predissociation level.

$$e + N_2(X\,^1\Sigma_g^+) \rightarrow e + N_2(b\,^1\Pi_u, v' = 4)$$

$$N_2(b\,^1\Pi_u, v' = 4) \rightarrow N(^4S) + N(^2D) + 0.693\ \text{eV}$$

dissociation energy of N_2 and the ionization potential of N. The product ions and atoms are also formed in excited states at higher energy thresholds. Inelastic collisions of energetic protons and other ions with atmospheric gases were discussed in Section 3.4.

4.4 Reactive collisions

This collision type involves the production of new species, such as the reactions $O^+ + N_2 \rightarrow NO^+ + N$ and $NO^+ + e \rightarrow N + O$. Many reactive collisions occur in the upper atmosphere and only the two examples cited above are needed to show that neutral atoms and molecules, ions, and electrons may take part in such collisions. The variety of important reactive collisions in atmospheric processes is large.

The complexity of reactive collisions is a challenge to quantum mechanical computations of the cross sections. An important difference in applying wave theory to reactive scattering is that the particles themselves are no longer conserved. In the usual mathematical description of a wave the imaginary part is associated with absorption or attenuation. The partial wave description of reactive scattering is therefore derived with a complex scattering potential

$$V(r) = \frac{\hbar^2}{2M_R}[U(r) + \mathrm{i}\,W(r)] \qquad (4.4.1)$$

in which the imaginary part accounts for possible absorption of particles. We noted in our earlier discussion that the phase shift of the partial waves depends on the scattering potential. Thus, the phase shift is also written as the sum of a real and imaginary part,

$$\eta_l = \xi_l + \mathrm{i}\mu_l \qquad (4.4.2)$$

and the total cross section for the lth partial wave is

$$\sigma_l(E) = \frac{4\pi\hbar^2}{2M_r E}(2l+1)\exp(-2\mu_l)2(\sin^2\xi_l + \sinh^2\mu_l) \qquad (4.4.3)$$

This is similar in form to Equation (4.2.12) for elastic scattering only, but has an additional term to account for the absorption of particles. When two particles collide there is a probability that the interaction may be elastic, inelastic, or reactive. Each has a probability specified by a cross section. Being concerned here with reactive collisions we may, for convenience, write an expression for the absorption cross section only for the lth partial wave

$$\sigma_l(E)_{\mathrm{abs}} = \frac{4\pi\hbar^2}{2M_R E}(2l+1)\beta_l \qquad (4.4.4)$$

in order to identify the opacity β_l as the probability that reactive scattering will occur. It is approximately related to the ratio of the elastic scattering differential cross section perturbed by reactive scattering to the pure elastic scattering differential cross section,

$$\beta \sim 1 - \sigma(E,\theta)/\sigma_e(E,\theta) \qquad (4.4.5)$$

The probability of a reactive collision may thereby be derived from elastic differential cross section measurements.

Theoretical computations of cross sections require accurate wave functions for the initial and final states as well as the interaction potential that characterizes the collision. The degree of complexity increases for collisions in which the interaction potential is no longer spherically symmetric, as is the case for ion–molecule reactions. Rotational excitation of the molecule may occur. There are several quantal formulations of reactive scattering that are well developed formally (see the bibliography to this chapter); however, application to processes of interest in the upper atmosphere is, as yet, limited. A formulation that has been applied to ion–molecule reactions, called the statistical theory of chemical reactions, leads to an expression for the reaction cross section that depends on so-called reaction channels, an entrance channel, γ', and an exit channel, γ. The channels are identified by the rotational quantum number, j, and the angular momentum quantum number, l. The total angular momentum quantum number of the channel is J. The cross section at relative kinetic energy $E_{\gamma'}$ for a reactive collision from channel γ' to γ is given by

$$\sigma(\gamma' \to \gamma) = \frac{4\pi\hbar^2}{2M_R E_{\gamma'}} \frac{1}{2j+1} \sum_J (2J+1) \frac{n_J(\gamma')n_J(\gamma)}{N_J} \tag{4.4.6}$$

where $n_J(\gamma)$ and $n_J(\gamma')$ are the number of states in channels γ and γ', respectively. In this formulation a complex is formed from the reactants which then decays into the products. The distribution of the final states is assumed to be independent of the channel by which the complex is formed. N_J is the total number of states that can be formed by the complex.

At large separation between ion and atom the interaction potential is dominated by polarization of the atom in the field of the ion. The potential then varies as the inverse fourth power of the distance between the collision partners,

$$V(r) = -\alpha e^2/2r^4 \tag{4.4.7}$$

where α is the polarizability of the molecule and e is the electronic charge. The reactive cross section for a polarization potential is

$$\sigma_R(E) = \pi(2\alpha e^2/E)^{1/2}\bar{\beta} \tag{4.4.8}$$

where $\bar{\beta}$ is a mean opacity defined in Equation (4.4.4) as the probability that reactive scattering occurs.

A fruitful approach to reactive collisions is through the potential energy surfaces, or curves, that characterize the electronic and vibrational states of the reactants, the products, and especially of the intermediate complex. Based on quantum mechanical concepts, the probability of a reaction depends on energy considerations, allowed symmetry and spin combinations, details of the potential curves such as repulsive states or the nuclear separation and dissociation energy of bound states, and curve crossings of states of the intermediate complex with the reactants and various products. Thus, the success of a theoretical analysis of reactive collisions depends very much on the accuracy and the details of the relevant potential curves and knowledge of the correlation between the states that contribute to the reaction.

If the dominant interaction potential in an ion–molecule reaction is due to a polarization force (Equation (4.4.7)), then the cross section for an exothermic reaction should decrease as the relative translational energy between the reactants increases. The charge transfer reaction, $N_2^+ + O_2 \to N_2 + O_2^+$, and the ion–atom exchange reaction, $N^+ + O_2 \to NO^+ + O$, qualitatively follow this trend, and the magnitude of the cross section is large, just as the simple theory predicts. However, the most important ion–atom exchange reaction in the ionosphere, the reaction that controls

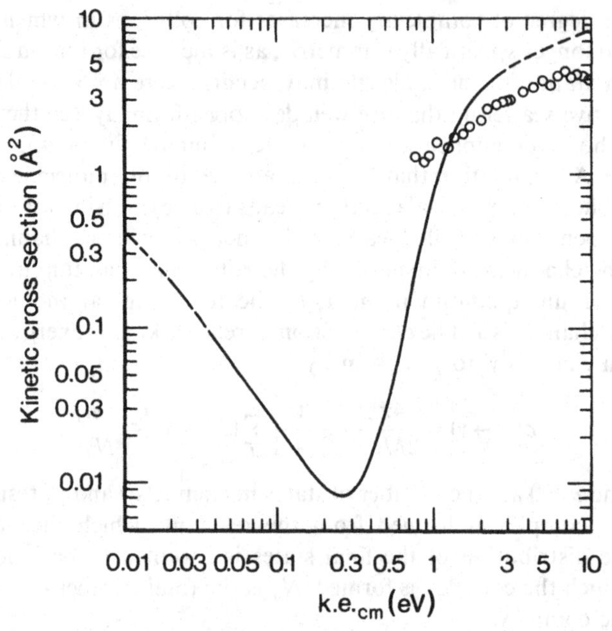

Fig. 4.4.1 Cross section for the ion-atom exchange reaction $O^+ + N_2 \rightarrow NO^+ + N$ as a function of the centre of mass kinetic energy of the collision partners. The solid and dashed curves were obtained in a flowing afterglow drift tube by D.L. Albritton, *et al.* (*J. Chem. Phys.*, **66**, 410, 1977) and the circles are results of a crossed beam experiment by J.A. Rutherford and D.A. Vroom (*J. Chem. Phys.*, **55**, 5622, 1971).

the ionization density in the F-region, has a cross section that bears no resemblance to the simple polarization interaction. The reaction is $O^+ + N_2 \rightarrow NO^+ + N$ and its cross section, shown in Figure 4.4.1, is about two orders of magnitude smaller at thermal energies than expected from simple theory and its magnitude increases with vibrational excitation of N_2, an aspect that simple theory does not consider. If this reaction were as rapid as predicted, then the F-region ionization would decay as rapidly as the E-region decays after sunset since the molecular ions that are formed recombine rapidly, as will be discussed below. If the F-region ions were to recombine the electron density would decrease to levels that would be insufficient to reflect radio waves, essentially destroying long-distance radio communication at night. Ionospheric ion chemistry is discussed in detail in Chapter 5 together with numerical values for various cross sections. The evidence that the reaction $O^+ + N_2 \rightarrow NO^+ + N$ is slow and that the magnitude depends on the vibrational distribution of ground state N_2 came from laboratory experiments at thermal and higher ion energies. The theoretical explanation, which followed the experimental results, draws on extensive and detailed analysis of the correlation between states that contribute to the reaction together with results derived from electron spectroscopy experiments. This important and illuminating case is discussed in a reference identified at the end of the chapter. The theoretical results clarify substantially the behaviour of the reaction $O^+ + N_2 \rightarrow NO^+ + N$, and the lesson which we learn from this example is to exercise great caution when attempting to predict the behaviour of complex reactions on the basis of inadequate information and simplified models. Understandably, theorists prefer examining reactive collisions between the simplest atomic and molecular systems while laboratory experimenters favour inert gases. In the upper atmosphere nature dictates the species with which we must deal, providing challenges for both theorists and experimenters.

The atmosphere differs from a laboratory environment and theory in yet another aspect. While in molecular beam laboratory experiments the relative energy of the colliding partners is well specified, reactive collisions in the atmosphere generally, but not always, occur at the temperature of the medium. Depending on the species involved, this could refer to the neutral gas temperature, the ion or electron temperature, the vibrational temperature of a specific molecule, or any combination of these appropriate to the reactants in the collision. It is useful, therefore, to define a quantity that takes account of the distribution in energy or velocity of the collision partners to characterize the probability of a reactive collision. Such a quantity is the reaction rate coefficient.

In a two-particle collision each species, a and b, may have its own velocity distribution, $f_a(\mathbf{v}_a)$ and $f_b(\mathbf{v}_b)$. If the cross section for a reactive collision from channel γ' to γ at relative particle speed v_{ab} is $\sigma(\gamma' \to \gamma, v_{ab})$ then the reaction rate coefficient is defined as follows

$$K(\gamma' \to \gamma) = \int v_{ab}\sigma(\gamma' \to \gamma, v_{ab}) f_a(\mathbf{v}_a) f_b(\mathbf{v}_b)\, d\mathbf{v}_a\, d\mathbf{v}_b \tag{4.4.9}$$

The cross section is specific to a reaction channel since more than one set of products may result from the collision. For example, the reaction $N^+(^3P) + O_2(X^3\Sigma_g^-, v = 0)$ yields three product channels, $NO^+ + O$, $O^+ + NO$ and $O_2^+ + N$ and the product species are found in several electronic and vibrational states that are discussed in Chapter 5. The relative production efficiency is a function of the relative collision energy, as shown in Figure 4.4.2.

The total reactive cross section, $\sigma_R(v)$, is defined as the sum over all internal states of the reactants and products,

$$\sigma_R(v) = \sum_{\gamma', \gamma} \sigma(\gamma' \to \gamma, v) \tag{4.4.10}$$

where the subscript has been dropped from the relative collision speed.

If the velocity distributions of the reactants are the equilibrium Maxwell–Boltzmann distribution at temperature T,

$$f_a(\mathbf{v}_a, T) = \left(\frac{m_a}{2\pi kT}\right)^{3/2} \exp\left(-\frac{m_a v_a^2}{2kT}\right) \tag{4.4.11}$$

with a similar expression for $f_b(\mathbf{v}_b, T)$, then the reaction rate coefficient is given by

$$K_R(T) = 4\pi\left(\frac{M_R}{2\pi kT}\right)^{3/2} \int v^3 \sigma_R(v) \exp\left(-\frac{M_R v^2}{2kT}\right) dv \tag{4.4.12}$$

(Problem 15). Experimental cross section measurements are frequently given as a function of the relative kinetic energy of the collision. With the change of variable $E = \frac{1}{2}M_R v^2$ the rate coefficient becomes

$$K_R(T) = \frac{1}{(kT)^{3/2}}\left(\frac{8}{\pi M_R}\right)^{1/2} \int_0^\infty E\sigma_R(E) \exp\left(-\frac{E}{kT}\right) dE \tag{4.4.13}$$

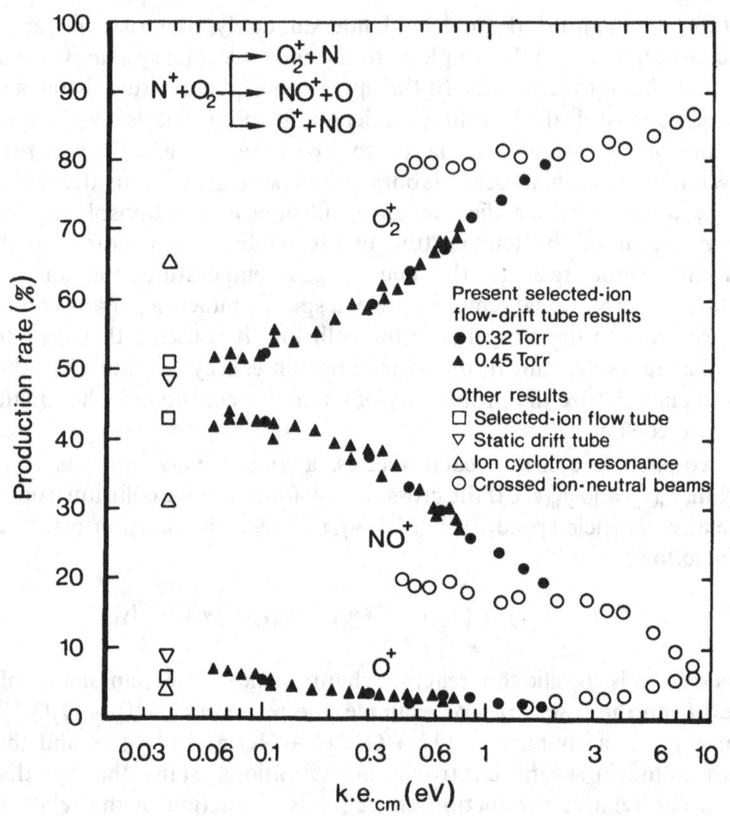

Fig. 4.4.2 Branching ratios of the reaction of N^+ ions with O_2 as a function of relative kinetic energy. Results of several experiments are included in the figure. (F. Howorka, *et al.*, *J. Chem. Phys.*, **73**, 758, 1980.)

The rate at which a reaction proceeds can therefore be determined if the collision cross section is known as a function of energy. However, it is essentially impossible to invert the computation; finding the energy dependence of the cross section would require an experimental accuracy not yet achieved in determining the temperature dependence of $K_R(T)$.

Very early in the development of chemical reaction kinetics and continuing with current experiments the macroscopic rate constant of a reaction is frequently described by the Arrhenius equation,

$$K_R(T) = A\exp(-E_{act}/kT) \tag{4.4.14}$$

which defines the activation energy of a reaction, E_{act}, and the quantity A called the preexponential factor. One form of the Arrhenius relationship is obtained with the simplest possible collision model in which the reactants are assumed to be hard spheres of diameter, d. The geometry of the collision is shown in Figure 4.4.3 which also identifies the impact parameter, b, the initial relative velocity vector, $\mathbf{v}$, and the velocity vector, $\mathbf{v}_c$, on impact along the line of centres. The energy on impact along the line of centres, E_c, cannot be less than a threshold value, E^*, which imposes conditions on both

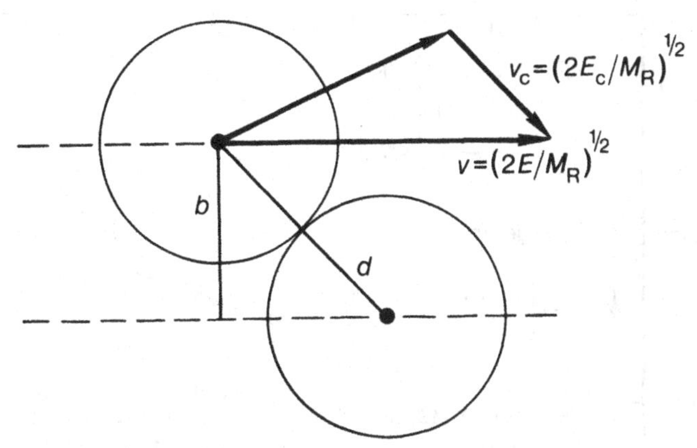

Fig. 4.4.3 Reactive scattering of hard spheres.

the initial relative energy and on the impact parameter, b. The relationship between these quantities is obtained from the figure.

$$b^2/d^2 = 1 - v_c^2/v^2 \tag{4.4.15}$$

The maximum value possible for the impact parameter, i.e., the spheres still touching, is the threshold energy which is then equal to the energy along the line of centers, $E^* = E_c = \frac{1}{2}M_R v_c^2$. The cross section as a function of relative impact energy, $E = \frac{1}{2}M_R v^2$ is given by

$$\begin{aligned} \sigma_R(E) &= 0, & E &< E^* \\ \sigma_R(E) &= \pi d^2(1 - E^*/E), & E &\geqslant E^* \end{aligned} \tag{4.4.16}$$

The reaction rate coefficient for the simple hard sphere collision model cross section is obtained by inserting Equation (4.4.16) into Equation (4.4.13) and evaluating the integral,

$$K_R(T) = \pi d^2\left(\frac{8kT}{\pi M_R}\right)^{1/2} \exp\left(-\frac{E^*}{kT}\right) \tag{4.4.17}$$

The threshold energy E^* corresponds to the activation energy, E_{act}, in the Arrhenius equation and the preexponential factor becomes a function of the temperature,

$$A = \pi d^2\left(\frac{8kT}{\pi M_R}\right)^{1/2} \tag{4.4.18}$$

The mean speed for a Maxwell–Boltzmann distribution is $\langle v \rangle = (8kT/\pi M_r)^{1/2}$ reducing the preexponential factor to $A = \pi d^2 \langle v \rangle$.

For example, the reaction rate coefficient for the production of nitric oxide in the thermosphere by the reaction $N(^4S) + O_2 \rightarrow NO + O$ is given by the Arrhenius form, $K_R(T) = 4.4 \times 10^{-12} \exp(-3220/T_n)\ \mathrm{cm^3\ sec^{-1}}$ with T_n the temperature of the neutral gas in degrees Kelvin. The temperature dependence of the preexponential factor is weak by comparison with the exponential factor.

We have already drawn attention to ion–molecule reactions at thermal energies as important processes in determining the ionization of the upper atmosphere. The cross section for the polarization potential appropriate to exothermic ion–molecule

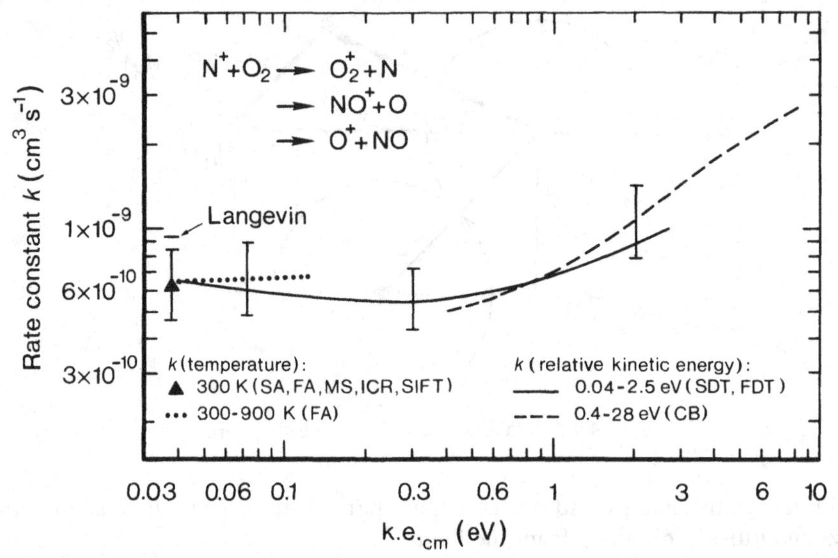

Fig. 4.4.4 Summary of the rate constant data that are available for the reaction of N^+ with O_2 as a function of relative kinetic energy. The temperature-dependent data were plotted here using the conversion k.e.$_{cm} = 3kT/2$. The investigations by experimental methods represented are static afterglow (SA), flowing afterglow (FA), mass spectrometer (MS), static drift tube (SDT), flow drift tube (FDT), ion cyclotron resonance (ICR), selected ion flow tube (SIFT), and crossed beams (CB). (F. Howorka, *et al.*, *J. Chem. Phys.*, **73**, 758, 1980.)

collisions is given by Equation (4.4.8). The reaction rate coefficient is obtained by using the cross section in Equation (4.4.13) to give

$$K_R(T) = 2\pi(\alpha e^2/M_R)^{1/2}\bar{\beta} \tag{4.4.19}$$

This result shows that the rate coefficient is independent of the temperature. The range of applicability of Equations (4.4.8) and (4.4.19) is limited to kinetic energies below 10 eV. The numerical value of K_R for the reaction $N^+ + O_2$ is shown in Figure 4.4.4 together with a summary of experimental results. At thermal energies the mean opacity, $\bar{\beta}$, for this reaction is about $\frac{2}{3}$.

The slow local decay rate of ionization in the F-region is due to the small rate coefficient of the processes that destroy the major ion, O^+. We have already discussed the ion–atom interchange reaction with N_2 and O_2. Another possible mechanism is radiative recombination, $O^+ + e \rightarrow O + h\nu$, the inverse reaction to photoionization. Applying the principle of microscopic reversibility the cross section for radiative recombination, $\bar{\sigma}_r(E_e)$, is related to the photoionization cross section, $\bar{\sigma}_I(\nu)$, by

$$\bar{\sigma}_r(E_e) = \bar{\sigma}_I(\nu)\frac{\tilde{\omega}_O}{\tilde{\omega}_{O^+}}\frac{(E_I)^2}{E_e}\frac{1}{2m_e c^2} \tag{4.4.20}$$

where E_e is the electron energy, m_e the electron mass, E_I the ionization energy, and $\tilde{\omega}_O$ and $\tilde{\omega}_{O^+}$ are the statistical weights of the atom and ion, respectively. The statistical weight of an energy level is $\tilde{\omega} = 2J + 1$ where J is the number of states in the level. This relationship shows that radiative recombination is a slow process (Problem 16).

Equally slow is the related process of dielectronic recombination, $O^+ + e \rightleftharpoons O^{**} \rightarrow O^* + h\nu$, which involves electron capture into an excited state of the atom while also exciting an atomic electron. The doubly excited state lies above the ionization limit in the continuum and is unstable. It can either revert to the ion by the process of auto-ionization or decay radiatively leaving the atom in a non-auto-ionizing state. With auto-ionization being the more likely fate of O^{**}, the dielectronic rate coefficient is generally small.

The dissociative recombination reactions, $NO^+ + e \rightarrow N^* + O$, $O_2^+ + e \rightarrow O^* + O$ and $N_2^+ + e \rightarrow N^* + N$, have large rate coefficients. One or both product atoms are generally in excited electronic states. The process is illustrated in Figure 4.4.5 by one possible dissociative recombination channel in NO^+. The electron is captured into a repulsive state of the excited intermediate NO^* molecule, which leads to dissociation. Similar situations are suggested by the potential curve diagrams for O_2^+–O_2 and N_2^+–N_2 given in Appendix 3. Quantal computations of dissociative recombination coefficients involve vibrational wave functions of the ion and the change in the internuclear distance with energy, the statistical weights of the electronic states of the ion and neutral molecule, the lifetimes for stabilization and radiationless transition of the repulsive molecular state, and the electron temperature. Equations are given in a reference listed with this chapter. Dissociative recombination cross sections or rate

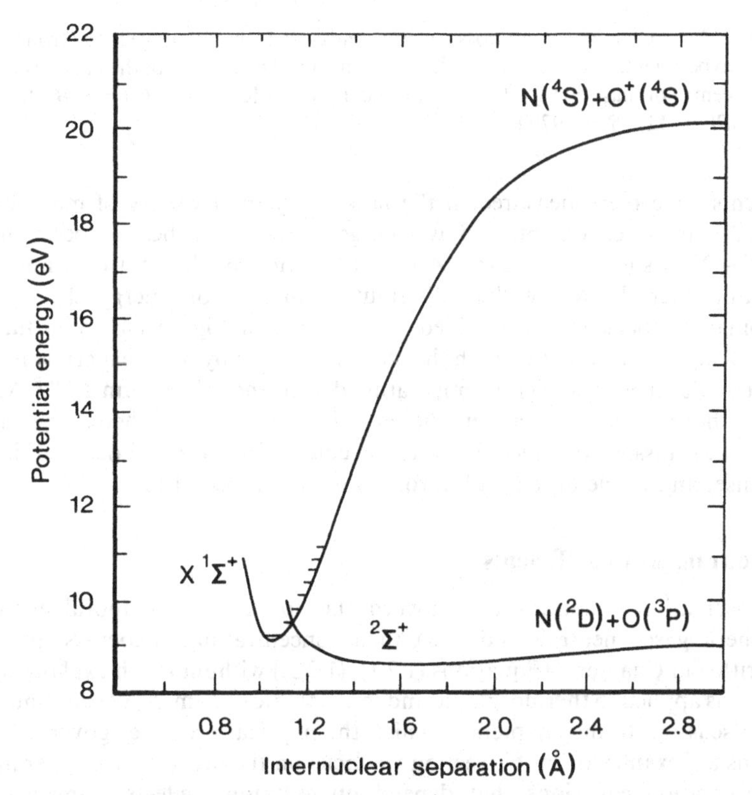

Fig. 4.4.5 Partial potential energy diagram for NO^+–NO showing the ground state ion and excited state neutral curves that contribute to dissociative recombination.

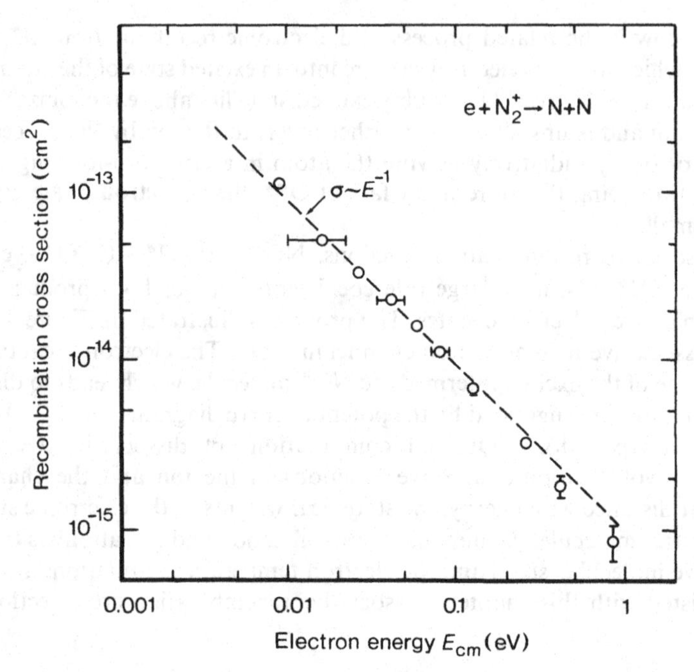

Fig. 4.4.6 e–N_2^+ recombination cross section measured by a merged electron ion–beam experiment. The dashed line shows an inverse dependence of the cross section on the centre of mass energy. (P.M. Mul and J. Wm. McGowan, *J. Phys. B: Atom. Molec. Phys.*, **12**, 1591, 1979.)

coefficients have been measured in the laboratory by a variety of methods. As an example, the cross section obtained by a merged electron–ion beam experiment for $N_2^+ + e \rightarrow N + N$ is shown in Figure 4.4.6 as a function of the centre of mass energy. Integrating over the Maxwellian distribution of electron energy the dissociative recombination coefficient is obtained. It is shown in Figure 4.4.7 as a function of electron temperature, together with the results obtained by other experiments. For the simplest model, theory predicts a temperature dependence of the form $T_e^{-0.5}$. Molecular ions are the dominant species in the E-region of the ionosphere. The large rate coefficient for dissociative recombination accounts for the rapid decay of ionization after sunset and in the absence of auroral sources of ionization.

4.5 The transport coefficients

The conservation equations that govern the spatial and temporal evolution of atmospheric gases (neutrals and ions), their concentrations, velocities, and energies were written in Chapter 1 (Equations (1.2.1)–(1.2.3)) without much explanation of the solutions as applied to thermospheric and ionospheric dynamics. We continue to defer such a discussion to the chapters in which the physical processes governed by these equations are examined. In this section we focus on the properties and parameters of the conservation equations that depend on collisional effects, namely diffusion, viscosity, heat conductivity and mobility. Collisions govern the transport of certain

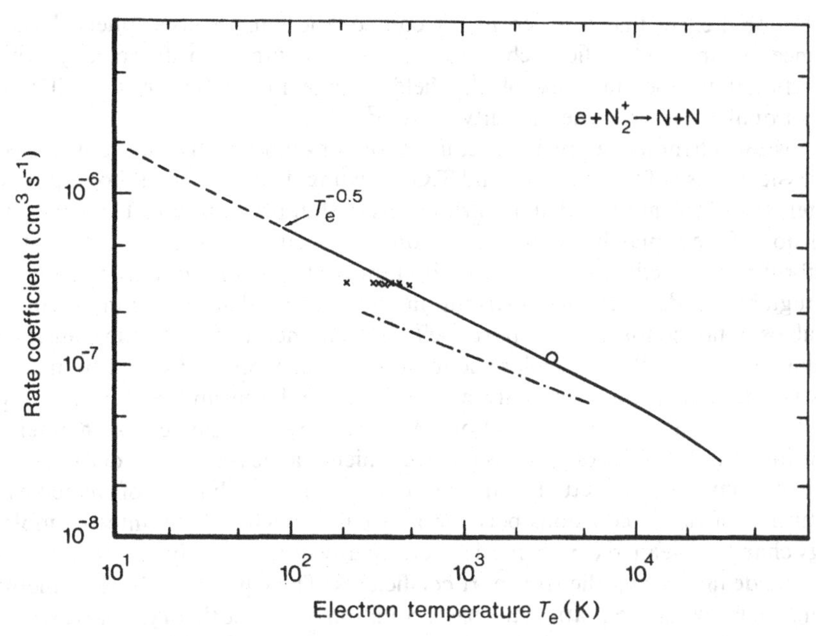

Fig. 4.4.7 Derived rate coefficient for e–N_2^+ dissociative recombination. The dashed line is an extrapolation to lower temperatures using a $T_e^{-0.5}$ dependence. Crosses, circle and the dot-dash curve are data derived from different experiments. (P.M. Mul and J. Wm. McGowan, *J. Phys. B: Atom. Molec. Phys.*, **12**, 1591, 1979.)

molecular quantities from one region of the atmosphere to another: diffusion the transport of mass, viscosity the transport of momentum, thermal conductivity the transport of energy, and mobility the transport of charge.

In the thermosphere different atmospheric species are distributed according to their own partial pressure and large density gradients exist as a function of altitude for any one species and between one species and a background gas. The density gradients cause molecular flow. The flux of molecules is proportional to the density gradient (Fick's law, $\Phi = -D\nabla n$), the flow direction being opposite to the gradient, and the proportionality factor, $D(\text{cm}^2\,\text{s}^{-1})$, is the molecular diffusion coefficient.

The wind speed in the thermosphere generally varies with altitude and gradients are a common occurrence. Thus, one layer of gas exerts a drag on adjacent layers that are moving at different speeds. On the molecular level this viscous force is the rate of momentum change which is proportional (and oppositely directed) to the speed gradient (Newton's law of viscous flow, $F = -\eta\nabla v$) and the proportionality factor is the coefficient of viscosity, η (gm cm^{-1} s^{-1}).

Substantial temperature gradients prevail in the lower thermosphere (Figure 1.1.1) causing adjacent layers of molecules to carry different amounts of kinetic energy. The layers are in contact, however, and transfer of energy, or heat flow, will occur. The heat flux is proportional and oppositely directed to the temperature gradient (Fourier's heat law, $Q = -\lambda\nabla T$) and the proportionality factor is the thermal conductivity, λ (erg cm^{-1} s^{-1} degree^{-1}).

In addition to the transport properties listed above charged particles are subject to an additional effect that leads to major consequences in the ionosphere. Under the influence of an electric field charged particles acquire a drift velocity which is proportional to the intensity of the field (Langevin relation $v_d = \mathscr{K}E$) and the proportionality factor is the mobility $\mathscr{K}$ ($cm^2\,V^{-1}\,s^{-1}$).

Transport phenomena are a consequence of non-uniform distributions in gases. In the classic treatise of S. Chapman and T.G. Cowling (listed in the bibliography to this chapter) the problem is treated at various levels of approximation. These involve the expansion of the actual distribution function of molecular velocities as a sum of a local equilibrium Maxwell–Boltzmann distribution and one or more correction terms. Although high order corrected distribution functions yield accurate kinetic equations, the values obtained for the transport coefficients are no better than the collision cross sections that specify the molecular encounters. Useful expressions have been obtained by assuming that the molecules are hard spheres, and a mean free path is computed from a quasi-equilibrium distribution. Alternatively, a mean collision interval or relaxation time between collisions is defined which is a measure of the decay rate of the difference between the actual and the equilibrium distribution of velocities. The assumption of elastic collisions between molecules precludes any internal molecular energy changes. Mean free path, hard sphere, or any other simplifying assumption need not be made in deriving the transport coefficients. The Chapman–Enskog method of solving the Boltzmann equation, the basic equation of kinetic theory, yields expressions for the transport coefficients in terms of so-called collision integrals which are a function of the differential scattering cross section. The binary collision integrals between species 1 and 2 are

$$\Omega_{1,2}^{(r,s)} = \frac{1}{2\pi^{1/2}} \int_0^\infty \exp\left(-\frac{M_R v^2}{2kT}\right)\left(\frac{M_R v^2}{2kT}\right)^{(2s+3)/2} \phi_{1,2}^{(r)}\,dv \qquad (4.5.1)$$

with

$$\phi_{1,2}^{(1)} = 2\pi \int_0^\pi (1 - \cos^r\theta)\sigma_{1,2}(v,\theta)\sin\theta\,d\theta \qquad (4.5.2)$$

where $\sigma_{1,2}(v,\theta)$ is the differential scattering cross section. The indices r and s denote different types of collision integrals appropriate to different transport coefficients. For example, the binary diffusion coefficient in the one-term expansion of the distribution function is given by

$$D_{12} = \frac{3kT}{16M_R n\Omega_{1,2}^{(1,1)}} \quad (cm^2\,s^{-1}) \qquad (4.5.3)$$

where n is the total molecular number density. The quantity $\phi_{1,2}^{(1)}$ is the diffusion cross section,

$$\phi_{1,2}^{(1)}(v) = 2\pi \int_0^\pi (1 - \cos\theta)\sigma_{1,2}(v,\theta)\sin\theta\,d\theta \quad (cm^2) \qquad (4.5.4)$$

which, on substituting Equation (4.2.12) for the differential cross section, gives

$$\phi_{1,2}^{(1)}(v) = \frac{4\pi\hbar^2}{M_R^2 v^2} \sum_{l=0}^{\infty} (l+1)\sin^2(\eta_l - \eta_{l+1}) \quad (cm^2) \qquad (4.5.5)$$

and

$$\Omega_{1,2}^{(1,1)} = \frac{1}{2\pi^{1/2}}\left(\frac{M_R}{2kT}\right)^{5/2}\int_0^\infty v^5 \exp\left(-\frac{M_R v^2}{2kT}\right)\phi_{1,2}^{(1)}(v)\,dv \qquad (\mathrm{cm^3\,s^{-1}}) \quad (4.5.6)$$

The collision integral can be evaluated if the differential scattering cross section is available.

The Chapman–Enskog solution theoretically predicted that diffusion can also arise from temperature gradients and did so before the effect was measured in the laboratory. It is introduced into the macroscopic molecular flow as a thermal diffusion factor, α_{12}^T, adding the term $-\alpha_{12}^T D_{12}\nabla \ln T$ to Fick's flux equation.

Mobility is a transport coefficient that applies to diffusion of charged particles, electrons, and ions. The earth's thermosphere is a weakly ionized gas so that the collision frequency between ions and neutrals exceeds the collision frequency between ions. Denoting the neutral species in a binary collision by the subscript n and the ion by i, the mobility of the charged particles in the neutral gas is given to a very good approximation (see references) by the so-called Einstein relation,

$$\mathscr{K}_{in} = eD_{in}/kT_i \qquad (\mathrm{cm^2\,V^{-1}\,s^{-1}}) \qquad (4.5.7)$$

where e is the electric charge of the ion. Ion and electron mobilities are the physical properties that are adopted in the derivation of the electrical conductivity in the ionosphere, presented in Chapter 8.

In the preceding section we noted that the interaction between ions and atoms at a separation greater than a cutoff distance is dominated by a polarization potential given by

$$V(r) = -\alpha e^2/2r^4 \qquad (4.4.7)$$

where α is the polarizability of the neutral species. The inverse fourth power variation of the interaction potential has the useful property that the transport coefficients can be computed without knowing the distribution function. Several references listed with this chapter discuss the quantal aspects of this problem such as the simple forms assumed by the phase shifts. The diffusion cross section obtained from quantal computations is

$$\phi_{1,2}^{(1)}(v) = 2.210\pi(\alpha e^2/M_R v^2)^{1/2} \qquad (4.5.8)$$

where v is the relative speed of the particles before collision and the numeric originates from the Bessel function of the second kind in the solution. The collision integral can be evaluated analytically,

$$\Omega_{1,2}^{(1,1)} = \tfrac{3}{16}2.210\pi(\alpha e^2/M_R)^{1/2} \qquad (4.5.9)$$

and the diffusion coefficient defined by Equation (4.5.3) becomes

$$D_{12} = \frac{kT}{2.210\pi M_R n}\left(\frac{M_R}{\alpha e^2}\right)^{1/2} \qquad (\mathrm{cm^2\,s^{-1}}) \qquad (4.5.10)$$

The ion mobility is given by the simple expression

$$\mathscr{K}_{12} = 1/2.210\pi n(\alpha M_R)^{1/2} \qquad (4.5.11)$$

with the mixed units of $\mathrm{cm^2\,V^{-1}\,s^{-1}}$.

The coefficient of viscosity in a single component gas composed of molecules of mass M, using the Chapman–Enskog one-term expansion, is given by

$$\eta = 5kT/8\Omega^{(2,2)} \qquad (\mathrm{gm\,cm^{-1}\,s^{-1}}) \qquad (4.5.12)$$

and the thermal conductivity is

$$\lambda = 75k^2T/32M\Omega^{(2,2)} \qquad (\text{erg cm}^{-1}\,\text{s}^{-1}\,K^{-1}) \qquad (4.5.13)$$

These two transport coefficients involve a cross section given by

$$\phi^{(2)}(v) = 2\pi \int_0^{\pi} \sigma(v,\theta)\sin^3\theta\,d\theta \qquad (\text{cm}^2) \qquad (4.5.14)$$

and a collision integral

$$\Omega^{(2,2)} = \frac{1}{2\pi^{1/2}}\left(\frac{M_R}{2kT}\right)^{7/2} \int_0^{\infty} v^7 \exp\left(-\frac{M_R v^2}{2kT}\right)\phi^{(2)}(v)\,dv \qquad (\text{cm}^3\,\text{s}^{-1}) \quad (4.5.15)$$

Collision integrals are functions analogous to the rate coefficients defined previously in that a cross section is weighted by a distribution function of molecular velocities. Differential scattering cross sections obtained experimentally may be used to obtain the transport cross sections, collision integrals, and transport coefficients by numerical integration of the equations given above. Actually, it is easier to measure some transport coefficients experimentally and use the theoretical relationships between them to obtain numerical values for those that may be more difficult to measure. Values obtained experimentally can then be compared with theoretically derived coefficients assuming an analytic function for the atomic, molecular or ionic interaction potential, of which several have been proposed. Much work has been done in this area over the years and we refer the reader to the bibliography to this chapter.

Transport coefficients for atmospheric gases and ions are listed in Appendix 6. We note that diffusion, viscosity, and thermal conductivity coefficients are functions of the density. The mobility is a function of density, mass, and the polarizability. Convenient empirical formulae have been developed to obtain the transport coefficients for any value of temperature, within a specified range, given a value at one temperature. For example, the viscosity η at temperature T(K), given the viscosity η_0 at T_0, is

$$\eta = \eta_0\left(\frac{T}{T_0}\right)^{3/2}\frac{T_0 + S - 273}{T + S - 273} \qquad (4.5.16)$$

where S is the so-called Sutherland's constant which has a value of 104.7 for N_2 in the temperature range between 293 and 1100 K.

The transport coefficients for mixtures of gases can be derived, in principle, by adopting the appropriate collision integrals, $\Omega_{i,j}$, and cross sections, $\phi_{i,j}$. Convenient formulae have also been developed (see references).

The thermal conductivity is simply related to the viscosity by Equations (4.5.12) and (4.5.13),

$$\lambda/\eta = 15\,k/4M \qquad (4.5.17)$$

in reasonable agreement with the experimental values listed in Appendix 6.

Both binary diffusion and self-diffusion occur in the atmosphere. Since the collision integrals are functions of the mass of the collision partners the coefficients of binary, D_{12}, and self-diffusion, D_{11}, are not expected to be identical, in addition to the possibility of different interaction potentials governing the encounters. Diffusion coefficients are a function of the particle concentration (in addition to the mass and temperature) and numerical values usually refer to standard conditions of 1 atm

($n = 2.7 \times 10^{19}\,\mathrm{cm}^{-3}$). The temperature dependence of the binary diffusion coefficients is given approximately by

$$\mathrm{d}(\ln D_{12})/\mathrm{d}(\ln T) = 2 - \tfrac{5}{2}C \qquad (4.5.18)$$

where the parameter C is in the range 0–$\frac{2}{15}$ depending on the molecular model adopted. It is zero for molecular encounters where the mutual repulsive force varies as the inverse fifth power of their distance apart, the same as polarization but with different sign. The coefficient of self-diffusion is related to the viscosity by the simple equation

$$nMD_{11}/\eta = 3A \qquad (4.5.19)$$

where the parameter A has values between 1.2 and 1.55, the latter appropriate to Maxwell molecules. As previously mentioned, experimentally obtained values of the transport coefficients provide insight to the appropriate model of molecular interaction potentials. Equations (4.5.12), (4.5.13) and those derived therefrom are valid for a single component gas.

The ion mobility is related to the binary diffusion coefficient by the Einstein relation (Equation (4.5.7)) but it should be pointed out that D_{12} now refers to the diffusion of an ion in a neutral gas. Typical values of ion mobility are given in Appendix 6. The process of charge transfer dominates the mobility of ions in their parent gas.

Bibliography

The topics in basic physics that underlie collision and reaction phenomena are the kinetic theory of gases and atomic and molecular physics. For the former topic we have drawn primarily on the development by Chapman and Enskog, presented in the classic monograph

S. Chapman and T.G. Cowling, *The Mathematical Theory of Non-Uniform Gases* (third edition) Cambridge University Press, Cambridge, 1970.

A different theoretical approach to the kinetic theory of gases is discussed by its originator

H. Grad, Principles of the kinetic theory of gases, *Handbuch der Physik* Vol. 12, Springer Verlag, Berlin, 1958, pp. 205–94.

A comprehensive book on the subject is

J.O. Hirschfelder, C.F. Curtiss and R.B. Bird, *The Molecular Theory of Gases and Liquids*, John Wiley, New York, 1954.

There are several additional texts on the kinetic theory of gases, for example, those by J.H. Jeans and L.B. Loeb. Of the many texts on statistical mechanics, we have found one that is especially useful for our purposes.

D.A. McQuarie, *Statistical Mechanics*, Harper and Row, New York 1976.

Our sketchy presentation of collision theory, giving final equations without derivation requires that the reader consult texts and review articles on various aspects of the subject. We list those books from which material has been drawn for this chapter.

D.R. Bates (Ed.), *Atomic and Molecular Processes*, Academic Press, New York, 1962.

J.B. Hasted, *Physics of Atomic Collisions*, Butterworth and Co., London, 1972.

E.W. McDaniel, *Collision Phenomena in Ionized Gases*, John Wiley, New York, 1964.

H.S.W. Massey, E.H.S. Burhop and H.B. Gilbody, *Electronic and Ionic Impact Phenomena*, Vols. 1, 2, 3, Clarendon Press, Oxford, 1969.

I.W.M. Smith, *Kinetics and Dynamics of Elementary Gas Reactions*, Butterworth and Co., London, 1980.

R.E. Johnson, *Introduction to Atomic and Molecular Collisions*, Plenum Press, New York, 1982.

A.R. Hochstim (Ed.), *Kinetic Processes in Gases and Plasmas*, Academic Press, New York, 1969.

E.W. McDaniel, V. Cermak, A. Dalgarno and E.E. Ferguson, *Ion-Molecule Reactions*, John Wiley, New York, 1970.

M.J. McEwan and L.F. Phillips, *Chemistry of the Atmosphere*, John Wiley, New York, 1975.

A five-volume treatise contains several chapters written by specialists in various aspects of collision phenomena relevant to upper atmosphere physics. The first volume is particularly useful.

H.S.W. Massey, E.W. McDaniel and B. Bederson, *Applied Atomic Collision Physics*, Vol. 1, *Atmospheric Physics and Chemistry*, Academic Press, New York, 1982.

Useful review articles frequently appear in an annual review series,

D.R. Bates and B. Bederson (Eds.) *Advances in Atomic and Molecular Physics*, Vols 1–20, Academic Press, New York, 1965–85.

References from which specific data have been extracted are given in the figure captions. An extended discussion of reactive collisions as seen through correlation of states is given by

J.J. Kaufman, Potential energy surface considerations for excited state reactions, Chapter 12 in *The Excited State in Chemical Physics*, Ed. J.W. McGowan, Vol. 28 in the series Advances in Chemical Physics, John Wiley, New York, 1975.

The above reference adopts the ionospherically important reaction $O^+ + N_2 \rightarrow NO^+ + N$ as an application of the general concepts. A more general discussion of ion–neutral interactions (and a lengthy list of references) is the review article

W. Lindinger, State selected ion-neutral interactions at low energies, in *Electronic and Atomic Collisions*, Eds. J. Eichler, I.V. Hertel and N. Stolterfoht, North Holland, Amsterdam, 1984, pp. 415–28.

A recent discussion of dielectronic recombination also provides references to previous work on the subject,

H. Nussbaumer and P.J. Storey, Dielectronic recombination at low temperatures, *Astron. Astrophys.*, **126**, 75, 1983.

Radiative recombination is discussed in Chapter 7 of a monograph already listed above,

D.R. Bates and A. Dalgarno, Electronic recombination, Chapter 7 in *Atomic and Molecular Processes*, Ed. D.R. Bates, Academic Press, New York, 1962.

and in a recent review article

D.R. Bates, Aspects of recombination, in *Advances in Atomic and Molecular Physics*, Vol. 15, Academic Press, New York, 1980.

Transport coefficients are discussed in several of the general references already listed above (see S. Chapman and T.G. C. Cowling) and in more specialized works,

E.A. Mason and E.W. McDaniel, *Transport Properties of Ions in Gases*, J. Wiley (Interscience) New York, 1985.

The basic references on ion mobilities are

A. Dalgarno, M.R.C. McDowell and A. Williams, The mobilities of ions in unlike gases, and A. Dalgarno, The mobilities of ions in their parent gases, *Phil. Trans. Roy. Soc. London*, **A250**, 411–39, 1958.

Atomic and molecular polarizabilities are reviewed and numerical values are given in the treatise by J.O. Hirschfelder *et al.* (listed above) and by

T.M. Miller and B. Bederson, Atomic and molecular polarizabilities, in *Advances in Atomic and Molecular Physics*, Vol. 13, Academic Press, New York, 1977, pp. 1–56.

Problems 9–16

Problem 9: Assuming that one of the particles in a binary collision is initially at rest, derive the relationships between the scattering angles and velocities in the centre of mass and laboratory coordinate systems.

Problem 10: Use classical mechanical scattering theory to show that the Coulomb potential leads to a differential cross section identical with the quantum mechanical result. Consult texts listed in the bibliography.

Problem 11: Compute the differential scattering cross section for 10 keV electrons scattered by nitrogen atoms in the range of 0–180°. Use the screened Coulomb potential in the Born approximation.

Problem 12: Compute the differential scattering cross section for 10 keV electrons scattered by N_2 in the range 0–180° assuming the most probable internuclear distance to be 1.1 Å (Appendix 3).

Problem 13: Sketch the phase function for elastic scattering of 1 keV, 10 keV and 100 keV electrons on N_2 from 0 to 180°. Use a linear ordinate and abscissa for scattering angles between 0 and 20°, noting the oscillatory nature of the phase function.

Problem 14: Compute the total elastic scattering cross section for electrons on N_2, O_2 and O between 1 keV and 100 keV electron energy using the Born approximation for a screened Coulomb potential. Use these results to extend to higher energies the cross section curves in Appendix 4.

Problem 15: Starting with the definition of the reaction rate coefficient, Equation (4.4.9), derive Equations (4.4.12) and (4.4.13) for the coefficients appropriate to the equilibrium distribution of the reactants. Also derive an expression for the reaction rate coefficient when the temperatures of the reactants are not the same.

Problem 16: Find the ratio of the radiative recombination cross section to the photoionization cross section for thermal electrons ($T_e = 10^3$ K) recombining with ground state O^+ ions.

5

Ion and neutral composition of the upper atmosphere

5.1 Formulation of the problem

The density and composition of the thermosphere are not uniform over the globe and variations at a given altitude are a function of solar UV illumination, auroral energy input and the effects of transport from one region to another. Variations in density and composition with latitude and longitude have been observed directly by neutral and ion mass spectrometers carried on board many satellites. Horizontal variations in density and composition are small, however, compared to variations with altitude. The ratio of the total density at the lower boundary of the thermosphere to the density at the upper boundary is about seven orders of magnitude which basically accounts for the large change in composition with altitude throughout the region. Consideration of the latitudinal and longitudinal variability will be deferred to Chapter 8 where we discuss three-dimensional thermospheric and ionospheric dynamics. In this chapter we focus on processes that operate locally or vary primarily with altitude, leading to one-dimensional equations. At the lower boundary of the thermosphere, the mesopause, the neutral atmosphere is essentially fully mixed but above this level the composition changes markedly with altitude. Even greater variability prevails in the ion composition.

The mass conservation equation (Equation (1.2.1)) is the starting point for our analysis. For each species j we have

$$m_j \frac{\partial n_j}{\partial t} + m_j \frac{\partial}{\partial z}(n_j v_j) = m_j P_j - m_j L_j \tag{5.1.1}$$

where $P_j(z)$ and $L_j(z)$ are the local production and loss rates of species j with mass m_j and concentration n_j, and the terms on the left side of the equation are the time rate of change of $n_j(z)$ at a fixed point and the divergence of the mass flux $m_j(\partial/\partial z)(\phi_j)$, in the vertical z direction with $v_j(z)$ the vertical velocity of the jth component. The total mass density is

$$\rho(z) = \sum_j m_j n_j(z) \tag{5.1.2}$$

and a mass averaged velocity is defined as

$$v(z) = (1/\rho(z))\sum_j m_j n_j(z) v_j(z) \tag{5.1.3}$$

Summing Equation (5.1.1) over all species j and using Equations (5.1.2) and (5.1.3) we obtain

$$\frac{\partial \rho(z)}{\partial t} + \frac{\partial}{\partial z}(\rho v) = 0 \tag{5.1.4}$$

Under steady state conditions the divergence of the total mass flux is zero and the mass flux is therefore a constant. Numerical estimates of the terms in the momentum balance equation (Equation (1.2.2)) show that the vertical acceleration term is quite small compared with the two major forces, the pressure gradient and the gravitational force. To a very good approximation these two forces balance so that

$$\partial p/\partial z = -\rho g \tag{5.1.5}$$

where p is the pressure and g is the gravitational acceleration. The atmosphere is assumed to obey the equation of state of an ideal gas which relates the pressure, density and temperature,

$$p = nkT \tag{5.1.6}$$

Defining a mean mass $\bar{m}$ by

$$\bar{m} = \sum_j m_j n_j / \sum_j n_j \tag{5.1.7}$$

gives

$$\rho g = \bar{m} n g \tag{5.1.8}$$

and

$$\partial p/\partial z = -p(\bar{m}g)/kT \tag{5.1.9}$$

Integration of this equation from a reference level z_0 where the pressure is p_0 yields the pressure at some arbitrary altitude z

$$p = p_0 \exp\left[-\int_{z_0}^{z} (\bar{m}g/kT)\,dz'\right] \tag{5.1.10}$$

Thus, the atmospheric pressure decreases exponentially with increasing altitude. Three quantities in the integrand vary with altitude, the mean mass $\bar{m}$, the gravitational acceleration g and the temperature T. In the thermosphere separation of species with different masses caused by the changes of pressure with height usually prevails over mixing and each constituent tends to a distribution with altitude governed by its own partial pressure.

5.2 Neutral gas diffusion

In our examination of the processes that govern the altitude profile and temporal variation of thermospheric and ionospheric species we first consider the transport term, $m_j(\partial/\partial z)(n_j v_j)$, in the continuity equation, Equation (5.1.1). Vertical motion of individual species relative to each other is due to diffusion but a parcel of air can also have a bulk mass averaged velocity, which is a vertical wind. While not large compared with horizontal components, vertical winds have been measured during periods of intense thermospheric heating, a topic that will be taken up in Chapter 8. Vertical winds counteract the diffusive separation of species with different masses, thereby influencing the composition of the neutral gas as a function of altitude. The diffusion velocity of species j is measured with respect to the mass averaged velocity, $v(z)$, and the transport term becomes

$$m_j \frac{\partial}{\partial z}(n_j v_j) = m_j \frac{\partial}{\partial z}[n_j(v_j - v)] + m_j \frac{\partial}{\partial z}(n_j v) \qquad (5.2.1)$$

In the preceding chapter diffusion was defined in terms of gradients in molecular density. The general diffusion equation appropriate to a non-equilibrium distribution in a gas has additional terms, namely a term due to a temperature gradient (already mentioned in Chapter 4) and a term due to the gradient in the total pressure. In the atmosphere it is the gravitational force that sets up the pressure gradients (Equation (5.1.5)). We consider, for the moment, diffusion of neutral species only, and will add subsequently electromagnetic forces which only act on the ions. The vertical diffusion velocity for species j in a multicomponent gas is given by

$$v_j(z) - v(z) = \sum_i \frac{n}{n_i} D_{ji} \left[\frac{\partial \ln n_j}{\partial z} + (1 + \alpha_{ji}^T) \frac{\partial \ln T}{\partial z} - \frac{m_j g}{kT} + \frac{\bar{m}}{kT} F \right] \qquad (5.2.2)$$

The notation has already been defined except for the quantity F which represents the sum of the pressure and gravitational forces per unit mass acting on the gas as a whole; α^T is the thermal diffusion coefficient. The subscript i refers to any other species besides j. Numerical values of the diffusion coefficients are usually given for binary diffusion, D_{ji}, and for self-diffusion, D_{jj}, (see Appendix 6) rather than for a multicomponent gas. In the first approximation of the Chapman–Enskog solution the multicomponent coefficients reduce to a summation of the binary coefficients providing an adequate description of diffusion in the atmosphere.

There is another mechanism that contributes to the transport properties in the atmosphere which is due to non-laminar flow of the gas. In the thermosphere turbulent flow is probably initiated by waves rather than by shocks; these waves can originate within the thermosphere or they can be propagated from below. A wide range of scale sizes exists, with turbulence tending to mix the atmospheric gases. Small scale turbulent mixing leads to eddy diffusion, eddy viscosity, and eddy heat conduction and a term to account for this effect, $D_E(\partial/\partial z)(\ln m_j n_j/\rho)$, is added to the equation for the diffusion velocity, Equation (5.2.2). The eddy diffusion coefficient, D_E, depends on the scale size and degree of turbulence, and observations of the phenomenon in the atmosphere are used for numerical estimates. Eddy diffusion becomes an important mass transport mechanism near the bottom of the thermosphere, below about 110 km, and it dominates over molecular diffusion below about 90 km where complete mixing prevails.

The time scale for changes in the density of the major neutral atmospheric species is generally long for diurnal and seasonal variations, and a quasi-steady state can be assumed to exist, except during twilight and periods of large geomagnetic and auroral storms. With this assumption, which is not always justified, the force F on the gas as a whole does not contribute to the vertical diffusion velocity and Equation (5.2.2) reduces to

$$v_j(z) - v(z) = \sum_i \frac{n}{n_i} D_{ji}\left[\frac{\partial \ln n_j}{\partial z} + (1 + \alpha_{ji}^T)\frac{\partial \ln T}{\partial z} - \frac{m_j g}{kT}\right] + D_E \frac{\partial}{\partial z}\left[\ln\left(\frac{m_j n_j}{\rho}\right)\right] \tag{5.2.3}$$

Above an altitude where eddy mixing is no longer effective an additional restrictive assumption is frequently made, that there is no net vertical mass flux. Thus, in a steady state when local production and loss rates balance the continuity equation reduces to an equation representing diffusive equilibrium,

$$\frac{\partial n_j}{\partial z} + (1 + \alpha_j^T)\frac{n_j}{T}\frac{\partial T}{\partial z} + \frac{n_j m_j g}{kT} = 0 \tag{5.2.4}$$

which has the solution

$$n_j(z) = n_j(z_0)\left[\frac{T(z_0)}{T(z)}\right]^{(1+\alpha_j)} \exp\left[-\int_{z_0}^{z} \frac{m_j(z')g(z')}{kT(z')}\,dz'\right] \tag{5.2.5}$$

Adopting an empirical representation of the variation of temperature with altitude, Equation (5.2.5) has been used to construct altitude profiles of neutral atmospheric constituents and the result is known as an analytic model atmosphere. The concentrations and temperature at the lower boundary, z_0, determine the absolute value of $n_j(z)$. Several satellites orbiting the Earth within the thermosphere have carried mass spectrometers capable of measuring concentrations of various atmospheric gases. The results of thousands of individual measurements over a range of altitudes (and geographic extent) have been used to construct empirical model atmospheres (Appendix 1). Empirical models have been helpful in understanding the processes that control atmospheric composition and density as well as testing various assumptions that are made to reduce the theoretical problem to soluble levels (Problem 17).

Any departure from complete hydrostatic equilibrium (Equation (5.1.5)) and diffusive equilibrium (Equation (5.2.4)) predicts mass motion which can only be correctly examined using the momentum conservation equations. Additionally, the equation of state of a gas relates the pressure, density, and temperature, so that even in diffusive equilibrium the density distribution cannot be established independently of the temperature distribution. Temperature is obtained by solving the energy conservation equation. We frequently need to remind ourselves that the physical and chemical processes that establish the composition, motion, and thermal characteristics of the atmosphere are coupled. Depending on the specific goal certain assumptions and simplifications are well justified but these may not be the same ones for different goals.

Only the transport term in the continuity equation, Equation (5.1.1), has been discussed so far while local production and loss rates of species j have been ignored. We pointed out earlier that the major molecular constituents of the thermosphere, N_2 and O_2, are products of the long-term evolution of the Earth's atmosphere and their concentration in the thermosphere is unaffected by local sources and sinks. O, the third major neutral constituent, is produced by photodissociation of O_2 (as described in Chapter 2) and destroyed by the three-body recombination reaction,

$O + O + M \rightarrow O_2 + M$, which becomes effective only at the base of the thermosphere, below about 110 km. Local production and loss of O have a negligible effect on the thermospheric O density compared with transport effects (Problem 18). The concentration of all three major neutral species is therefore controlled by transport processes or dynamic effects. Vertical transport, diffusion and mixing have been treated in this section but we defer to Chapter 8 a discussion of horizontal transport.

Local sources and sinks are very important in determining the concentrations of the minor neutral thermospheric species N and NO, of all the ionic species, O^+, O_2^+, N_2^+, NO^+, N^+, H^+ and He^+ and of the ambient electrons. The continuity equation, Equation (5.1.1), for these species therefore includes production, P_j, and loss, L_j, terms. Before taking up the complete continuity equations we first look at the transport terms for charged particles.

5.3 Ion diffusion

Ion motion occurs in response to density gradients, temperature gradients and the gravitational field, analogous to the neutral species but, in addition, charged particles experience a force due to electric and magnetic fields. An additional term appears in the equation for the vertical diffusion velocity (Equation (5.2.2)), the Lorentz force term

$$D_{ij}\frac{q_j}{kT}\left[E_z + \frac{1}{c}(\mathbf{v}_j \times \mathbf{B})_z\right] \tag{5.3.1}$$

in cgs–emu units. The quantity $D_{ij}q_j/kT$ was identified as the ion mobility, $\mathscr{K}_{ij}$, by the Einstein relation (Equation (4.5.7)), where q_j is the electric charge of the ion and D_{ij} is the binary diffusion coefficient of ion j in a neutral gas i.

The derivation of the diffusion coefficient by the Chapman–Enskog theory assumes that various species in the gas mixture have a common temperature, T. This assumption is implicit in the definition of the collision integral, Equation (4.5.1). It is justified for evaluating the transport properties of the neutral gases in the thermosphere. Temperature sensors carried by satellites since the beginning of space exploration have shown, however, that the temperature of the ions and the electrons can be substantially higher than the neutral gas temperature. In retrospect, this is to be expected, and we will discuss temperatures at length in Chapter 6 dealing with the energetics of the upper atmosphere. The effect of unequal electron, ion, and neutral gas temperatures may be included in the binary diffusion coefficient and the mobility by defining a reduced temperature

$$T_R = M_R\left(\frac{T_2}{m_2} + \frac{T_1}{m_1}\right) \tag{5.3.2}$$

where M_R is the reduced mass (Equation (4.2.3)). T_R replaces T in the collision integral (Equation (4.5.1)) and the diffusion coefficient (Equation (4.5.3)). The collision term for momentum exchange in the transport equations derived by the so-called 13-moment approximation of Grad makes use of the reduced temperature concept. Additional reading on this topic is listed in the bibliography.

There are two components to the Lorentz force term (Expression (5.3.1)); one, due to an electric field, E_z, is parallel to the ion drift direction, v_z, the other due to the vertical velocity component of ions drifting in a magnetic field is due to ion motion orthogonal to v_z. Considering, for the moment, only the direction of the vectors, the vertical ion velocity reaches a maximum at the magnetic equator where the magnetic field is horizontal and for a horizontal ion velocity that is orthogonal to the magnetic field direction. Approaching the poles the geomagnetic field vector becomes vertical and the $(\mathbf{v}_j \times \mathbf{B})_z$ component vanishes. The contribution of this term to the vertical ion drift depends not only on the direction of the vectors but also on the magnitude of the horizontal ion velocity, i.e., the ion convection pattern. Ion convection velocities are largest in the auroral oval and polar cap at geomagnetic latitudes above about 60° where the magnetic field vector becomes increasingly more vertical. The electric field component of the Lorentz force is very small in the ionosphere due to the high mobility of electrons along the magnetic field inhibiting the formation of potential gradients.

With the aim of studying the vertical structure of ion composition in the thermosphere, we examine the forces that influence their motion. The large difference in mass between the positive ions and electrons gives rise to a corresponding difference in the gravitational force terms in the respective equations for the diffusion velocity. Separation of the oppositely charged species is prevented, however, by the development of a polarization field forcing the ions and electrons to drift as a single gas. This maintains bulk charge neutrality in the ionosphere,

$$\sum_i n(\mathrm{i}) = n(\mathrm{e}) \tag{5.3.3}$$

where $n(\mathrm{i})$ and $n(\mathrm{e})$ are the ion and electron concentrations and the summation is over all species of ions. The term applied to this drift motion parallel to the magnetic field vector is ambipolar diffusion and the coefficient is defined by Fick's law, which in Chapter 4 defined the binary diffusion coefficient,

$$\Phi = -D_\mathrm{a}\nabla n \tag{5.3.4}$$

To be a useful parameter the ambipolar diffusion coefficient, D_a, must be obtainable from experimentally measurable or theoretically derivable parameters, the binary diffusion coefficients and the ion mobilities. The ion and electron fluxes are governed by the density gradients and the polarization field which tends to slow down the electrons and speed up the ions,

$$\Phi_i = -D_\mathrm{in}\nabla n + \mathscr{H}_\mathrm{in}En \tag{5.3.5}$$

$$\Phi_\mathrm{e} = -D_\mathrm{en}\nabla n - \mathscr{H}_\mathrm{en}En \tag{5.3.6}$$

where E is the polarization field, n is the ion or the electron density, D_in and D_en are the binary ion–neutral and electron–neutral diffusion coefficients and $\mathscr{H}_\mathrm{in}$ and $\mathscr{H}_\mathrm{en}$ the corresponding mobilities. Eliminating the polarization field between Equations (5.3.5) and (5.3.6) and writing the result in the form of Equation (5.3.4) we find that

$$D_\mathrm{a} = \frac{D_\mathrm{in}\mathscr{H}_\mathrm{en} + D_\mathrm{en}\mathscr{H}_\mathrm{in}}{\mathscr{H}_\mathrm{in} + \mathscr{H}_\mathrm{en}} \tag{5.3.7}$$

Due to their smaller mass, the mobility of electrons parallel to the magnetic field is substantially larger than the ion mobility, $\mathscr{H}_\mathrm{en} \gg \mathscr{H}_\mathrm{in}$. Using the Einstein relation, $D = kT\mathscr{H}/e$, the ambipolar diffusion coefficient reduces to

$$D_\mathrm{a} = D_\mathrm{in}(1 + T_\mathrm{e}/T_\mathrm{R}) \tag{5.3.8}$$

where T_R is the reduced temperature for the ions and neutrals

$$T_R = M_R\left(\frac{T_i}{m_i} + \frac{T_n}{m_n}\right) \tag{5.3.9}$$

with M_R the reduced mass and T_e the electron temperature (Problem 19). We will learn in Chapter 6 that the electron temperature is generally higher than the ion temperature, both being higher than the neutral kinetic temperature. Therefore, the ambipolar diffusion coefficient is usually more than twice as large as the ion–neutral binary diffusion coefficient and transport of ionization by ambipolar diffusion is defined as operating parallel to the magnetic field vector.

The diffusion coefficient and the mobility are derived from the collision integrals (Equation (4.5.1)) that depend on the diffusion cross sections (Equation (4.5.4)). It is therefore the interaction between charged particles and neutrals that governs these transport coefficients. We illustrated this concept in Chapter 4 by noting that an induced dipole polarization potential (Equation (4.4.7)) dominates the interaction between an ion and a neutral at large separation, with more complex potential functions required to take account of the short-range interaction. In the F-region of the ionosphere the dominant ion, O^+, diffuses in the dominant neutral species, O, and the interaction of an ion in its parent gas is primarily by resonance charge transfer,

$$O^+ + O \rightarrow O + O^+ \tag{5.3.10}$$

The interaction potential still has the inverse fourth power dependence of dipole polarization but the diffusion cross section must be modified to take account of the identical nuclear fields of the colliding particles. Resonance charge transfer is discussed in more detail in Section 6.4.2.

The occurrence of several species of ions and electrons in the ionosphere considerably complicates the formulation of the transport properties, the resulting equations and the numerical solutions. Ions can interact with each other in ion–ion collisions, with electrons in ion–electron collisions, and electrons can collide with other electrons. Electrostatic forces between charged particles are specified by the long-range Coulomb potential (described in Chapter 4) which decreases much more slowly with increasing distance between particles compared to all other intermolecular forces. A given ion may therefore find itself in the field of several ions; collisions are no longer strictly binary and collective or plasma interactions need to be considered.

The relative magnitude of the ion–neutral collision frequency and the Coulomb collision frequency determines whether or not collisions between charged species need to be included in the ion diffusion equation. The collision frequency (s^{-1}) is a convenient parameter commonly used in ionospheric studies that is defined in terms of the collision integral for diffusion, $\Omega_{i,j}^{(1,1)}$ (Equation (4.5.1)),

$$\nu_{ij} = \tfrac{16}{3}\frac{n_j}{m_i} M_R \Omega_{i,j}^{(1,1)} \tag{5.3.11}$$

In the E-region of the ionosphere the degree of ionization only reaches about one part per million and the ion–neutral collision frequency is much larger than the Coulomb collision frequency. However, in the F-region there may be one ion per thousand neutrals which is sufficient to make the Coulomb collision frequency comparable to or even larger than the ion–neutral collision frequency.

Diffusion coefficients may be expressed in terms of the collision frequency; for example, the ambipolar diffusion coefficient is given by

$$D_a = \frac{kT_R}{m_i \nu_{in}}\left(1 + \frac{T_e}{T_R}\right) \tag{5.3.12}$$

where ν_{in} is the ion–neutral collision frequency.

We have shown that ambipolar diffusion governs field-aligned transport of the positive ions and the electrons, constraining both to one velocity, the ambipolar diffusion velocity. There are several ion species in the ionosphere, the molecular ions NO^+ and O_2^+ being the major species in the E-region (even though N_2 is the major neutral constituent), the atomic ion O^+ in the F-region, with a gradual transition to He^+ and H^+ at the highest levels of the ionosphere. Ambipolar diffusion couples the transport of all ion species and electrons. Denoting the electron, ion and neutral drift velocities by v_e, v_i and v_n, respectively, the electron diffusion velocity with respect to the diffusion velocity of the neutrals, assumed to be the same for all species, is given by,

$$\mathbf{v}_e - \mathbf{v}_n = -\frac{1}{n_e m_e \sum_n \nu_{en}}(\nabla p_e - n_e m_e \mathbf{g} + n_e e\mathbf{E}) \tag{5.3.13}$$

where the sum of the collision frequencies ν_{en} is over the neutral species, p_e is the scalar pressure and $\mathbf{E}$ is the polarization electric field. Drift motion is assumed to be parallel to the magnetic field vector. The analogous diffusion velocity equation for each ion is

$$\mathbf{v}_i - \mathbf{v}_n = -\frac{1}{n_i m_i \sum_n \nu_{in}}(\nabla p_i - n_i m_i \mathbf{g} - n_i e\mathbf{E}) \tag{5.3.14}$$

Solving Equation (5.3.13) for $e\mathbf{E}$ and putting the result into Equation (5.3.14), introducing the equations of state $p_e = n_e kT_e$ and $p_i = n_i kT_i$, and neglecting the small terms in m_e/m_i we obtain the diffusion velocity

$$\mathbf{v}_i - \mathbf{v}_n = -\frac{1}{m_i \nu_i}\left[\frac{kT_i}{n_i}\nabla n_i + \frac{kT_e}{n_e}\nabla n_e + k\nabla(T_i + T_e) - m_i \mathbf{g}\right] \tag{5.3.15}$$

where we have written ν_i for $\sum_n \nu_{in}$. In the upper F-region O^+ ions are the dominant charged species and, to a good approximation, $n(O^+) \approx n(e)$. The O^+ velocity is then given by

$$\mathbf{v}_i - \mathbf{v}_n = -D_a\left[\frac{\nabla n_i}{n_i} + \frac{\nabla(T_i + T_e)}{T_i + T_e} - \frac{m_i \mathbf{g}}{k(T_i + T_e)}\right] \tag{5.3.16}$$

where the ambipolar diffusion coefficient, D_a, is given by Equation (5.3.8). The quantity $(T_i + T_e)/2$ is frequently called the effective plasma temperature, T_p (Problem 20).

Equations (5.3.13)–(5.3.16) were derived with the tacit assumption of a vertical magnetic field, but the geomagnetic field is actually inclined at an angle I measured from the horizontal. The field-aligned component of v_i is therefore $v_i \sin I$ and its vertical component is $v_i \sin^2 I$. A horizontal neutral wind can also induce a vertical plasma drift, the magnitude depending on the wind direction. This topic will be discussed in Chapter 8 and we neglect the effect here in writing the vertical ambipolar diffusion velocity for a major ion,

$$v_z(\mathrm{O}^+) = -D_a \sin^2 I \left[\frac{\partial \ln n(\mathrm{O}^+)}{\partial z} + \frac{\partial \ln T_p}{\partial z} + \frac{m_{\mathrm{O}^+} g}{2kT_p} \right] \tag{5.3.17}$$

Thermal diffusion adds a term in the square brackets,

$$\alpha_{\mathrm{O}^+}(\partial \ln T_R)/\partial z \tag{5.3.18}$$

where α_{O^+} is the thermal diffusion factor, an experimentally obtainable quantity (Appendix 6).

5.4 Sources and sinks

In the two preceding sections we discussed the contribution of transport to the continuity equation, Equation (5.1.1), or, more accurately, we derived the vertical diffusion velocity for neutral species and for ions. The transport term in the continuity equation is actually the divergence of the flux, $\partial/\partial z(n_j v_j)$, which accounts, in part, for the numerical complexity of the term. In this section the sources and sinks of various ions and minor neutral species are investigated. The contributions of photoionization and dissociation and of electron impact ionization and dissociation were brought to light in Chapters 2 and 3, respectively, while the consequences of heavy ion impact were discussed in Section 3.4. These are the direct sources that account for the initial production of ions and atoms but measurements by rocket and satellite borne mass spectrometers in the thermosphere have shown that the ion and minor neutral composition is very different from that suggested by the primary production processes. For example, even though NO is a minor neutral species, the major ionic species in the E-region is NO^+. Chemical/ionic reactions account for the conversion of species, acting as sources for some and loss processes for others.

5.4.1 Collisional sources of ions and minor neutral species

The chemical equations that symbolically specify photoionization of N_2, O_2 and O are given in Chapter 2 (Equations (2.3.1), (2.3.2), and (2.3.3)) while dissociative ionization is described by Equations (2.3.6) and (2.3.7). The analogous processes for photoelectrons and auroral electrons are the following,

$$\mathrm{N_2 + e_p \rightarrow N_2^+ + e_p + e_s} \tag{5.4.1}$$

$$\mathrm{N_2 + e_p \rightarrow N^+ + N + e_p + e_s} \tag{5.4.2}$$

$$\mathrm{O_2 + e_p \rightarrow O_2^+ + e_p + e_s} \tag{5.4.3}$$

$$\mathrm{O_2 + e_p \rightarrow O^+ + O + e_p + e_s} \tag{5.4.4}$$

$$\mathrm{O + e_p \rightarrow O^+ + e_p + e_s} \tag{5.4.5}$$

The two electrons produced in ionization reactions are the scattered primary auroral or photoelectron, e_p, and the ejected secondary electron, e_s. These are intrinsically indistinguishable of course, but the scattered primaries are for the most part more energetic than the secondaries and the angular distribution of scattered primaries favours the forward direction while secondaries are ejected, more or less, isotropically. Secondaries may have sufficient energy to cause additional ionization with the production of tertiary electrons, and so on. Electron impact ionization and dissociative cross sections are listed in Appendix 4. The ion production rate of species j at altitude z

by electron impact is given by an equation similar to Equation (2.3.4) which describes photoionization.

$$\sum_l P_j^l(z) = n_j(z) \int_{E_{th(l)}}^{E_{max}} I(z, E)\sigma_j^l(E, l)\,\mathrm{d}E \qquad (\text{ion cm}^{-3}\,\text{s}^{-1}) \tag{5.4.6}$$

$n_j(z)$ is the number density of species j, and $\sigma_j^l(E, l)$ is the ionization cross section for various electronic states l of the ion. The integration is carried from the threshold energy for a given ionic state, $E_{th}(l)$, to the maximum electron energy in the spectral distribution. The differential electron intensity is

$$I(z, E) = 2\pi \int_0^\pi I(\mu, z, E) \sin\mu\,\mathrm{d}\mu \tag{5.4.7}$$

discussed at length in Chapter 3. Excited states of the ions are identified in the energy level diagrams in Appendix 3. The $A\,^2\Pi_u$ and $B\,^2\Sigma_u^+$ states of N_2^+ are the upper states for transitions that produce observable band spectra. In addition to the ground $X\,^2\Pi_g$ state of O_2^+ only the first excited state, a $^4\Pi_u$, is generally included in the large family of ionospheric ions. Both excited states of the ground configuration of O^+ are non-negligible ionospheric ions, the $O^+(^2D)$ and $O^+(^2P)$ states, although by far the major atomic ion is ground state $O^+(^4S)$. Equation (5.4.6) is also applicable to dissociative ionization processes, given the appropriate cross section, $\sigma_j^D(E)$.

In this chapter our interest focuses on the composition of the thermosphere and ionosphere. A steady state does not exist in these regions but different species of ions and neutrals have a wide range of lifetimes. For example, the excited $B\,^2\Sigma_u^+$ state of N_2^+ radiates in an allowed transition to the ground state of the ion (previewed in Section 3.3, Equations (3.3.2) and (3.3.3)) and the lifetime of the excited state is of the order of 10^{-8} s, while the ground state of O^+ ions, $O^+(^4S)$, is destroyed by chemical reaction and its lifetime in the F-region is several hours. Short-lived excited states are usually not included in ionospheric composition studies but they are most important in optical studies of thermospheric spectral emissions, the dayglow and the aurora, the subject of Chapter 7.

Electron impact also causes molecular dissociation. We discussed this collision process in Section 4.3 and here identify this source by the symbolic chemical equations

$$O_2 + e_p \rightarrow O + O^* + e_p \tag{5.4.8}$$

$$N_2 + e_p \rightarrow N + N^* + e_p \tag{5.4.9}$$

noting that a fraction of the O and N are produced in excited states. Since N_2 dissociates via predissociation levels, the probability of N atoms being produced in excited states is large, well over half the total.

To complete the discussion of direct production sources of ionospheric ions and thermospheric minor neutrals we take note of the contribution from energetic proton precipitation, principally of auroral origin. The reactions described in Section 3.4 yield the ions N_2^+, O_2^+ and O^+. The contribution to the H population in the thermosphere by proton bombardment is negligibly small by comparison with the terrestrial source which is primarily photochemical.

In summary, the ionospheric ions produced by direct sources are N_2^+, O_2^+,

$O_2^+(a\,^4\Pi_u)$, $O^+(^4S)$, $O^+(^2D)$, $O^+(^2P)$ and N^+, and the thermospheric minor neutral species are $N(^4S)$, $N(^2D)$, $N(^2P)$ and $O(^1D)$.

Radio wave probing of the ionosphere and rocket borne detectors have shown that at night there is always a small residual ionization density in the E-region, in spite of the rapid recombination of the major molecular ions. Ion mass spectrometers have identified this night time layer to consist of metallic ions, mostly Mg^+, Fe^+ with some Na^+, Al^+, Ca^+ and Ni^+. The neutral species and their oxides are produced by meteor ablation, followed by photoionization during the day and by charge transfer with the non-metallic ions. Persistence of the metallic ions at night is due to their slow rate of recombination with electrons and with the major neutral species. Indeed, the lifetime and abundance of metallic ions is probably controlled by dynamic processes, eddy diffusion and winds, which may also account for the small vertical extent of the layers.

5.4.2 Chemical reactions as sources and sinks

The physics and chemistry of reactive collisions are described in Section 4.4. In reactive collisions the original collision partners are destroyed and new species are produced so that chemical sources and sinks cannot be treated separately. To avoid getting lost in the plethora of reactions that occur in the thermosphere we focus here on the different types only, and give some examples, while organizing the full set by individual species in Appendix 5.

Atomic and molecular ions are denoted by A^+, B^+ or C^+ and AB^+, BC^+, or AC^+ respectively and neutral atoms and molecules by A, B, C and AB, BC, AC. Electrons are represented by e.

(1) Ion–atom interchange,

$$A^+ + BC \rightarrow AB^+ + C \quad \text{or} \quad AB^+ + C \rightarrow BC^+ + A \tag{5.4.10}$$

is the chemical reaction that produces NO^+ at the expense of other ions, making it the dominant ion in the E-region. The reaction

$$O^+ + N_2 \rightarrow NO^+ + N(^4S) + 1.10\,\text{eV}, \tag{5.4.11}$$

already discussed in Section 4.4, destroys O^+ while producing NO^+ ions and N atoms with an excess energy of 1.10 eV. This amount is the difference between the energy released in a reaction, in this case the ionization energy of O (13.61 eV), and formation of the NO molecule (6.51 eV), and the energy absorbed in a reaction, in this case dissociation of N_2 (9.76 eV), and ionization of NO (9.25 eV). The excess energy may go into internal excitation of the products or it may go into translational kinetic energy which is shared by the products in inverse proportion to their masses, required by the conservation of momentum. While governing the abundance of ionospheric ions and minor neutral species, Reaction (5.4.11) and all other chemical-ionic reactions only weakly perturb the density of the major neutral atmospheric constituents in the thermosphere, N_2, O_2 and O, which are controlled principally by dynamic and thermodynamic inputs. Ionic reactions contribute to dissociation of O_2 into O in the F-region. Another important ion–atom interchange reaction is

$$N_2^+ + O \rightarrow NO^+ + N(^2D) + 0.70\,\text{eV} \tag{5.4.12}$$

which is also a source of NO^+ and produces N atoms in the excited ^2D state and is a sink of N_2^+ ions.

(2) Atom–atom interchange,

$$A + BC \rightarrow AB + C \tag{5.4.13}$$

converts minor neutral species in the thermosphere as in the reaction

$$N(^2D) + O_2 \rightarrow NO + O(^3P) + 3.76\,eV \tag{5.4.14}$$

which converts one 'odd nitrogen' species, $N(^2D)$ to another, NO. The term 'odd nitrogen' is applied to all molecules containing a single nitrogen atom, e.g., N, NO, NO_2, NO_3, etc. Atom–atom interchange is the chemical reaction that destroys 'odd nitrogen' in the thermosphere,

$$N + NO \rightarrow N_2 + O \tag{5.4.15}$$

converting the minor species into two major species.

(3) Charge transfer,

$$A^+ + BC \rightarrow BC^+ + A \tag{5.4.16}$$

has a high probability of occurrence if the reaction is exothermic. An example is the reaction

$$O^+(^2D) + N_2 \rightarrow N_2^+ + O + 1.33\,eV \tag{5.4.17}$$

where the O^+ ion is in the excited ^2D state which is 3.31eV above the ground state (shown in the energy level diagram in Appendix 3). Ground state thermal $O^+(^4S)$ ions would make the reaction endothermic and the occurrence probability or cross section is vanishingly small.

The ionization potentials of O and H are almost equal (Table 2.2.2) allowing charge transfer between these species to occur readily in both directions

$$O^+ + H \rightleftarrows H^+ + O \tag{5.4.18}$$

The relative magnitude of the rates of the forward and reverse reactions is a function of the statistical weights of the reactants and the ion and neutral temperature. Charge transfer with O atoms is the only chemical loss process for H^+ ions in the thermosphere.

(4) Dissociative recombination,

$$AB^+ + e \rightarrow A + B \tag{5.4.19}$$

is the principal loss process for ionization in the thermosphere. An example of this process is

$$NO^+ + e \rightarrow N + O + 0.38\,eV \quad \text{or} \quad 2.75\,eV \tag{5.4.20}$$

which destroys an electron–ion pair and produces two atoms. The N atom may be in the excited ^2D state in which case the excess energy of the reaction is 0.38 eV, or it may remain in the ground ^4S state with a total excess energy of 2.75 eV. The results of laboratory measurements of this reaction indicate that 78% of the product N atoms are in the ^2D state. Sufficient energy is also available for the product O atoms to be in the excited $O(^1D)$ state; however, this is unlikely due to such a reaction not conserving electron spin (Problem 21). We have already learned in Chapter 4 that the probability of a reaction, the cross section, depends not only on energy considerations but also on symmetry and spin combinations and details of the interaction potentials. Most cross sections and reaction rate coefficients required in thermospheric ion–neutral chemistry are measured in the laboratory, and the formulae given in Appendix 5 are empirical analytic functions to fit the experimental results, or simply numerical values.

(5) Collisional deactivation and radiative cascading: the list of ions and minor neutrals that are produced in the thermosphere includes several excited species, e.g., $O^+(^2D)$ and $N(^2D)$. Only excited species with long radiative lifetimes can build up concentrations sufficient to take part in the chemical reactions described above. The destruction process, which generally limits the concentration of these metastable species, is collisional deactivation or *quenching*. Using a superscripted asterisk to denote an excited state, the reaction is

$$A^* + B \rightarrow A + B \quad \text{or} \quad A^* + e \rightarrow A + e \tag{5.4.21}$$

to show that atoms, molecules and electrons can cause quenching. An example of the process is

$$O^+(^2D) + e \rightarrow O^+(^4S) + e \tag{5.4.22}$$

The radiative lifetime of $O^+(^2D)$ ions is about 3.6 h but the lifetime against quenching is some tens of seconds in the F-region. If the radiative lifetime is not excessively long, metastable species are also destroyed by radiation,

$$A^* \rightarrow A + h\nu \tag{5.4.23}$$

For example, the $O^+(^2P)$ ion has a radiative lifetime of about 5 s; in the upper F-region, where quenching collisions are relatively infrequent, the decay is by radiation

$$O^+(^2P) \rightarrow O^+(^2D) + h\nu(7320\,\text{Å}, 7330\,\text{Å}) \tag{5.2.24}$$

The reaction is also a source of $O^+(^2D)$ ions. Line and band emissions are discussed in Chapter 7.

While most ions and minor neutrals have direct as well as chemical sources we note that the latter are the only source of thermospheric NO and the principal source of NO^+. Photoionization and electron impact ionization of NO are minor sources of the ion due to the low abundance of the parent neutral molecule. Appendix 5 reveals the complex coupling of the ion and minor neutral chemistry in the thermosphere. There is also a wide range of magnitudes in the rate coefficients for different reactions and it is tempting to neglect those that have very small rate coefficients. The rate at which a reaction proceeds, however, is the product of the rate coefficient and the concentration of the reactants and it is the latter quantity that changes drastically with altitude. Thus, a reaction that is important in the E-region may be negligible in the F-region and vice-versa. Since diffusive transport and mixing couples different height levels, all the reactions should be included in deriving the ionospheric and atmospheric structure.

It is conceptually helpful to summarize the salient aspects of the ion and minor neutral chemistry of the thermosphere by a flow chart. This is shown in Figure 5.4.1. Only the major reactions are included and no distinction is made between ground state and excited species. N_2 and O_2 are the parent molecules and the O atoms that are produced by photodissociation are long lived in the thermosphere. The three species are ionized and molecules are dissociated. Atomic ions are converted to molecular ions which recombine with ambient electrons to form O and N. The O atoms must diffuse downwards before three-body recombination becomes effective in producing O_2. N atoms react with O_2 to form NO which can be destroyed by photodissociation. However, the ultimate destruction mechanism is NO + N, two 'odd nitrogen' species that produce N_2. The flow chart shows that energetic photons and particles produce a

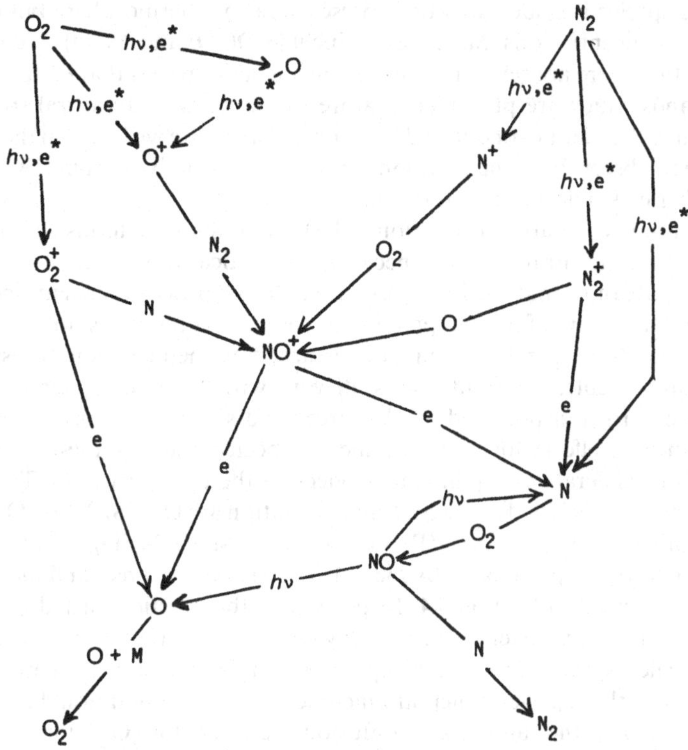

Fig. 5.4.1 Flow chart of the reactions in the thermosphere that account for the major sources and sinks of ions and minor neutral species.

variety of ions and neutrals from a parent gas consisting of N_2 and O_2 molecules. If the ionizing and dissociating sources are turned off the system will return to the original parent gas. Time constants for many processes in the chain are sufficiently long in the thermosphere, however, to prevent complete relaxation of the system.

5.5 Composition of the thermosphere and ionosphere

There are 14 species of ions and minor neutrals, some in metastable states, in the thermosphere that are produced at non-negligible rates and that have a sufficiently long lifetime to build up measurable concentrations at some altitude. Mass spectrometers cannot differentiate between ground state and excited atoms or ions so that the abundance of most metastable species must be inferred indirectly or derived from model computations that make use of results of laboratory experiments. For example, a large fraction of O_2^+ ions produced by electron impact ionization of O_2 are in the metastable O_2^+ (a $^4\Pi$) state which has a radiative lifetime of about 10^4 s and decays by reactive collisions and not by radiation. Vibrationally excited molecules are produced at an appreciable rate and chemical reactions with vibrationally excited species need to be considered separately from ground state reactions. Spectral emission from some metastable species is evidence of their presence in the thermosphere but cannot give their actual concentrations. Such species include O(^{1}D) and N(^{2}D). We will learn in Chapter 7 that there are several additional metastable species that emit characteristic lines or bands which are prominent features in dayglow and auroral spectra. For a description of the thermospheric and ionospheric structure we focus on the species that contribute to observable concentrations or that play an active part in the ionospheric–thermospheric chemical reaction scheme.

The continuity equation (Equation (5.1.1)) contains four terms: the time rate of change of the concentration n_j of species j, the vertical flux gradient of n_j, the local production rate of n_j, and the local loss rate of n_j. As previously mentioned, the one-dimensional treatment of the composition of the thermosphere assumes that the major neutral species, N_2, O_2 and O, remain unperturbed by chemistry and transport but the ion and minor neutral concentrations all vary with time, and their variation with altitude is due to transport and local sources and sinks. Local production and loss exceed transport effects for many of the 14 species and, for these, the continuity equation is considerably simplified by neglecting the transport term. The species for which transport is included in the continuity equation are $O^+(^4S)$, NO^+, O_2^+, N(^{4}S) and NO. The other species, N_2^+, $O^+(^2P)$, $O^+(^2D)$, N^+, N(^{2}P), N(^{2}D), H^+, $O_2^+(a\,^4\Pi)$, and O(^{1}D) are controlled principally by local processes, i.e., by direct and indirect sources and sinks described in Section 5.4. Inspection of the reactions listed in Appendix 5 reveals that there are reaction products which do not appear as source reactants elsewhere. Such species do not participate in chemical reactions because they decay radiatively to other species which are included in the computational scheme.

In Section 5.3, the ambipolar diffusion velocity for O^+ ions was derived (Equation (5.3.17)) under the assumption that it is the major ion and therefore $n(O^+) \approx n(e)$. It was noted that this simplification is only applicable in the upper F-region and the general expression (Equation (5.3.15)) which contains both the ion density and the electron density must be used if the continuity equation for $O^+(^4S)$ is to apply throughout the ionosphere. The assumption of a small vertical wind compared to the vertical ion drift is generally valid. The altitude profile of $O^+(^4S)$ ions is specified by the continuity equation that includes transport (Equation (5.3.15)) and local sources and sinks (Equations (2.3.4), (5.4.6) and Appendix 5). $O^+(^4S)$ is abbreviated to O^+ for convenience,

$$\begin{aligned}
\frac{\partial n(O^+)}{\partial t} = \frac{\partial}{\partial z}\Bigg\{ & n(O^+) D_a \sin^2 I \left[\frac{T_i}{T_R + T_e} \frac{\partial \ln n(O^+)}{\partial z} + \frac{T_e}{T_R + T_e} \frac{\partial \ln n(e)}{\partial z} \right. \\
& \left. + \frac{1}{T_R + T_e} \frac{\partial (T_i + T_e)}{\partial z} \right] - \frac{m_{O^+} g}{k(T_R + T_e)} \Bigg\} + n(O) \int_{\lambda_{th}}^{\lambda_{min}} I(z,\lambda) \sigma_O^i(\lambda) P_{O^+}^i(\lambda, {}^4S)\, d\lambda \\
& + n(O_2) \int_{\lambda_{th}}^{\lambda_{min}} I(z,\lambda) \sigma_{O_2}^{di}(\lambda) p_{O_2}^{di}(\lambda, {}^4S)\, d\lambda + n(O) \int_{E_{th}}^{E_{max}} I(z,E) \sigma_O^I(E, {}^4S)\, dE \\
& + n(O_2) \int_{E_{th}}^{E_{max}} I(z,E) \sigma_{O_2}^{di}(E, {}^4S)\, dE + \gamma_{13} n(O^{+\,2}D) n(O) + \alpha_5 n(O^{+\,2}D) n(e)
\end{aligned}$$

$$+\gamma_{24}n(O^{+}\,{}^{2}D)n(N_2)+\gamma_{25}n(O^{+}\,{}^{2}P)n(O)+\alpha_7 n(O^{+}\,{}^{2}P)n(e)$$
$$+A_3 n(O^{+}\,{}^{2}P)+\gamma_{19}n(N_2^{+})n(O)+\gamma_{20}n(H^{+})n(O)+\gamma_{27}n(N^{+})n(O_2)$$
$$-\gamma_2 n(O^{+})n(O_2)-\gamma_1 n(O^{+})n(N_2)-\gamma_{21}n(O^{+})n(NO)$$
$$-\gamma_{12}n(O^{+})n(H)-\gamma_{26}n(O^{+})n(N\,{}^{2}D) \qquad (5.5.1)$$

where the γs and αs are reaction rate coefficients and A specifies a radiative transition probability; numerical values are given in Appendix 5. Two terms in Equation (5.5.1) have not yet been explicitly identified. These are the production of $O^{+}(^{4}S)$ ions by dissociative photoionization (cross section $\sigma^{di}_{O_2}(\lambda)$) of O_2 and dissociative electron impact ionization of O_2 (cross section $\sigma^{di}_{O_2}(E, {}^{4}S)$). There are four direct sources of $O^{+}(^{4}S)$ and nine chemical sources and there are five chemical loss reactions. The parameters that appear in the $O^{+}(^{4}S)$ continuity equation are the major species concentrations, $n(N_2)$, $n(O_2)$, $n(O)$, as well as the concentrations of H, H^{+}, $O^{+}(^{2}D)$, $O^{+}(^{2}P)$, N_2^{+}, N^{+}, $N(^{2}D)$ and NO, plus $n(e)$.

Reaction rate coefficients, k, cross sections, σ, the neutral, ion and electron temperatures, T_n, T_i, T_e, the dip angle, I, the gravitational acceleration and the ambipolar diffusion coefficient are required. Reaction rate coefficients may be temperature- (and therefore altitude-) dependent. The dip angle is a function of the location of the station at which the profile is computed. The photon intensity $I(z, \lambda)$ and the energetic electron intensity $I(z, E)$ are required as a function of altitude and the relevant cross sections and branching ratios are obtained from laboratory or theoretical investigations. Thresholds for the direct ionization sources, λ_{th} and E_{th}, can be obtained from energy level diagrams (Appendix 3) or Table 2.2.2. Integration is carried out to photon wavelengths at which $I(z, \lambda)$ no longer contributes to the integrand and to electron energies at which $I(z, E)$ likewise becomes negligibly small. Occasionally, energetic proton precipitation into the thermosphere becomes comparable to, or exceeds, electron precipitation in which case a direct source of $O^{+}(^{4}S)$ due to proton impact is included in Equation (5.5.1).

The $O^{+}(^{4}S)$ continuity equation contains the largest number of terms of all the 14 continuity equations and the reader may already have noted that even more terms are generated upon carrying out the differentiation with respect to z in the flux term. This reality is symptomatic of the difference between elegant theoretical formulations and numerical modelling of the real world.

An ion for which transport is unimportant and that has no chemical sources is $O_2^{+}(a\,{}^{4}\Pi)$; for this the continuity equation becomes,

$$\frac{\partial n(O_2^{+}\, a^{4}\Pi)}{\partial t} = n(O_2)\left[\int_{\lambda_{th}}^{\lambda_{min}} I(\lambda, z)\sigma^{i}_{O_2}(\lambda)P^{i}{}_{O_2^{+}}(\lambda, a^{4}\Pi)\,d\lambda + \int_{E_{th}}^{E_{max}} I(z, E)\sigma^{i}_{O_2}(E, a^{4}\Pi)\,dE\right]$$
$$- n(O_2^{+}\, a^{4}\Pi)[\gamma_6 n(N_2)+\gamma_7 n(O)+\alpha_4 n(e)+A_8] \qquad (5.5.2)$$

Inspection of Figure 5.4.1 shows that NO^{+} ions only have chemical sources and sinks, avoiding the complexity of direct sources. This is offset, however, by the transport term which must be included in the NO^{+} continuity equation (Problem 22).

The 14 continuity equations represent a set of coupled differential equations that

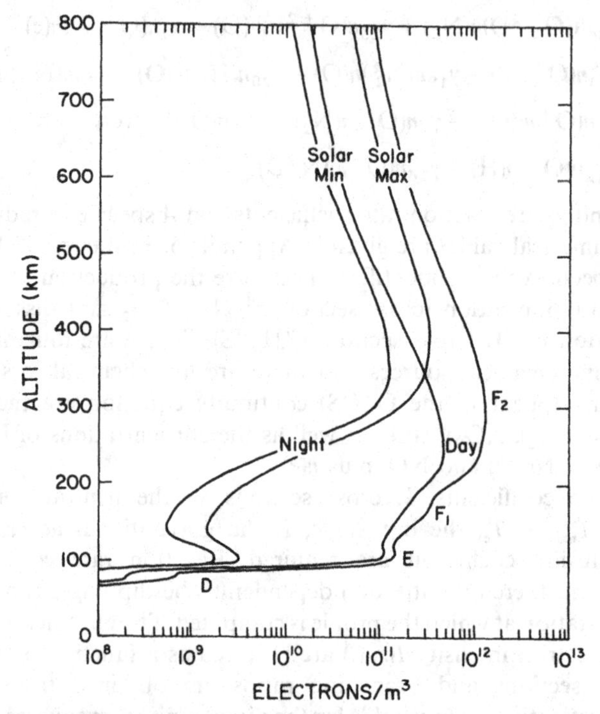

Fig. 5.5.1 Typical midlatitude electron density profiles for sunspot maximum and minimum, day and night. Different regions of the ionosphere are labelled D, E, F_1 and F_2. (A.D. Richmond, The ionosphere, Chapter 7 in *The Solar Wind and the Earth*, Eds. S.-I. Akasofu and Y. Kamide, Terra Scientific Publ. Co., Tokyo, 1987.)

must be solved simultaneously. All require initial conditions and those that include a transport term also require boundary conditions. Chemical equilibrium is generally assumed at the lower boundary except for NO which is transported downward into the mesosphere. A flux condition is generally applied at the upper boundary but the actual value is difficult to obtain experimentally. Incoherent scatter radar measurements of the height profile of electron density have been used to infer the gradient and the flux, assuming that the O^+ density is equal to the electron density and that the H^+ and He^+ ion densities are small by comparison.

Charge neutrality requires that the electron density be equal to the sum of the positive ion densities, negative ions being only a trace species above about 85 km. Electron density profiles vary with geographic or geomagnetic location, with season, with magnetic activity, with solar activity and diurnally. Typical altitude profiles of the electron density are shown in Figure 5.5.1 for day and night and for solar maximum and solar minimum conditions.

The physical and chemical processes that govern the composition of the ionosphere and thermosphere are reasonably well understood and the equations required to determine the altitude profiles of ions and minor neutrals can be formulated. There is uncertainty, however, in establishing the appropriate boundary conditions. Rate coefficients for some chemical reactions are not as certain as is desirable. Finally, obtaining computer based numerical solutions to the coupled equations is a non-trivial problem, especially since the continuity equations are coupled to the energy equations through the electron and ion temperatures.

Bibliography

Ionospheric phenomenology and a theoretical framework for discussing the physical processes at work are given in

H. Rishbeth and O.K. Garriott, *Introduction to Ionospheric Physics*, Academic Press, New York, 1969.

S.J. Bauer, *Physics of Planetary Ionospheres*, Springer Verlag, Heidelberg, 1973.

Several chapters in the following two volume monograph deal with topics on thermospheric and ionospheric structure and composition

P.M. Banks and G. Kockarts, *Aeronomy*, Academic Press, New York, 1973.

Development of an analytic model of the neutral atmosphere is given by

J.C.G. Walker, Analytic representation of upper atmosphere densities based on Jacchia's static diffusion models, *J. Atmosph. Sci.*, **22**, 462, 1965.

In a recent paper the one-dimensional, field-aligned, ion diffusion equations are derived in detail, and references to previous work are given,

J. Grochulska, Diffusion equations for the major ions in the mid-latitude ionosphere and plasmasphere, *J. Atmos. Terr. Phys.*, **47**, 423, 1985.

The method of moments of the distribution function for deriving the transport equations is given by

H. Grad, Principles of the kinetic theory of gases, *Handbuch der Physik*, Vol. 12, Springer Verlag, Berlin, 1958, p. 205.

Problems 17–22

Problem 17: Construct an analytic model atmosphere using Equation (5.2.5) adopting values for the concentrations at the lower boundary (120 km) and for the temperature profile given by the empirical MSIS model listed in Appendix 1 for solar maximum conditions. Compare the results with the empirical model.

Problem 18: Atomic O is produced by photodissociation of O_2 and is lost by downward transport to altitudes (below 110 km) where three-body recombination becomes a local loss process. Compute the production rate of O at 170 km by solar UV radiation (under conditions of your choice) and the loss rate by vertical diffusion. Eddy diffusion is unimportant at this altitude but thermal diffusion should be included. Assume $\alpha^T = 0.38$. Also, assume that the mass averaged vertical velocity is zero.

Problem 19: Derive Equation (5.3.8) for the ambipolar diffusion coefficient starting from its definition.

Problem 20: Derive the equation for the vertical diffusion velocity of O^+ ions in the F-region assuming zero vertical neutral wind.

Problem 21: (a) Check the following reactions for exothermicity and for electron spin conservation.

$$NO^+(X^1\Sigma^+) + e \rightarrow N(^2D) + O(^3P) + \Delta E$$

$$NO^+(X^1\Sigma^+) + e \rightarrow N(^4S) + O(^1D) + \Delta E$$

(b) What are the statistical weights of the ground state reactants and products in the charge transfer reaction

$$O^+ + H \rightleftarrows H^+ + O$$

Problem 22: Write the one-dimensional, time-dependent continuity equation for NO^+ in the ionosphere, including local sources and sinks as well as transport. Do not solve.

6

Temperatures in the upper atmosphere

6.1 Introduction

The subject of this chapter is plural because the neutral gas, the ions and the electrons are generally all at different temperatures in the upper atmosphere. Although each component has energy sources and sinks the temperatures are not independent since, in a collision dominated, partially ionized gas, energy is shared amongst the various constituents. We focus principally on the kinetic temperatures of various species but note that excitation of internal energy can be characterized by vibrational and rotational temperatures in molecules. Rotational relaxation is rapid, and the distribution of rotational lines in a band may be used to infer the kinetic temperature of the gas. Vibrational relaxation is relatively slow so that the vibrational temperature of a molecular species may differ from the kinetic temperature of the gas.

Temperature is the observable parameter in the energy balance of the thermosphere. It is important, therefore, to understand the physical processes that underlie the energetics of the region and thereby control the altitude profiles of the several temperatures. In addition, temperature influences processes other than those associated with energetics. We have already pointed out in Chapter 5 that several reaction rate coefficients are temperature-dependent, so that the composition of the thermosphere and the ionosphere is influenced by the thermal structure. The dynamic behaviour of the thermosphere is also influenced by the temperature structure through the equation of state of a gas which relates temperature to pressure, and by the temperature dependence of the transport coefficients, discussed in Chapter 4.

A typical neutral temperature profile from the ground to the exobase was sketched in Figure 1.1.1. The pronounced minimum at the mesopause (80–85 km) indicates that the thermosphere is energetically decoupled from the mesosphere (not entirely, of course), has its own heat sources and that heat flows downward toward the minimum. At the mesopause and below the ion and electron temperatures are both equal to the neutral temperature, but in the thermosphere the altitude profiles of the three temperatures may be quite different. In this chapter we investigate the processes that determine the temperature profiles: these are sources, sinks and transport of energy.

The principal source of energy in the thermosphere is the absorption of solar UV radiation. At high latitudes there is a substantial, but variable, contribution from the dissipation of electric currents driven by electric fields of magnetospheric origin, and from the absorption of energetic particles associated with the aurora. Dissipation of gravity waves that propagate upward from the troposphere is a minor source of neutral gas heating.

6.2 UV photon and energetic electron energy sources

Turning firstly to absorption of solar radiation, the following questions need to be addressed. How much solar UV energy is absorbed at various levels throughout the thermosphere, how much goes into each constituent, what fraction of the absorbed energy goes into heating of the neutrals, ions, and electrons, and what are the various paths by which the energy is channelled into heating? Several types of inelastic or reactive collisions are responsible for absorbing solar UV photons in the atmosphere. The principal ones have already been discussed in Chapter 2: they are photoionization (Section 2.3) and photodissociation (Section 2.4). In Chapter 2 we focused on the type of process involved and the products of the reaction, i.e., the production of ion pairs by photoionization and the production of atoms by molecular dissociation. Noting only that every reaction has a threshold energy that must be available from the photon, we did not concern ourselves with the overall energy budget of solar UV absorption, which is taken up in this chapter.

Visible light, of course, penetrates to the Earth's surface (we see by it!). Solar radiation at wavelengths less than 3100 Å is absorbed by various atmospheric constituents. The altitude at which vertically incident solar radiation is decreased to e^{-1} of its value outside the atmosphere is shown in Figure 6.2.1 as a function of wavelength. The data

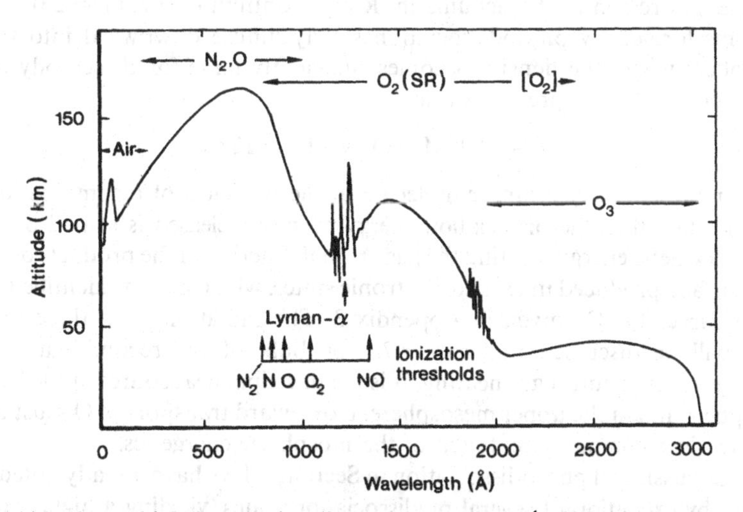

Fig. 6.2.1 Altitude at which the solar irradiance decreases to e^{-1} of its value outside the Earth's atmosphere for vertical incidence. The principal absorbing species are identified. In the wavelength region labelled 'air' all constituents contribute to attenuation. Wavelengths corresponding to the ionization thresholds of important species are marked by arrows. (Adapted from L. Herzberg in *Physics of the Earth's Upper Atmosphere*, Eds. C.O. Hines *et al.*, Prentice Hall, New Jersey 1965.)

required to construct the curve shown in this figure (below 2000 Å) are given in Chapter 2. The ordinate is the altitude where the optical depth is unity (Equation (2.2.1)). Solar photons at wavelengths beyond about 1750 Å are not strongly absorbed in the thermosphere. O_3 and O_2 in the stratosphere become the principal absorbers. We already noted in Chapter 2 that solar H-Ly-α radiation (1216 Å) penetrates to the mesosphere and X-rays ($\lambda < 10$ Å) are absorbed in the mesosphere and stratosphere (Figure 2.2.5). Thus, we need only to concern ourselves with photons at wavelengths less than about 1800 Å.

Photodissociation of O_2 is the dominant absorption process in the wavelength region 1750–1300 Å by the reaction

$$O_2 + h\nu(\lambda \leqslant 1748\,\text{Å}) \rightarrow O_2(B\,^3\Sigma_u^-) \rightarrow O(^3P) + O(^1D) + \text{k.e.} \qquad (6.2.1)$$

The excited B $^3\Sigma_u^-$ state is identified in the potential diagram of O_2, shown in Appendix 3. Since the dissociation energy of O_2 is 5.12 eV and the excitation energy of $O(^1D)$ is 1.97 eV, photons with at least 7.09 eV energy are required. The excess energy of photons at shorter wavelengths ($\lambda < 1748$ Å) goes into kinetic energy or heating of the reaction products, the O atoms. Although the $O(^1D)$ atoms can decay by radiation the state is metastable with a radiative lifetime that is so long that, in the altitude range where most of the dissociation occurs, the excited atoms are deactivated to the ground state by collisions. Elastic collisions result in sharing the 1.97 eV kinetic energy. Inelastic collisions lead to products and states that may radiate, in which case the energy is lost from the thermosphere, but much of the energy is eventually dissipated, heating the neutral gas. Reaction (6.2.1) describes dissociation in the Schumann–Runge continuum ($\sim 1300 < \lambda < 1750$ Å) shown in Figure 2.2.3. Photodissociation of O_2 occurs at a large rate because the solar flux and the absorption (dissociation) cross section are large in the wavelength region of the Schumann–Runge continuum (Problem 23).

O atoms formed by photodissociation slowly diffuse downward into the lower thermosphere where the density becomes sufficiently large for three-body recombination to occur at an appreciable rate,

$$O + O + M \rightarrow O_2 + M + 5.12\,\text{eV} \qquad (6.2.2)$$

where M represents any atom or molecule. If the products of recombination are in the ground state then the dissociation energy which is released is shared between the products as kinetic energy, heating the gas. A small fraction of the products of Reaction (6.2.2) is, in fact, produced in excited electronic states, which may be identified from the potential curves for O_2 given in Appendix 3. The radiation from these states (the airglow) will be discussed in Chapter 7. The bulk of the recombination energy, however, goes into neutral gas heating. It is the principal heat source at the base of the thermosphere and in the upper mesosphere. Downward transport of O must therefore be included in a complete treatment of thermospheric energetics.

In our discussion of photodissociation in Section 2.4 we have already noted that N_2 dissociates by excitation of several predissociation states, yielding a highly structured band absorption cross section (Figure 2.2.2). The dissociation products have both internal excitation energy and excess kinetic energy which are indirect and direct heat sources, respectively. Photodissociation of N_2 is a minor, yet direct, source of neutral gas heating by comparison with that of O_2. Vertical transport of the dissociation products, the N atoms, and subsequent exothermic chemical reactions identify the process as a non-local heat source that contributes substantially to thermospheric heating. Appendix 5 lists various reactions and the available kinetic energy.

Electron impact dissociation is a high latitude heat source associated with auroral precipitation,

$$e_p + O_2 \rightarrow O^* + O + e_p \tag{6.2.3}$$

$$e_p + N_2 \rightarrow N_2^* + e_p; N_2^* \rightarrow N^* + N + \text{k.e.} \tag{6.2.4}$$

where e_p identifies a primary energetic auroral electron and the asterisk refers to an excited state. Details of electron impact collision processes are discussed in Section 4.3, highlighting the difference between O_2 and N_2. The excited states in which the dissociation products are formed are potential sources of energy through subsequent chemical reactions and collisional deactivation. The energetic electrons undergo many collisions before becoming part of the ambient electron population and a large fraction of their energy ends up in neutral gas heating.

Photoionization of the major neutral thermospheric constituents absorbs most of the photons at wavelengths less than 1000 Å. The ionization thresholds are listed in Table 2.2.2 and the cross sections are given in Appendix 2. Electron impact ionization initiates a sequence of processes that channel a substantial fraction of the initial electron energy into kinetic energy of the ionospheric electrons and ions and of the thermospheric neutral gas.

Photoionization destroys the photon and produces a photoelectron and a positive ion,

$$M + h\nu(\lambda < \text{threshold}) \rightarrow M^+ + e_s \tag{6.2.5}$$

Auroral primary electrons likewise produce an electron–ion pair,

$$e_p + M \rightarrow e_p + M^+ + e_s \tag{6.2.6}$$

and undergo energy degradation and angular scattering. Differences between photoelectrons and auroral electrons emerged in Chapters 2 and 3. In this chapter we combine their roles as energy or heat sources for the ambient ionized and neutral gas. To gain an overview of the physical and chemical processes that take part in channelling the energy towards heating, a flow chart becomes helpful (Figure 6.2.2).

Precipitating heavy ions, H^+ and O^+, usually contribute only a small fraction of the total energy deposited in the thermosphere. Occasionally, however, the auroral proton energy flux accounts for all the particle input which, at night, then becomes the dominant heat source. Electrons ejected by heavy particle ionization become indistinguishable from auroral secondary electrons or photoelectrons, following similar energy degradation and heating channels.

Referring to the flow chart, Figure 6.2.2, we note that the initial interaction between the energetic photons or particle fluxes and the atmosphere leads primarily to ionization and dissociation of the neutral gas. A small amount of energy is expended in direct impact excitation, some of which is lost by radiation but an appreciable part is dissipated as heat by collisional deactivation. Ionized and dissociated species undergo a variety of chemical/ionic reactions described in Section 5.4. These reactions are exothermic and the excess energy may go into internal excitation of the products or into kinetic energy, i.e., heating of the neutrals and ions. Photoelectrons and secondary or

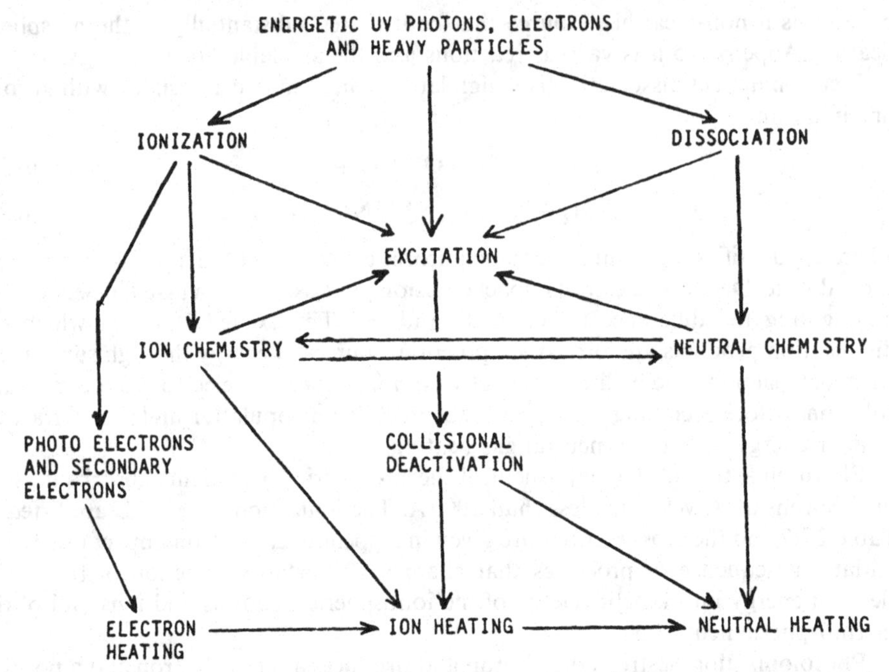

Fig. 6.2.2 Physical and chemical processes by which energy carried into the upper atmosphere by UV photons, electrons and ions is channelled into heating of the plasma and neutral gas.

ejected auroral electrons produced in ionization reactions are degraded by a variety of inelastic collisions, including ionization, dissociation and excitation of internal states and sharing of energy with the ambient electron gas.

In connection with our discussion in Chapter 3 on electron transport in the atmosphere we dwelt at length on the variety of inelastic processes that contribute to energy degradation of the electron stream. Equation (3.2.8) specifies the rate of energy transfer from the energetic or suprathermal electrons to the ambient electrons. This process is the principal heat source for ionospheric electrons, and the heating rate is given by

$$Q(\mathrm{e}, z) = n_{\mathrm{e}}(z) \int_{E_{\mathrm{c}}} I(E, z) L(E, \mathrm{e})\, \mathrm{d}E \qquad (\mathrm{eV\,cm^{-3}\,s^{-1}}) \qquad (6.2.7)$$

where $L(E, \mathrm{e})$ is the stopping cross section of the ambient electrons defined by Equation (3.2.8), $I(E, z)$ is the intensity of energetic electrons at altitude z integrated over all angles, and $n_{\mathrm{e}}(z)$ is the electron density. The lower limit of integration, E_{c}, is the crossover energy at which the thermal electron intensity is equal to the photoelectron or secondary electron intensity. A practical upper limit of integration is about 10 eV because inelastic collisions with the more abundant neutral species are the dominant energy loss processes for electrons with energy greater than a few eV. Interaction between suprathermal and ambient electrons is primarily by Coulomb collisions, but excitation of waves (Cerenkov radiation) also contributes to energy loss. The complete expression for the loss rate or stopping cross section is given in a reference listed at the end of the chapter. The complex expression is closely approximated by a simpler empirical equation,

$$L(E, \mathrm{e}) = \frac{3.37 \times 10^{-12}}{E^{0.94} n_e(z)^{0.03}} \left[\frac{E - E_e}{E - 0.53 E_e} \right]^{2.36} \quad (\mathrm{eV\,cm}^2) \qquad (6.2.8)$$

where $E_e = kT_e$ and E is the suprathermal electron energy, both in units eV.

What is the justification for adopting an electron temperature in the stopping cross section, given the definition of the concept of a temperature in terms of a Maxwellian distribution of energy? It is that the suprathermal photoelectrons and auroral electrons account for less than 0.01% of the total electron density at any altitude, while the bulk of the ambient ionospheric electron population does have a Maxwellian energy distribution at some temperature, T_e.

Computing the ion and neutral heating rates requires detailed book-keeping of the rates at which various exothermic reactions occur and the excess kinetic energy associated with each reaction. Several examples are listed in Section 5.4.2 and we repeat two to illustrate the procedure. The ion–atom interchange reaction,

$$O^+ + N_2 \rightarrow NO^+ + N(^4S) + 1.10\,\mathrm{eV} \qquad (5.4.11)$$

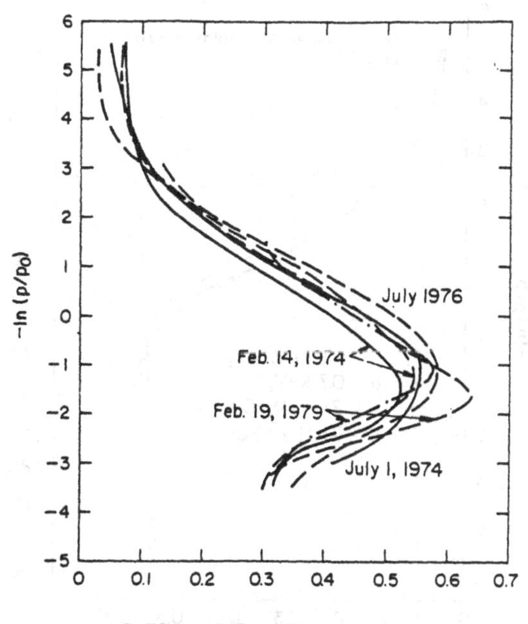

Fig. 6.2.3 The neutral gas heating efficiency for solar UV energy deposition in the thermosphere. Adopting pressure as a coordinate instead of altitude substantially decreases the difference in heating efficiences between solar maximum and minimum conditions at a specified level in the atmosphere. The reference pressure level is $p_0 = 5 \times 10^{-4}\,\mu$b. Results of different model computations are shown. (Adapted from R.G. Roble and B. Emery, *Planet. Space Sci.*, **31**, 597, 1983.)

has an excess kinetic energy of 1.10 eV which is shared between the NO^+ ion and the neutral N atom in inverse proportion to their masses. The reaction is a source of both ion and neutral heating. The atom–atom interchange reaction,

$$N(^2D) + O_2 \rightarrow NO + O(^1D) + 1.74\,eV \qquad \text{(Appendix 5)}$$

is a neutral gas heat source and, in addition, the $O(^1D)$ excited atoms may be collisionally deactivated,

$$O(^1D) + M \rightarrow O(^3P) + M + 1.97\,eV \qquad (6.2.9)$$

supplying an additional 1.97 eV of energy to the gas. M represents any deactivating species, most likely N_2 molecules. In Appendix 5 a list of the more important chemical/ionic reactions is given, including their exothermicity. While simple in concept, the task of computing the heating rate of the ion and neutral gas is complex and lengthy. We adopt results from the research literature (referenced with this chapter) and determine a neutral and ion heating efficiency from the ratio of the local rate of heating to the local energy deposition rate by UV photons or energetic electrons. Although the efficiency is somewhat dependent on the neutral atmosphere (altitude distribution of density and composition), on the solar activity cycle or on the energy spectrum of the primary auroral electrons, departures from an average value are not large and the curves shown in Figures 6.2.3 and 6.2.4 represent well the results of

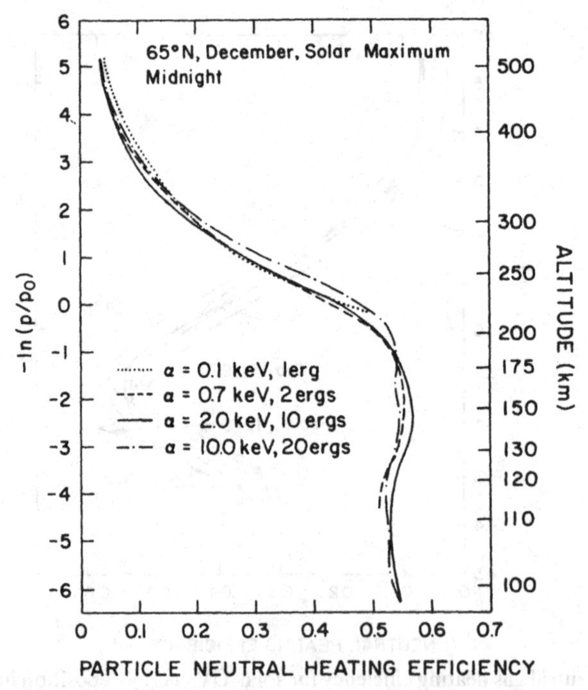

Fig. 6.2.4 Generalized log pressure profiles of the neutral gas heating efficiency for auroral electron fluxes with Maxwellian energy spectra of several characteristic energies, α, and a range of energy deposition rates. The altitude scale applies to the specific conditions of season, time, position and solar activity. (Adapted from M.H. Rees *et al.*, *J. Geophys. Res.*, **88**, 6289, 1983.)

current (1987) research. The heating efficiency is defined in terms of local effects only but we have already learned in Chapter 5 that vertical transport of ions and minor neutral species is an important process in establishing their height distributions in the atmosphere. The local heating rate represents energy deposited in the ion and neutral gas.

The fraction of energy that goes into heating the ambient electrons is a function of the intensity of suprathermal electrons and is more variable than the neutral or ion heating efficiencies. Between 1 and 10% of the energy deposition rate may go into electron heating. Although small, this is sufficient to ensure that the electron temperature is always higher than the neutral gas temperature. Under photon and particle energy input only, it is also higher than the ion temperature, which is higher than the neutral temperature. The flow chart in Figure 6.2.2 indicates that the electron gas heats the ion gas which heats the neutral gas. The circumstances under which this situation may be different are described in the following section.

6.3 Electric field energy sources

Under the action of an electric field, charged particles drift relative to one another and relative to neutral particles. Collisions between species limit the drift velocities and convert some of the drift energy into thermal energy. If we assume that the force per unit mass on the ions due to collisions with neutrals is proportional to the velocity difference between the two species, then the rate of frictional heating is

$$Q_E = n_i m_i \nu'_{in}(v_i - v_n)^2 \qquad (\mathrm{erg\,cm^{-3}\,s^{-1}}) \tag{6.3.1}$$

where n_i is the ion density, m_i is the ion mass, and ν'_{in} the ion–neutral energy transfer collision frequency. The collision frequency ν'_{in} in Equation (6.3.1) is related to the collision frequency ν_{in} defined by Equation (5.3.11) by $m_i \nu'_{in} = M_R \nu_{in}$, where M_R is the reduced mass. Friction simultaneously heats the neutral gas. Under the assumption stated above the ion–neutral, ν'_{in}, and neutral-ion, ν'_{ni}, collision frequencies satisfy the condition

$$n_i m_i \nu'_{in} = n_n m_n \nu'_{ni} \tag{6.3.2}$$

and the neutral heating rate may be derived from the ion heating rate.

In our discussion of ion diffusion in Section 5.3 we found that the relative drift velocity under the influence of an electric field, **E**, is given by (Equation (5.3.14))

$$\mathbf{v}_i - \mathbf{v}_n = e\mathbf{E}/m_i \nu_{in} \tag{6.3.3}$$

The rate of frictional heating is therefore

$$Q_E = (eE)^2 n_i / M_R \nu_{in} \tag{6.3.4}$$

or, in terms of the collision integral for diffusion,

$$Q_E = (eE)^2 \frac{3}{16} \frac{n_i m_i}{n_n M_R^2} \frac{1}{\Omega^{(1,1)}_{i,n}} \tag{6.3.5}$$

These equations are valid in the absence of a magnetic field. An additional force acts on the ions in the Earth's ionosphere where there is a magnetic field. The second term in the expression for the Lorentz force (Equation (5.3.1))

$$\mathbf{F} = e[\mathbf{E} + (1/c)\mathbf{v}_i \times \mathbf{B}] \tag{6.3.6}$$

causes ions that do not travel parallel to **B** to gyrate about the magnetic lines of force with an angular frequency

$$\omega_i = eB/m_i c \tag{6.3.7}$$

modifying the relative motion of ions and neutrals. The consequences on ionospheric currents will be explored in Chapter 8 while in this chapter we note that the frictional heating rate is modified by the ion gyrofrequency,

$$Q_E = (eE)^2 \sum_{i,n} \frac{n_i}{M_R} \frac{\nu_{in}}{\omega_i^2 + \nu_{in}^2} \tag{6.3.8}$$

where the summations over i and n apply if several ion and neutral species are present.

An equation analogous to Equation (6.3.8) applies to heating of the electron gas. For a given magnitude of electric field, electron heating is substantially smaller than ion heating in the F- and E-regions of the ionosphere but becomes comparable in the *D*-region (Problem 24).

What is the source of the electric field? In our discussion of ion drift in Section 5.3 the electric field was a polarization field due to charge separation between ions and electrons. The effect is self-limiting, however, and polarization fields are small. In the collision dominated region of the ionosphere the magnitude of electric fields parallel to the magnetic field is always small since the mobility of electrons along **B** is large. Electric fields orthogonal to **B**, however, can be sustained because the mobilities of electrons and ions are much smaller transverse to **B**, as will be discussed in Chapter 8.

There are external and internal sources of electric fields in the ionosphere. Large scale orthogonal electric fields in the ionosphere have their origin externally in the magnetosphere, the two regions being linked by the highly conducting magnetic lines of force. The large scale magnetospheric electric field is generated by the interaction between the solar wind and the geomagnetic field, causing electrons and ions to drift in different directions, thereby setting up a large potential difference between regions of the magnetosphere.

We will learn in Chapter 8 that a neutral wind is a perennial feature of the thermosphere while ion drifts or currents flow in the ionosphere. Collisions between the two species couple their velocities and a large neutral wind can drag ions across the geomagnetic field generating an effective electric field,

$$\mathbf{E}_i = (1/c)\mathbf{v}_i \times \mathbf{B} \tag{6.3.9}$$

that contributes to heating the ion and neutral gas. A thermospheric neutral wind can thereby become an internal electric field source while the solar wind is an external source. The respective generating mechanisms are very different.

Orthogonal electric fields are a major source of ion (and neutral) heating causing a substantial increase in the ion temperature which may, indeed, exceed the electron temperature as well as the neutral temperature. The relationship $T_n < T_i > T_e$, frequently observed by satellite borne detectors and inferred from incoherent scatter radar measurements, serves as evidence for large scale electric fields in the high latitude ionosphere. In the absence of electric field heating $T_e > T_i$.

6.4 Energy exchange by collisions

Energy can be shared locally between species at different temperatures. We have already tacitly assumed and will state explicitly here that a kinetic temperature, T_e, is associated with the ambient electrons, T_i with all the ions regardless of species, and T_n with all the neutral gas constituents. This assumption is valid for the neutrals and electrons since both types have Maxwellian velocity distributions. (The suprathermal photoelectrons and auroral electrons account for a very small fraction of the total population.) Different species of positive ions gain and lose energy at somewhat different rates, however, and may thereby acquire slightly different kinetic temperatures. In addition, under the influence of a strong electric field the directed component of the ion velocity may become comparable to the thermal speed yielding a non-Maxwellian velocity distribution. References to this chapter expand on these interesting phenomena which here will not be investigated further.

The most common situation in the Earth's upper atmosphere regarding the relative magnitude of the three temperatures is $T_e > T_i > T_n$: collisions between electrons and ions or neutrals cool the electrons and heat the ions or neutrals, and collisions between ions and neutrals cool the former and heat the latter. The rate at which energy is transferred by particles of type 1 at temperature T_1 in elastic collisions with particles of type 2 at temperature T_2 is given by,

$$L_{1,2} = -2n_1 \frac{m_1}{m_1 + m_2} \nu_{12} \frac{3}{2} k(T_1 - T_2) \tag{6.4.1}$$

where ν_{12} is the momentum transfer collision frequency already defined in Chapter 5,

$$\nu_{12} = \frac{16}{3} \frac{n_2}{m_1} M_R \Omega^{(1,1)}_{1,2} \tag{5.3.11}$$

in terms of a collision integral, Ω, similar to that given by Equation (4.5.6), but with the substitution

$$\frac{2kT}{M_R} \rightarrow 2k\left(\frac{T_1}{m_1} + \frac{T_2}{m_2}\right) \tag{6.4.2}$$

to take account of the temperature difference between the interacting Maxwellian gases. To evaluate the collision integral requires the collision cross section. As already discussed in Chapter 4, theoretical values of cross sections depend crucially on knowing the interaction potential. Whenever available, reliable, experimentally derived cross sections are used for solving atmospheric problems.

6.4.1 Energy exchange collisions involving electrons

The elastic collision cross sections between electrons and the major atmospheric species are shown in Appendix 4 as functions of electron energy. The rate of energy transfer between electrons and neutral species n by elastic collisions is given by

$$L_{\text{elastic}}(\text{e,n}) = -16 n_e n_n \frac{m_e}{m_n} \Omega_{\text{e,n}}^{(1,1)} k(T_e - T_n) \tag{6.4.3}$$

with the reasonable assumption that $m_e \ll m_n$. The collision integral, $\Omega_{\text{e,n}}^{(1,1)}$ is defined by Equations (4.5.6) and (6.4.2) and must be solved numerically if the experimental velocity-dependent cross sections are adopted. Evidently, the energy transfer rate depends on the neutral species with which the electron collides. Energy transfer rates between electrons and the neutral species N_2, O_2, O, He and H by elastic collisions are listed in Appendix 6, a result adopted from the recent research literature (see references to this chapter).

Due to the small mass ratio in Equation (6.4.3) elastic collisions are inefficient in transferring energy from electrons to neutrals. More effective processes are the inelastic collisions. These were first introduced in Section 3.2, in connection with the energy degradation of photoelectrons and auroral electrons. Three types of inelastic collisions are described in Section 4.3: ionization, excitation of electronic states and molecular dissociation. The threshold energy for most of these processes is in the energy range of several electron volts to tens of electron volts, and the electrons responsible for them can hardly be identified with the thermal population, even in the high energy tail. Their number density is too small to contribute to the thermal balance of the electron gas. Inelastic collisions that are effective in cooling the electron gas are those for which the threshold energies lie within the thermal distribution. These are rotational and vibrational excitation in molecules, excitation of ground state fine structure levels in O and very low lying electronic states.

Equation (6.4.3) applies to the energy transfer rate by elastic collisions only. For inelastic collisions between electrons and neutrals a general expression that is applicable if both species have a Maxwellian distribution at temperatures T_e and T_n, respectively, is given by,

$$\begin{aligned} L_{\text{inelastic}}(\text{e,n}) &= 2\pi \left(\frac{1}{\pi k T_e}\right)^{3/2} \left(\frac{2}{m_e}\right)^{1/2} \frac{n_e n_n}{\sum_j \tilde{\omega}_j \exp(-E_j/kT_n)} \\ &\times \sum_{i=0}^{m-1} \sum_{j=i-1}^{m} \tilde{\omega}_i (E_j - E_i) \exp(-E_i/kT_n) \\ &\times \left\{ \exp\left[\frac{E_j - E_i}{kT_e T_n}(T_e - T_n)\right] - 1 \right\} \\ &\int_{E_j - E_i}^{\infty} E' \sigma_n(i \to j, E') \exp\left(-\frac{E'}{kT_e}\right) \mathrm{d}E' \end{aligned} \tag{6.4.4}$$

in which $\tilde{\omega}_j$ is the statistical weight of the jth state required to evaluate the partition function

$$\sum_j \tilde{\omega}_j \exp(-E_j/kT_n) \tag{6.4.5}$$

E_j and E_i are the energies of the jth and ith state, with $E_j - E_i > 0$, and $\sigma_n(i \to j, E)$ is the energy-dependent excitation cross section of the nth species from state i to j. Equation

(6.4.4) applies to both excitation and deexcitation. The two processes are related by the principle of detailed balance, derivable from quantum mechanics, which asserts that the transition probabilities for a process and its reverse are equal. The density of the neutral species is n_n and m represents the number of energy states. Other symbols have previously been defined. As written, Equation (6.4.4) is positive when $T_e > T_n$, representing energy loss by electrons through inelastic collisions with neutrals.

Rotational excitation of molecules by electron impact requires a change in the electron's angular momentum. For low energy electrons, however, the probability that an electron has non-zero angular momentum is appreciable only at large electron–molecule distances. Therefore, rotational excitation is due primarily, though not entirely, to long-range interactions. The homonuclear atmospheric molecules N_2 and O_2 have a permanent quadrupole moment (but no dipole moment) and the leading term in the long-range interaction potential is due to coupling between the electron and the molecular charge distribution,

$$V(r) = -qe/r^3 \tag{6.4.6}$$

where the quadrupole moment of the molecule is

$$q = q_0 e a_0^2 \tag{6.4.7}$$

with q_0 an experimentally determined coefficient, which is close to unity for N_2 molecules, and a_0 the Bohr radius. Possible coupling to excited electronic levels adds a polarization term to the interaction potential (already discussed in Section 4.4) which varies as r^{-4}. Experimental measurements of rotational excitation cross sections are sparse (see references). The rotational excitation cross section has, however, been computed using the Born approximation. The angular dependence of the quadrupole interaction potential leads to the selection rule $J' = J \pm 2$ for transition between rotational levels, and the cross section is given by

$$\sigma_{\text{rot}}(J \pm 2; J) = \frac{K'}{15K} \pi a_0^2 \left[8q_0^2 + \frac{\pi q_0(\alpha_\parallel - \alpha_\perp)}{3} \left(\frac{3K^2 + K'^2}{K} \right) + \frac{\pi^2(\alpha_\parallel - \alpha_\perp)^2}{32}(K^2 + K'^2) \right] \frac{(J + 1 \pm 1)(J \pm 1)}{(2J + 1)(2J + 1 \pm 1)} \tag{6.4.8}$$

with

$$q_0 = q/ea_0^2, \; K = (2M_R E)^{1/2}/\hbar \tag{6.4.9}$$

The initial and final wave numbers (in atomic units) of the scattered electron are K and K', respectively, πa_0^2 is the atomic cross section (8.7974×10^{-17} cm^2) and $\alpha_\parallel$ and $\alpha_\perp$ are the longitudinal and transverse polarizabilities of the molecule, listed in Appendix 6 (Problem 25).

The quantity of interest is the rate of energy loss by electrons through rotational excitation and deexcitation,

$$L_{\text{rot}}(\text{e}, \text{n}) = n_e v \sum_J n_j [\sigma(J \to J + 2)(E_{J+2} - E_J) - \sigma(J \to J - 2)(E_J - E_{J-2})] \tag{6.4.10}$$

where v is the electron velocity, n_J is the concentration of molecules in the Jth rotational level and E_J is the rotational energy of the Jth level,

$$E_J = BJ(J+1) \tag{6.4.11}$$

The rotational constant is

$$B = h^2/8\pi^2 I \tag{6.4.12}$$

and I is the moment of inertia of the molecule about the axis through the center of mass and orthogonal to the internuclear axis.

In a neutral gas that has a Maxwellian distribution at temperature T_n the rotational population is given by

$$n_n(J) = n_n \frac{(2J+1)\exp(-E_J/kT_n)}{\sum_J (2J+1)\exp(-E_J/kT_n)} \tag{6.4.13}$$

and the energy loss rate becomes

$$L_{rot}(e,n) = \left(\frac{2E}{m_e}\right)^{1/2} n_e n_n \sum_J \left\{ \frac{(2J+1)\exp(-E_J/kT_n)}{\sum_J (2J+1)\exp(-E_J/kT_n)} \times [\sigma(J+2;J)(E_{J+2}-E_J) - \sigma(J-2;J)(E_J - E_{J-2})] \right\} \tag{6.4.14}$$

where E is the electron energy and n_n the neutral molecular gas density. If it is also assumed that the electron gas has a Maxwellian distribution at temperature T_e, then the averaged loss rate is given by

$$L_{rot}(e,n) = 2\pi \left(\frac{1}{\pi k T_e}\right)^{3/2} \left(\frac{2}{m_e}\right)^{1/2} \frac{n_e n_n}{\sum_J (2J+1)\exp(-E_J/kT_n)} \times \int_0^\infty E \exp\left(-\frac{E}{kT_e}\right) \sum_J (2J+1)\exp\left(-\frac{E_J}{kT_n}\right) \times [\sigma(J+2,J)(E_{J+2}-E_J)\sigma(J-2;J)(E_J - E_{J-2})]\,dE \tag{6.4.15}$$

Various approximations to the energy loss rate have been made to avoid the complexity of evaluating the expression for the wide range of T_e and T_n in the thermosphere under varying geophysical conditions. A useful approximation is given in terms of the temperature difference $(T_e - T_n)$,

$$L_{rot}(e,n) = n(e)n(n)(G\bar{v})_{rot} k(T_e - T_n) \tag{6.4.16}$$

Using a set of rotational cross sections for N_2 that are consistent with laboratory experiments bearing on transport coefficients and the N_2 rotational energy levels,

$$L_{rot}(e, N_2) = n(e)n(N_2)(G\bar{v})_{rot} k(T_e - T_n) \qquad (\text{erg cm}^{-3}\,\text{s}^{-1}) \tag{6.4.17}$$

for which the rate coefficient for energy loss, $(G\bar{v})_{rot}$, is shown in Figure 6.4.1 as a function of the neutral gas temperature T_n (Problem 26).

Rotational cross section data for slow electrons colliding with O_2 molecules are not as well established as those for N_2. Measurements are hampered by the ease of low energy electron attachment to O_2 to form a negative ion, while the low vibrational

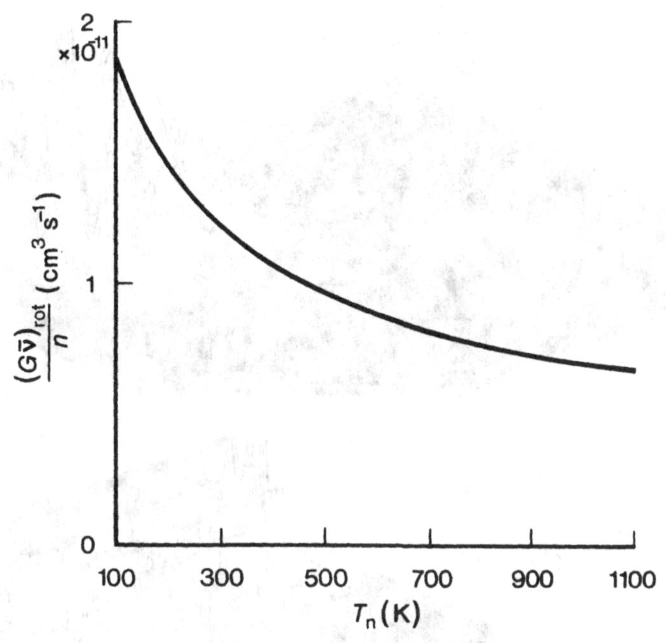

Fig. 6.4.1 Energy loss parameter, $(G\bar{v})_{rot}/n$, for rotational excitation in N_2 as a function of the neutral gas temperature. (Adapted from A. Dalgarno and R. J. Moffett, *Proc. Nat. Acad. Sci. India*, **A33**, Pt. IV, 511, 1963.)

excitation threshold, 0.195 eV, precludes reliable identification of rotational energy loss. In addition, O_2 has two excited electronic levels at 0.98 eV and 1.62 eV which further cloud the rotational excitation measurements. While it is desirable to understand the individual energy loss processes, we need to remind ourselves that the energy balance and temperature in the thermosphere are governed by the overall effect of all the reactions. Formulae for the electron energy loss rate due to rotational excitation of O_2 that are adopted in research on the energy balance in the atmosphere are based on current understanding of this complex problem. The experimentally derived effective quadrupole moment is about $1.8ea_0^2$ compared to about ea_0^2 for N_2 while the rotational constant, B, is 1.45 cm^{-1} for O_2 and 2.01 cm^{-1} for N_2.

Energy loss by collisional excitation of vibrational states in N_2 and O_2 is an important process in establishing the electron energy balance in the thermosphere. The excitation cross sections of the various vibrational levels in N_2 are unexpectedly large, around 2 eV, and have considerable structure (Figure 6.4.2). Inspection of the energy level diagram of N_2 (Appendix 3) shows that there is a temporary negative ion, $N_2^-(^2\Pi_g)$, that captures the electron and has a lifetime that is at least comparable to the vibrational period. Formation of the negative ion is followed by auto-detachment of the electron, leaving the neutral molecule in a vibrationally excited state. Qualitatively, the structure in the cross section is due to resonance energies,

$$E_{nv} = E_n + (v + 1/2)\,h\nu_{vib} \tag{6.4.18}$$

where E_n is the energy of the electronic state of the molecular ion and v is the vibrational

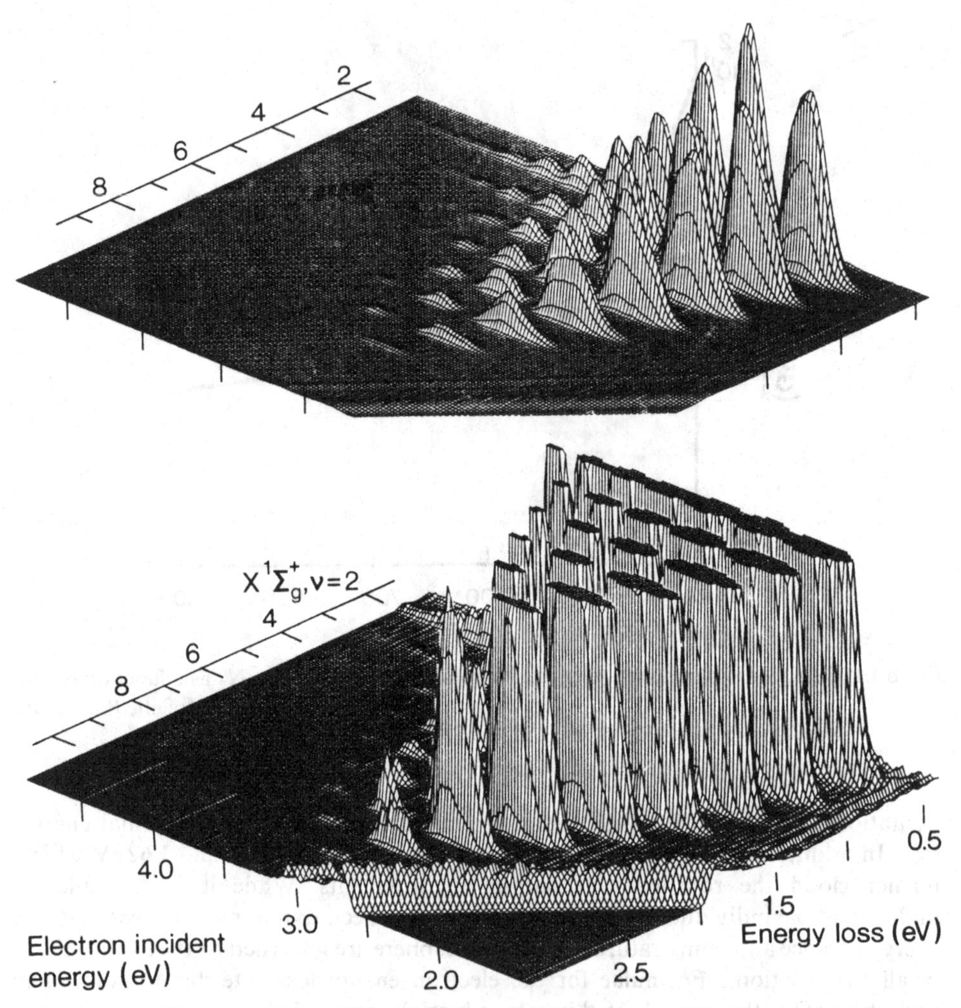

Fig. 6.4.2 Laboratory measurement of the differential cross section for excitation of the $v = 2$ to $v = 10$ vibrational levels in ground state N_2 by electron impact. The maximum cross section has a value of about $2.7 \times 10^{-16}\,cm^2$. The diagonal ridge in the expanded ordinate presentation is due to a background of electrons close to zero energy. (F. Currell, *et al.*, *XIV ICPEAC*, collected abstracts, 1985.)

quantum number. The vibrational energy is $h\nu_{vib}$. Detailed quantum mechanical computations predict the observed structure and magnitude of the cross section (see references). The energy loss rate by the inelastic collision

$$e + N_2(v=0) \rightarrow e + N_2(v>0) \tag{6.4.19}$$

is given by

$$L_{vib}(e, N_2) = n(e)n(N_2)u_v \Delta E(0,v)\sigma(0 \rightarrow v, E) \tag{6.4.20}$$

where u is the electron velocity, $\Delta E\,(0,v)$ the internal energy change and $\sigma(0 \rightarrow v, E)$ the energy-dependent excitation cross section to level v. This equation is analogous to Equation (6.4.10) but does not include deexcitation collisions, the possibility that vibrationally excited molecules transfer energy back to the electron gas. Although possible in the weakly ionized thermosphere, the current view is that vibrational deactivation occurs by neutral and not electron collisions, contributing to neutral heating.

Equation (6.4.20) appears deceptively simple but the energy-dependent structure in each cross section and the number of vibrational levels that need to be included, at least to $v = 8$, result in a large computational task. The vibrational spacing is about 0.29 eV and each vibrational level contributes to the overall energy loss. Assuming that the electron gas and the neutral gas have Maxwellian distributions at temperatures T_e and T_n, respectively, the cooling rate of electrons by vibrational excitation of N_2 has been computed (see references) and the result has been fitted by a cumbersome analytic formula that we give in Appendix 6.

The vibrational excitation of O_2 for electron energies less than about 4 eV occurs principally through the intermediate formation of the $O_2^-(^2\Pi_g)$ ion. The cross sections for vibrational levels $v = 1$–4 have been measured in a laboratory cross-beam apparatus. Excitation of the electronic states $^1\Delta_g$ and $^1\Sigma_g^+$ also contributes to electron cooling, as already pointed out, and the excitation cross sections at low energy are shown in Figure 6.4.3. Both electronic states must be included when computing

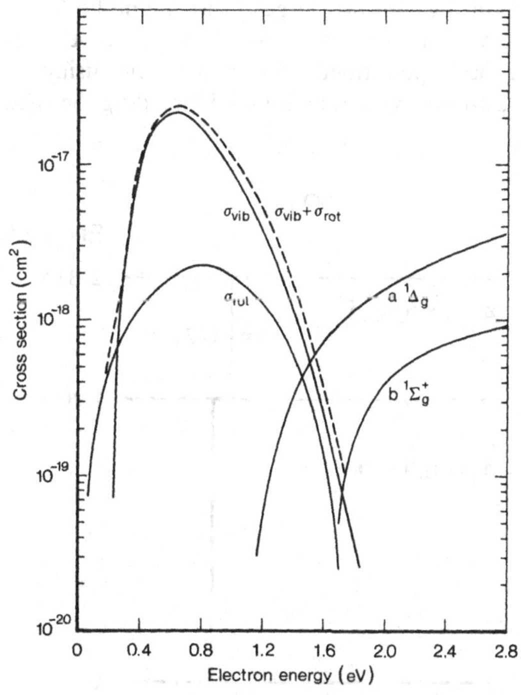

Fig. 6.4.3 Cross sections for the excitation by electron impact of vibrational and rotational levels and of low lying electronic states in O_2 as a function of electron energy (data compiled by A.V. Phelps, 1981).

electron energy loss rates in the thermosphere. An equation analogous to Equation (6.4.20) gives the vibrational loss rate, supplemented by excitation of the electronic states. The neutrals and the electrons are assumed to have a Maxwellian energy distribution at temperatures T_n and T_e, respectively, and the loss rate can thereby be given in terms of these parameters. The numerical results have been fitted to an approximate analytic expression given by

$$L(\mathrm{e},\mathrm{O}_2) = n_e n(\mathrm{O}_2) 7.45 \times 10^{-13} \exp\left(f\frac{T_e - 700}{700T_e}\right) \cdot \left[\exp\left(\frac{3000}{T_e} - \frac{3000}{T_n}\right) - 1\right] \quad (\mathrm{eV\,cm^{-3}\,s^{-1}}) \qquad (6.4.21)$$

with $f = 3.902 \times 10^3 + 4.38 \times 10^2 \tanh[4.56 \times 10^{-4}(T_e - 2400)]$, and temperatures in degrees Kelvin (Problem 27).

The ground state of atomic oxygen is $\mathrm{O}(^3\mathrm{P}_{2,1,0})$. The energy level diagram, the transition probabilities and the wavelengths of the radiations are shown in Figure 6.4.4. Noting that the energy of the fine structure levels above the ground state, $J = 2$, is in the electron thermal regime, it is likely that the $J = 1$ and 0 levels are excited by electron impact,

$$\mathrm{e} + \mathrm{O}(^3\mathrm{P}_J) \rightarrow \mathrm{e} + \mathrm{O}(^3\mathrm{P}_{J'}) \qquad (6.4.22)$$

and the process becomes an energy loss mechanism for the electron gas if the cross section for the reaction is sufficiently large. There are no laboratory measurements of the O fine structure excitation cross section; however, several quantum mechanical computations have been performed, the most recent using a multiconfiguration representation of the atomic wave function and including the effects of the long-range

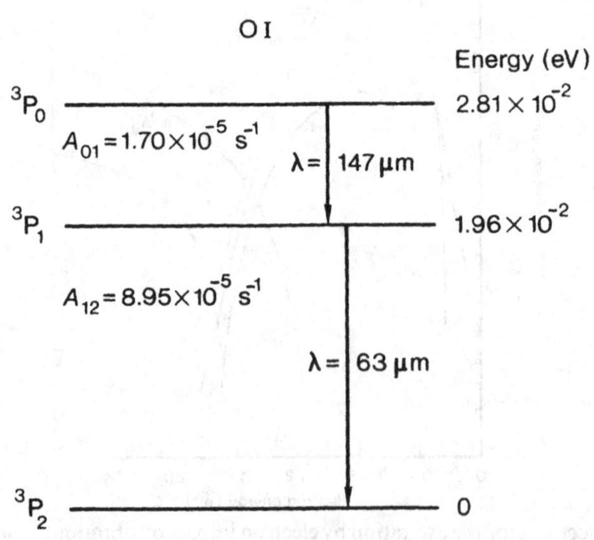

Fig. 6.4.4 Energy level diagram of the fine structure in the ground state of O. Transition probabilities are given for the radiation between levels.

Table 6.4.1. [a]*Collision strengths of O, $\Omega(J, J')$*

T_e(K)	$\Omega(2,1)$	$\Omega(2,0)$	$\Omega(1,0)$
500	0.0027	0.0006	0.0008
1000	0.0076	0.0022	0.0018
2000	0.0174	0.0054	0.0039
5000	0.0474	0.0148	0.0112
10000	0.0987	0.0292	0.0265
20000	0.2066	0.0536	0.0693

[a]Adapted from M. Le Dourneuf and R.K. Nesbet, *J. Phys.* **B9**, L241, 1976.

polarization interaction in the collision. The cross section for the transition from level J to J' is

$$\sigma(J \rightarrow J';\, E) = \frac{\pi a_0^2}{\tilde{\omega}_J}\Omega(J, J')\frac{13.604}{E} \tag{6.4.23}$$

where πa_0^2 is the area of the first Bohr orbit, $\tilde{\omega}$ is the statistical weight $(2J+1)$, the energy of 1 Rydberg is 13.604 eV, E is in eV, and $\Omega(J, J')$ is the collision strength for the transition ^3P_J to $^3P_{J'}$. The last quantity is the result of the quantum mechanical computations. The most widely accepted current (1987) values are listed in Table 6.4.1 as a function of electron temperature ($E = kT_e$).

The rate of change of electron energy through excitation and deexcitation of the fine structure levels of O is given by

$$L_{FS}(e, O) = n(e)v \sum_{J=0}^{2} n_J(O) \sum_{J \neq J'} (E_J - E_{J'})\sigma(J \rightarrow J';\, E) \tag{6.4.24}$$

where v is the electron velocity and the number density of atoms in the Jth level follows a Boltzmann distribution,

$$n_J(O) = n(O)\frac{(2J+1)\exp(-E_J/kT)}{\sum_{J=0}^{2} (2J+1)\exp(-E_J/kT)} \tag{6.4.25}$$

equilibrium being established by collisions with the neutral gas. The electron gas has a Maxwellian distribution at temperature T_e and the energy loss rate becomes

$$L_{FS}(e, O) = 2\pi\left(\frac{1}{\pi k T_e}\right)^{3/2}\left(\frac{2}{m_e}\right)^{1/2} n_e n_J(O) \int_0^{\infty} E\exp\left(-\frac{E}{kT_e}\right) \times \sum_{J' \neq J} (E_J - E_{J'})\sigma(J \rightarrow J';\, E)\,dE \tag{6.4.26}$$

Energy loss by excitation of the fine structure level of O is an important electron cooling process in the low energy regime where other inelastic collisions are no longer effective (Problem 28).

The electron gas in the high latitude F-region can attain a high temperature caused by a large rate of local, aurorally associated, electron energy deposition, as given by

Equation (6.2.7). At middle and high latitudes the electron gas gains heat that is conducted into the ionosphere from an even hotter electron gas in the magnetosphere. At a high electron temperature an appreciable fraction of the Maxwellian electron population has sufficient energy to excite the lowest electronic state in atomic oxygen, $O(^1D)$, thereby cooling the electron gas. The energy-dependent excitation cross section for the reaction

$$e + O(^3P) \rightarrow e + O(^1D) \quad (6.4.27)$$

has recently been measured in the laboratory and the results, which agree rather well with theoretical computations, are shown in Figure 6.4.5.

The electron energy loss rate is given by

$$L_{O\,^1D}(e, O) = n_e n(O) 2\pi \left(\frac{1}{\pi k T_e}\right)^{3/2} \left(\frac{2}{m_e}\right)^{1/2} \left\{ \exp\left[-\frac{1.97}{k}\left(\frac{1}{T_n} - \frac{1}{T_e}\right)\right] - 1 \right\}$$

$$\times 1.97 \int_{1.97}^{\infty} E\sigma(O\,^1D) \exp\left(-\frac{E}{kT_e}\right) dE \quad (\mathrm{eV\,cm^{-3}\,s^{-1}}) \quad (6.4.28)$$

where 1.97 eV is the excitation energy of the $O(^1D)$ state, as shown in the energy level diagram given in Appendix 3. Since $T_e > T_n$ the right side of the equation is negative and represents a loss rate for the electron gas. We will learn in Chapter 7 that the process described by Equation (6.4.28) is one of several sources of $O(^1D)$ atoms that radiate in the well-known nebular lines at 6300 Å and 6364 Å (Problem 29).

There is one additional process by which electrons exchange energy, i.e., collisions with the ambient ion gas. Such collisions assume importance as the degree of ionization

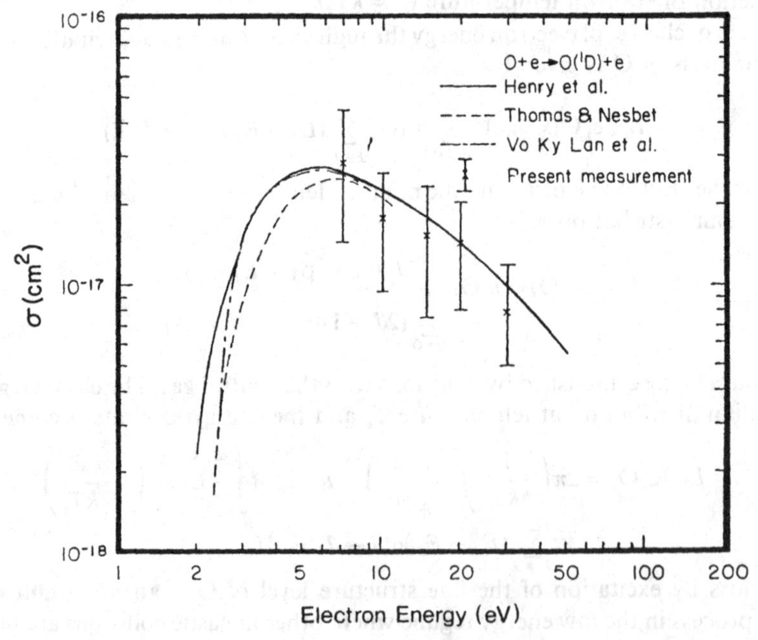

Fig. 6.4.5 Cross sections for the excitation of $O(^1D)$ by electron impact on ground state O. The experimental results, shown as crosses, are compared with several theoretical studies. (T.W. Shyn and W.E. Sharp, *J. Geophys. Res.*, **91**, 1691, 1986.)

in the thermosphere increases; this occurs with increasing altitude. Under the most common circumstances $T_e > T_i$ and electron–ion collisions represent energy loss for the electron gas. Under conditions of large ion heating rates resulting from large orthogonal electric fields in the ionosphere, ion–electron collisions may serve as an energy source for the electron gas and an energy sink for the ion gas.

Elastic collisions between electrons and ions are described by the screened Coulomb potential (Equation (4.2.8)). The inverse of the parameter λ_0 in Equation (4.2.7) is a measure of the effective range of the Coulomb interaction between two charged particles. It has been called the Debye shielding length, λ_D, and is defined in a binary mixture of electrons and singly charged ions by

$$\frac{1}{\lambda_D^2} = 4\pi\frac{e^2}{k}\left[\frac{n_e}{T_e} + \frac{n_i}{T_i}\right] \tag{6.4.29}$$

where the colliding species have a Maxwellian distribution at temperatures T_e and T_i. The screened Coulomb potential yields a differential scattering cross section (Equation (4.2.8)) that is used to evaluate the momentum transfer cross section (Equation (4.5.4)),

$$\phi_{1,2}^{(1)}(v) = (4\pi e^4/M_R^2 v^4)\ln\Lambda \tag{6.4.30}$$

where

$$\Lambda = (M_R v^2/e^2)\lambda_D \tag{6.4.31}$$

The relative velocity between collision partners is v and M_R is the reduced mass. The rate at which energy is transferred between electrons and singly charged ions is therefore (Problem 30)

$$L_{\text{elastic}}(\text{e}, \text{i}) = \frac{4(2\pi m_e)^{1/2}}{m_i} n_e n_i e^4 \frac{\ln\Lambda}{(kT_e)^{3/2}} k(T_e - T_i) \tag{6.4.32}$$

Summarizing this subsection, several types of collision contribute to energy exchange between the electron gas and other atmospheric species. Due to the large changes with altitude of the atmospheric density and composition, different processes become important at various levels, requiring adequate values for all in order to compute reliable temperature profiles.

6.4.2 Energy exchange between ions and neutrals

Two processes contribute to the exchange of energy between ions and neutrals. One process is elastic non-resonant collisions, for example O^+–N_2 or N_2^+–O collisions. The encounter is dominated at low temperatures by an induced dipole or polarization interaction described in Section 4.4. The interaction potential, which is given by Equation (4.4.7), leads to a collision integral that is independent of temperature, Equation (4.5.9). At the higher temperatures of the upper F-region, attractive forces due to quantum mechanical effects lead to a weakly temperature-dependent collision. Since high temperature cross section data are currently not available, we adopt the induced dipole interaction potential in deriving the energy exchange rate by non-resonant collisions (Problem 31)

$$L_{\text{NR}}(\text{i}, \text{n}) = 3n_i n_n \frac{2.21\pi}{m_i + m_n}(\alpha M_R e^2)^{1/2} k(T_i - T_n) \tag{6.4.33}$$

where α is the polarizability defined in Section 4.4. Numerical values of the polarizabilities of atmospheric neutral species are listed in Appendix 6.

The other process that contributes to exchange of energy between ions and neutrals is symmetrical resonant charge transfer. It involves collisions between ions and their parent neutrals, for example

$$O^+ + O \rightarrow O + O^+ \tag{5.3.10}$$

which has already been mentioned in Section 5.3, or

$$N_2^+ + N_2 \rightarrow N_2 + N_2^+ \tag{6.4.34}$$

While a transfer of charge occurs in the collision, each particle tends to retain its original kinetic energy so that the kinetic energy associated with the ion before the collision appears as kinetic energy of the neutral after the collision. Since the cross section is large, the process efficiently transfers kinetic energy from ions to neutrals.

The cross section for symmetric charge transfer is determined largely by long-range forces between ion and atom so that asymptotic collision theories and the impact parameter approximation (Figure 4.4.3) yield satisfactory results. However, the impact velocity must exceed a threshold below which the cross section varies rapidly as a function of impact parameter. In addition, the impact velocity must not be so large that transfer of electron momentum occurs. The temperature range in the thermosphere lies within the above lower and upper bounds. The cross section for symmetric charge transfer is given by

$$\phi_{\mathrm{SCT}} = 2\pi \int_0^\infty b \sin^2\Gamma(b)\,\mathrm{d}b \tag{6.4.35}$$

where b is the impact parameter and

$$\pm\Gamma(b) = \pm(\eta_l^+ - \eta_l^-) \tag{6.4.36}$$

The η_ls are the lth order phase shifts of the partial wave scattering analysis discussed in Section 4.2, but here we distinguish between the phase shift due to a scattering potential $V^+(r)$, symmetric in the nuclei, and $V^-(r)$, antisymmetric in the nuclei. In a reference given at the end of this chapter it is shown that the momentum transfer or diffusion cross section, $\phi_{1,2}^{(1)}$ (Equation (4.5.5)), is very nearly twice the resonant charge transfer cross section,

$$\Phi_{\mathrm{i,n}}^{(1)} = 2\Phi_{\mathrm{SCT}} \tag{6.4.37}$$

This follows from the observation that the transfer of momentum in elastic scattering is nearly the same as that in inelastic scattering while only a charge is transferred in the latter process.

An effective momentum transfer cross section is defined by averaging the differential cross section over the ion and neutral velocity distribution functions, assumed to be Maxwellian,

$$\phi_{\mathrm{SCT}}(\mathrm{i,n}) = \int_0^\infty f_{\mathrm{i}}(v) f_{\mathrm{n}}(v) \phi_{\mathrm{i,n}}^{(1)}(v)\,\mathrm{d}v_{\mathrm{i}}\,\mathrm{d}v_{\mathrm{n}}. \tag{6.4.38}$$

At low impact velocity the induced dipole polarization interaction becomes dominant. This is illustrated in Figure 6.4.6 where $\phi_{\mathrm{SCT}}(\mathrm{i,n})$ is given as a function of $T_{\mathrm{i}} + T_{\mathrm{n}}$.

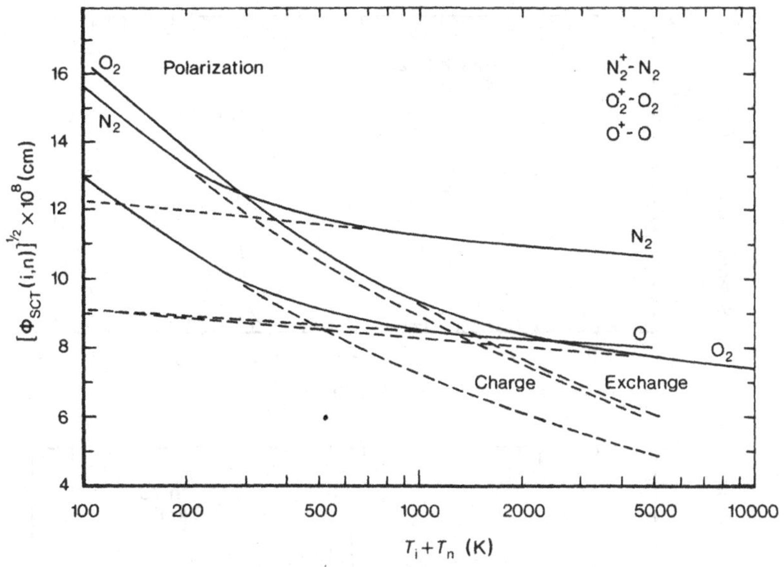

Fig. 6.4.6 Momentum transfer cross sections between the major atmospheric neutral species and their ions. Symmetric charge exchange is shown by the dashed curves and polarization interaction by the solid curves. Cross sections are given as a function of the sum of $T_i + T_n$. (Adapted from P.M. Banks and G. Kockarts, *Aeronomy*, Part A, Academic Press, New York, 1973.)

Polarization interactions in N_2^+–N_2 and O^+–O charge exchange may be neglected at thermospheric temperatures but contribute to the O_2^+–O_2 cross section.

The energy transfer rate between ions and neutrals by resonant charge exchange is

$$L_{SCT}(i, n) = -\tfrac{3}{2} n_i \nu_{SCT} k(T_i - T_n) \qquad (\text{erg cm}^{-3}\,\text{s}^{-1}) \tag{6.4.39}$$

where ν_{SCT} is the average collision frequency for charge exchange. The collision frequency is defined by Equation (5.3.11) in terms of the collision integral. In Appendix 6 average ion-neutral charge exchange collision frequencies are listed for thermospheric ion and neutral species.

6.4.3 Energy loss by the neutral gas

Since $T_n < T_i$ and T_e, collisional energy exchange between the neutral gas and thermal ions and electrons can only result in neutral heating, how does the neutral gas lose energy? The neutral temperature gradient is positive with increasing altitude in the thermosphere assuring downward conduction of heat throughout the region. While this accounts in part for the observed neutral temperature profile there is a mechanism that acts as a local energy sink: it is radiation. To be effective, collisionally excited states must decay by radiation rather than by collisions with other species which would merely pass the energy locally from one to the other. In addition, the radiation must not be reabsorbed within the thermosphere if it is to extract energy from the region. The airglow and the aurora are visible evidence that energy does escape the thermosphere and that lines, bands, and continua that originate in neutral species are an energy sink. Optical phenomena are discussed in detail in Chapter 7 and we only note here that, in spite of

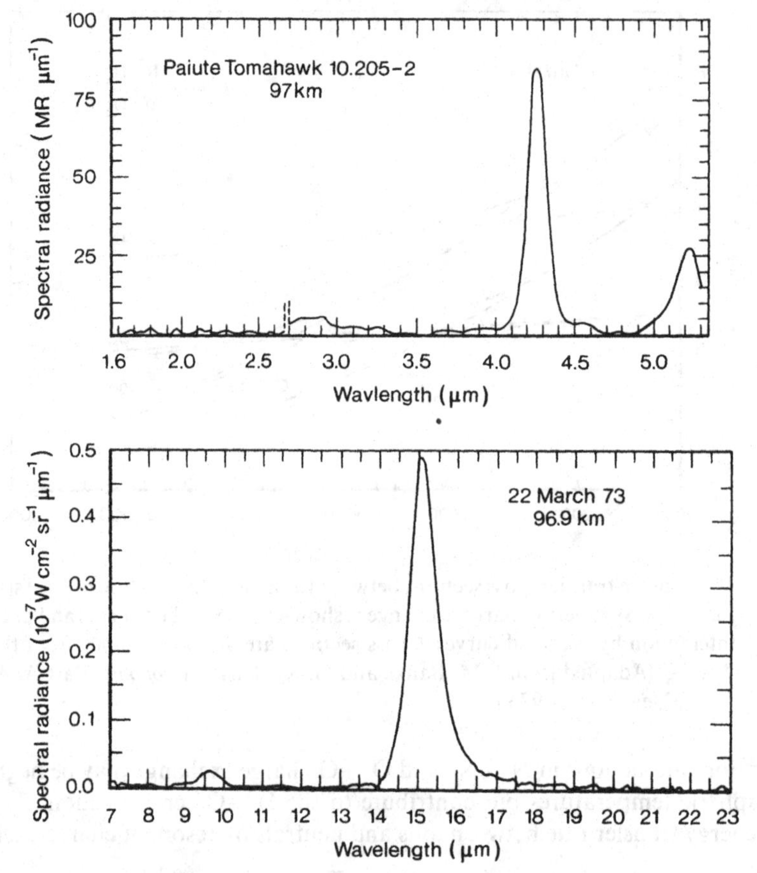

Fig. 6.4.7 Energy loss by radiation of the neutral gas in the IR spectrum. The radiation was measured by a rocket borne spectrometer viewing upward at 97 km. The spectral features are attributed to CO_2, O_3, NO and NO^+ ($1\ \mu m = 10^4$ Å). (A.T. Stair, *et al.*, *Atmospheres of Earth and the Planets*, Ed., B.M. McCormac, D. Reidel Publ. Co., Dordrecht, 1975, p 335.)

the brilliance of auroral displays, the airglow and auroral emissions in the visible and UV regions of the spectrum account for a very small fraction only of the total radiative loss in the thermosphere. Not surprisingly, the important radiative loss occurs in the IR part of the spectrum.

The strongest IR emissions detected by rocket borne spectrophotometers under a variety of geophysical conditions are at 2.8, 4.3, 5.3, 9.6, 15 and 63 μm (Figure 6.4.7). The 63 μm radiation corresponds to the fine structure level transition $^3P_2$–$^3P_1$ in O, shown in Figure 6.4.4. The companion transition at 147 μm is too weak to be detected. While substantial excitation occurs by electron impact (Equation (6.4.20)), followed by photon emission, the atmosphere is optically thick to the 63 μm radiation below about 150 km and very little escapes the thermosphere below 100 km. The 63 μm radiation is therefore not an effective radiative energy sink. The 9.6 μm radiation corresponds to the ν_3 fundamental vibration band of ozone. The radiation is a cooling mechanism at some levels in the middle atmosphere but has no influence in the thermosphere, above 100 km.

The feature observed around 5.3 μm is attributed to emission from the $\Delta v = 1$ sequence of ground state NO($X^2\Pi$). While resonance absorption of solar radiation and of earthshine account for most of the excitation below 120 km, in the absence of auroral energetic particle input, collisional excitation becomes dominant above this level. Under auroral bombardment collisional excitation prevails throughout the thermosphere. The excitation mechanisms are: (1) exchange of vibrational energy with other virbationally excited molecules, (2) collisional excitation by O or exchange between NO and O, and (3) the chemical reaction, principally in aurora,

$$N(^4S, ^2D, ^2P) + O_2 \rightarrow NO(v > 0) + O \tag{6.4.40}$$

Since NO is a minor constituent, the atmosphere is optically thin to the 5.3 μm emission and the radiation escapes from the thermosphere at the top and at the bottom. The 5.3 μm band sequence is the brighest NO emission feature and detailed computations (see references) have demonstrated that this radiation is an effective cooling mechanism for the thermosphere that must be included in the energy balance of the neutral atmosphere. The energy loss rate by radiation in the 5.3 μm band of NO is given by

$$L_{rad}(NO) = n(NO, v = 1)A_{10}h\nu \tag{6.4.41}$$

where the photon energy is $h\nu = 1876\,cm^{-1}$ and A_{10} is the radiative transition probability, $\sim 13.3\,s^{-1}$. The difficulty arises in determining the population of NO, $v = 1$ molecules since there are three potential sources of vibrationally excited NO and the rate at which various vibrational levels are populated is different for each process. Moreover, collisional deactivation of the $v = 1$ level competes with radiative transitions. O appears to be most effective and a collisional quenching rate coefficient $k(O) = 6.5 \times 10^{-11}\,cm^3\,s^{-1}$ has been adopted in the recent literature.

The features at 15 and 4.3 μm are attributed to the CO_2 (0v0) (bending) and CO_2 (00v) (asymmetric stretch) bands respectively. However, the $\Delta v = 1$ sequence of NO^+ ions also emits at 4.3 μm. Comparison of observations with synthetic spectra is necessary to separate the contributions of the two emitters. Vibrational energy exchange between N_2 and CO_2

$$N_2(v > 0) + CO_2 \rightleftharpoons N_2 + CO_2(001) \tag{6.4.42}$$

accounts for the former identification, while the chemical reaction

$$N^+ + O_2 \rightarrow NO^+(v > 0) + O \tag{6.4.43}$$

accounts for the latter. In aurora, both N_2 ($v > 0$) and NO^+ ($v > 0$) are produced at a large rate and rocket observations show a strong enhancement of the 4.3 μm feature. We indicated earlier in this chapter that collisional deactivation of vibrationally excited nitrogen is a neutral gas heat source. In view of the alternatives just described, some of the energy is radiated out of the thermosphere in the CO_2 bands.

6.5 Energy balance in the thermosphere

In addition to sources and sinks of energy, the temperature profiles in the atmosphere are influenced by energy transport processes. The energy balance equation contains the transport terms regardless of whether its derivation follows the Chapman–Enskog procedure of summational invariants or Grad's system of moment equations. Several references listed at the end of this chapter give extended discussions, including comparison of the methods. Since we are primarily interested in the physical processes and their mutual relationship, we write the energy balance equation for any (unlabelled) species as follows,

$$\frac{\mathcal{N}}{2}k\frac{D(nT)}{DT} = -\tfrac{5}{2}nkT\nabla\cdot\mathbf{v} - \nabla\cdot\mathbf{q} - \bar{\bar{\mathbf{P}}}:\nabla\mathbf{v} + Q - L \tag{6.5.1}$$

Each term specifies a rate of change of energy per unit volume. $\mathcal{N}$ is the number of degrees of freedom, 3 for a monoatomic gas and 5 for N_2 and O_2. The substantial derivative taken along the trajectory of the moving gas is

$$D/Dt = \partial/\partial t + \mathbf{v}\cdot\nabla \tag{6.5.2}$$

where the second term on the right side represents heat advection. In Equation (6.5.1) the terms on the right represent: (1) cooling (heating) due to expansion (contraction) of the gas under the influence of the scalar pressure $p = nkT$, (2) the divergence of the heat flow vector, (3) viscous heating due to stress in the fluid, (4) local energy sources, and (5) local and radiative energy sinks. The relative importance of various terms depends on the type of gas, neutral or ionized, and is also a function of altitude in the atmosphere.

Viscous heating is a consequence of relative motion between parcels of gas in a fluid, the quantity $\bar{\bar{\mathbf{P}}}$ representing the stress or pressure tensor. The colon in the viscous heating term signifies the double or scalar product. Viscous heating influences the energy balance in the neutral gas but is not an important source of ion or electron heating.

The thermal flow vector, $\mathbf{q}$, represents transport of molecular kinetic energy resulting from a temperature gradient. For ionized species heat is also carried by a current. Thus,

$$\mathbf{q} = -\lambda\nabla T - \beta\mathbf{J} \tag{6.5.3}$$

The quantity λ is the coefficient of thermal conductivity defined for a simple gas by Equation (4.5.13) in terms of a collision integral. This parameter will be discussed later in this section. The quantity β is the thermoelectric coefficient which is given by the approximate expression

$$\beta = \frac{5}{2}kT_{e,i}/|e| \tag{6.5.4}$$

where the absolute value of the electric charge assures applicability to both electrons and ions. The current density is defined by the difference in the flow of positive and negative charges,

$$\mathbf{J} = e(n_i\mathbf{v}_i - n_e\mathbf{v}_e) \tag{6.5.5}$$

The ionosphere consists almost entirely of singly charged ions and electrons and charge neutrality prevails, $n_e = n_i$, in the fluid.

The principal energy sources are solar UV photons, energetic charged particles of magnetospheric origin and orthogonal electric fields, as set forth in Sections 6.2 and 6.3. Exchange reactions are an energy source for one species and a sink for the collision partner, as discussed in Section 6.4, where radiative energy sinks are also presented.

Some insight into the viscous heating term is obtained by decomposing the scalar product into its elements,

$$\bar{\bar{\mathbf{P}}}:\nabla\mathbf{v} = P_{xx}\frac{\partial v_x}{\partial x} + P_{xy}\frac{\partial v_y}{\partial x} + P_{xz}\frac{\partial v_z}{\partial x}$$

$$+ P_{yx}\frac{\partial v_x}{\partial y} + P_{yy}\frac{\partial v_y}{\partial y} + P_{yz}\frac{\partial v_z}{\partial y}$$

$$+ P_{zx}\frac{\partial v_x}{\partial z} + P_{zy}\frac{\partial v_y}{\partial z} + P_{zz}\frac{\partial v_z}{\partial z} \tag{6.5.6}$$

The pressure tensor contains diagonal and off-diagonal terms. The diagonal terms,

$$P_{xx} = p - \frac{2}{3}\eta\left(2\frac{\partial v_x}{\partial x} - \frac{\partial v_y}{\partial y} - \frac{\partial v_z}{\partial z}\right) \tag{6.5.7}$$

include the scalar pressure p and viscous stress components, while the off-diagonal terms,

$$P_{xy} = -\eta\left(\frac{\partial v_x}{\partial y} + \frac{\partial v_y}{\partial x}\right) \tag{6.5.8}$$

have stress components only. Other terms are obtained by commuting the subscripts and the scalar pressure is given by

$$p = \frac{1}{3}(P_{xx} + P_{yy} + P_{zz}) \tag{6.5.9}$$

Equations (6.5.7) and (6.5.8) result from the second approximation of the Chapman–Enskog solution to the Boltzmann equation for which the reader is directed to a reference given at the end of this chapter. The coefficient of viscosity, η, is defined by Equation (4.5.12) in terms of the collision integral $\Omega^{(2,2)}$. Adopting a very simple collision model which assumes that neutral molecules are rigid elastic spheres without force fields, we obtain the collision integral for a single species gas

$$\Omega_1^{(2,2)} = 2\sigma_1^2(\pi kT/m_1)^{1/2} \tag{6.5.10}$$

where σ_1 is the molecular diameter and m_1 the mass. The coefficient of viscosity is therefore

$$\eta_1 = \frac{5}{16}\frac{1}{\sigma_1^2}\left(\frac{m_1 kT}{\pi}\right)^{1/2} \tag{6.5.11}$$

The coefficient of viscosity has been obtained experimentally for several neutral gases over a wide temperature range and the results are represented by the empirical equation

$$\eta_j = A_j T^B \qquad (\text{gm cm}^{-1}\,\text{s}^{-1}) \tag{6.5.12}$$

in the temperature range between 200 K and 2000 K. The parameter B has the value 0.69 for all major atmospheric species and the coefficient A_j takes the values shown in Table 6.5.1 (Problem 33). Eddy viscosity, mentioned in Section 5.2, is the mechanism associated with heating of the gas by the conversion of laminar flow to turbulence of varying scale sizes. The effect is unimportant in the thermosphere where only molecular viscosity contributes to the energy balance.

Table 6.5.1. *Numerical coefficients in Equations (6.5.12) and (6.5.14)*

j	A_j	C_j
N_2	3.43×10^{-6}	56
O_2	4.03×10^{-6}	56
O	3.90×10^{-6}	75.9
He	3.84×10^{-6}	299
H	1.22×10^{-6}	379

The coefficient of thermal conductivity was introduced in Equation (6.5.3) without detail. The discussion of viscosity simplifies the exposition of the thermal conductivity in view of the relationship for monoatomic gases given in Section 4.5,

$$\lambda = \eta \frac{15}{4} \frac{k}{M} \tag{4.5.17}$$

where M is the mass of the single species gas. With the rigid elastic sphere model assumed in the derivation of the viscosity one obtains for the coefficient of thermal conductivity

$$\lambda_1 = \frac{75}{64} \frac{1}{\sigma_1^2} \left(\frac{k^3 T}{\pi m_1} \right)^{1/2} \tag{6.5.13}$$

If intermolecular force fields are included in the collisions, the transport coefficients η and λ assume different (and more complex) forms than those given by Equations (6.5.11) and (6.5.13). A simple empirical equation satisfies the thermal conductivity coefficient in a neutral gas. It has the form

$$\lambda_j = C_j T^B \qquad (\mathrm{erg\,cm^{-1}\,s^{-1}\,K^{-1}}) \tag{6.5.14}$$

analogous to the coefficient of viscosity. B equals 0.69, as before, and the coefficient C_j assumes the values given in Table 6.5.1. An approximate expression for the coefficient of thermal conduction in a multicomponent gas is

$$\lambda = \frac{\sum_j C_j n_j}{\sum_j n_j} T^{0.69} \qquad (\mathrm{erg\,cm^{-1}\,s^{-1}}\,K^{-1}) \tag{6.5.15}$$

where n_j is the concentration of the jth constituent. An analogous expression may be used for the viscosity coefficient in a multicomponent gas.

6.5.1 Energy balance in the neutral gas

Collecting the terms that describe processes in the energy balance of the neutral gas in the thermosphere, Equation (6.5.1) becomes

$$\begin{aligned} &\frac{\mathcal{N}}{2} k \left[\frac{\partial (n T_n)}{\partial t} + \mathbf{v} \cdot \nabla (n T_n) \right] \\ &= -\frac{5}{2} n k T_n \nabla \cdot \mathbf{v} + \nabla \cdot (\lambda \nabla T_n) - \bar{\mathbf{P}} : \nabla \mathbf{v} + Q - L_{\mathrm{rad}} \end{aligned} \tag{6.5.16}$$

The thermal structure varies in all directions, latitude, longitude and altitude, and as a function of time. Vertical temperature gradients are, however, much larger than horizontal gradients and the concept of a global mean temperature profile has been useful. On a global mean basis heating by solar UV radiation, energetic particles and Joule dissipation is essentially balanced by downward thermal conduction and radiative cooling. Asymmetries, or non-uniformity in the heat sources, lead to compressional heating and cooling by adiabatic expansion as parcels of fluid move horizontally and vertically. Intense local heating might occur in a narrow, long, auroral arc leading to large velocity shear or stress in the small scale structure; viscous energy dissipation then becomes important in maintaining the energy balance.

The global mean temperature near the ground is a sensitive parameter indicative of climate changes on Earth resulting from the balance between the heating and cooling. Similarly, modelling the thermospheric mean temperature profile tests our overall knowledge of the global scale heat sources and sinks. Changes in global heating rates over a cycle of solar activity (due to changes in the solar UV flux discussed in Chapter 2) lead to corresponding changes in the global mean temperature. We will learn in Chapter 8 that in modelling thermospheric dynamics and thermodynamics theoretically it is convenient to specify the temperature in terms of departures from the global mean.

The global mean energy equation then reduces to the one-dimensional time-dependent form

$$\frac{\mathcal{N}}{2}nk\frac{\partial T_n}{\partial t} = \frac{\partial}{\partial z}\left(\lambda\frac{\partial T_n}{\partial z}\right) + Q_n - L_{rad} \qquad (6.5.17)$$

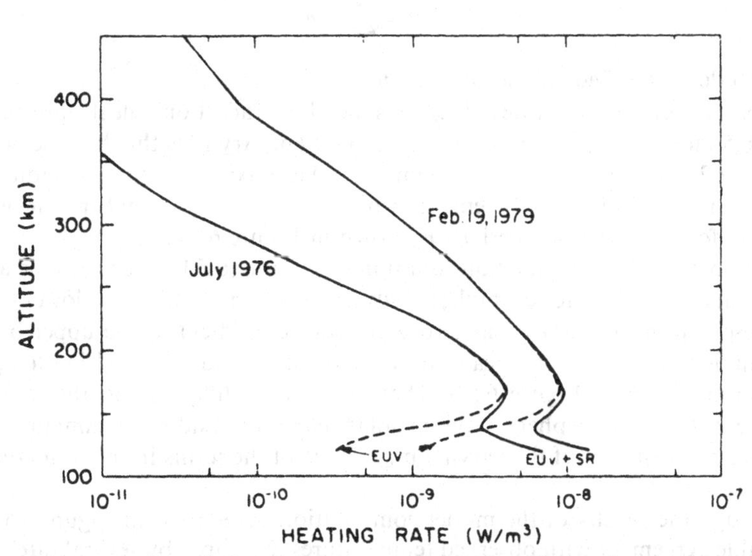

Fig. 6.5.1 Altitude profiles of the global mean solar UV radiation heating rates for solar minimum (July, 1976) and solar maximum (February, 1979) conditions. The dashed curves refer to the wavelength region below 1050 Å, the ionization region, while the solid curves include the entire ultraviolet spectrum. SR refers to the Schumann–Runge region. ($1\ \mathrm{W\,m^{-3}} \equiv 10\ \mathrm{erg\,cm^{-3}\,s}$) (R.G. Roble and B.A. Emery, *Planet. Space Sci.*, **31**, 597, 1983.)

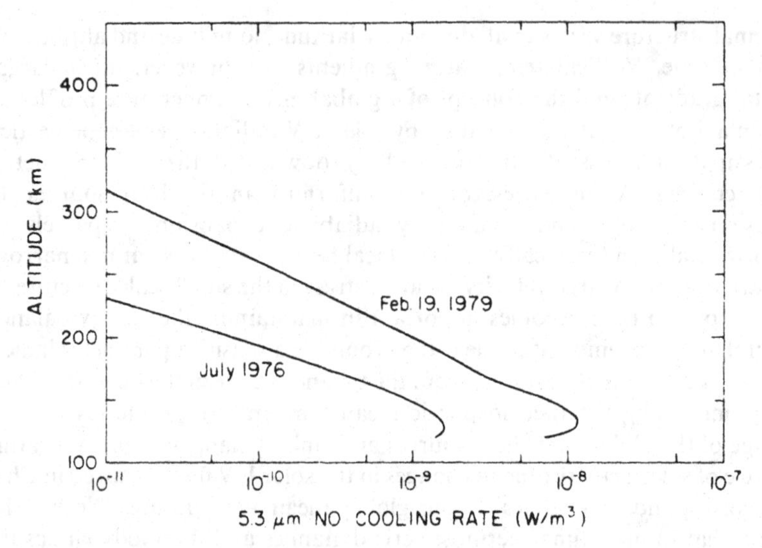

Fig. 6.5.2 Altitude profiles of the global mean cooling rate by radiation of the 5.3 μm vibrational transition in NO molecules. July, 1976 and February, 1979 represent periods of solar minimum and solar maximum activity, respectively. (R.G. Roble and B.A. Emery, *Planet. Space Sci.*, **31**, 597, 1983.)

with

$$\mathcal{N} n = \sum_j \mathcal{N}_j n_j \tag{6.5.18}$$

The altitude is specified by the coordinate z.

Global mean heating rates due to solar UV radiation effects are shown in Figure 6.5.1 for the two periods in July 1976 and February 1979 that have been adopted in Chapter 2 to represent solar minimum and solar maximum activity conditions. The mean radiative cooling rate by emission of the 5.3 μm vibrational transition in NO molecules for the same two periods is shown in Figure 6.5.2.

Solution of the heat conduction equation (Equation (6.5.17)) requires the adoption of boundary conditions, for example, specifying the temperature at the lower boundary from observations and assuming zero temperature gradient at the upper boundary. The equation is solved numerically to a steady state and the resulting temperature profiles are shown in Figure 6.5.3. There is a large difference in the temperature structure of the thermosphere between solar maximum and solar minimum activity which is a consequence of the varying magnitude of the terms in the heat conduction equation.

Although the results of the model computations presented in Figure 6.5.3 are in reasonable agreement with observed temperatures measured by several satellites over many years and a wide geographic coverage, the reader should not view the detailed numbers as being definitive; rather, they are illustrative of the relative importance of different processes that influence the temperature structure of the thermosphere.

The global mean structure of the thermosphere and ionosphere has recently (1987) been modelled to predict not only the neutral temperature for solar maximum and

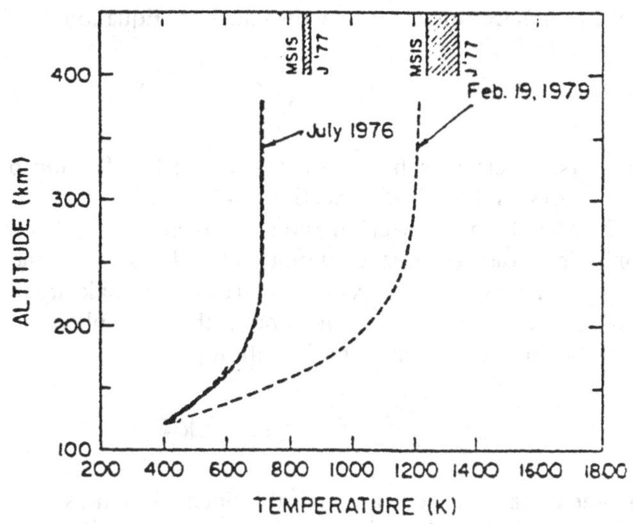

Fig. 6.5.3 Altitude profiles of the global mean temperature computed with a global thermospheric model for periods of solar minimum (July, 1976) and solar maximum (February, 1979) activity. Temperatures predicted for the corresponding periods by two empirical atmospheric models, the MSIS-83 and the Jacchia-77, are also shown. (Adapted from R.G. Roble and B.A. Emery, *Planet. Space Sci.*, **31**, 597, 1983.)

solar minimum conditions but also the plasma temperatures and the number densities of neutral and ion species. Self-consistent solutions of the one-dimensional, time-dependent equations for the neutral gas energy, the neutral and ion composition and the electron and ion energy were obtained. Auroral processes are included in the model as these were required to bring the calculated global mean structure into agreement with empirical models based on observational data. The relevant research paper is listed in the bibliography.

6.5.2 Energy balance in the electron and ion gas

Thermospheric electron and ion temperatures are measured remotely from the ground using incoherent scatter radars and *in situ* by instruments carried on board rockets and satellites. The picture that has emerged from numerous measurements acquired over the past quarter century shows diurnal, seasonal, latitudinal, and solar activity variability with random enhancements attributable to auroral activity. Analysis of such a complex global plasma temperature pattern would seem to require, at first sight, solving multidimensional, time-dependent equations. Upon closer examination of observations and analyses, however, the major and minor influences may be identified. First, a large collisional energy exchange rate insures the dominance of local sources and sinks in the E- and lower F-regions and reduces the role of energy transport over much of the ionosphere. Secondly, horizontal transport by advection, adiabatic heating and cooling and by conduction is small by comparison with vertical heat transport. Thirdly, viscous heating or cooling is very small. Electrons are not heated by chemical reactions. Applied to an electron gas the energy equation, Equation (6.5.1), becomes,

$$\frac{3}{2}k[\partial(n_e T_e)/\partial t + \mathbf{v}_e \cdot \nabla(n_e T_e)] = -\frac{5}{2}kT_e n_e \nabla \cdot \mathbf{v}_e - \nabla \cdot \mathbf{q} + Q_e - L_e \qquad (6.5.19)$$

where Q_e represents the electron heat sources described in Section 6.2 and L_e the electron cooling processes described in Section 6.4.

In Chapters 2 and 3 the photoelectron and auroral electron fluxes were treated as sources of ionization, dissociation, excitation, etc. These suprathermal electron streams, together with a much larger flux of thermal electrons make up the field-aligned currents that have been measured in and above the ionosphere. The ambipolar diffusion velocity for electrons is given by Equation (5.3.13),

$$\mathbf{v}_e = -\frac{1}{n_e m_e \nu_e}[k\nabla(n_e T_e) + n_e|e|\mathbf{E}] \qquad (6.5.20)$$

where it is assumed that the neutral vertical velocity is almost always small by comparison with the field-aligned electron drift velocity and that the gravitational force on electrons is also small by comparison with the other terms. In a weakly ionized gas, the electrons undergo collisions with ions as well as neutrals and the sum of the collision frequencies is simply labelled ν_e.

The mobility of electrons parallel to the magnetic field is larger than the ion mobility in inverse proportion to their masses and the parallel current is therefore determined by the lighter electrons,

$$\mathbf{J} = |e|n_e\mathbf{v}_e = \frac{|e|}{m_e \nu_e}(kn_e\nabla T_e + kT_e\nabla n_e + n_e|e|\mathbf{E}) \qquad (6.5.21)$$

Equation (6.5.21) shows that a current flows not only in response to an electric field but also to the temperature and density gradients that are found in the ionosphere.

In the general formulation of the conservation equations (Chapter 1), various observable parameters in a system are derived from the Boltzmann equation by taking velocity moments of the distribution function. One of the so-called higher moments describes the heat flow,

$$\mathbf{q} \approx \tfrac{1}{2}mn\overline{C^2\mathbf{C}} \qquad (6.5.22)$$

where C is the random or intrinsic velocity relative to the mean motion of the gas. Carrying out the averaging computation over the distribution function (detailed in a reference listed at the end of this chapter) we obtain, assuming a Maxwellian distribution,

$$\mathbf{q} = -\frac{5}{2}\frac{kT_e}{m_e \nu_e}(2kn_e\nabla T_e + kT_e\nabla n_e + n_e|e|\mathbf{E}) \qquad (6.5.23)$$

The intrinsic electron heat flow occurs in response to the same forces as the current, i.e., an electric field and temperature and density gradients. Differential flow between positive ions and electrons determines the polarization field **E** which may be eliminated between Equations (6.5.21) and (6.5.23) to give the heat flow vector in the electron energy equation,

$$\mathbf{q} = -\frac{5}{2}\frac{k^2 T_e n_e}{m_e \nu_e}\nabla T_e - \frac{5}{2}\frac{kT_e}{|e|}\mathbf{J} \tag{6.5.24}$$

This equation defines the coefficient of thermal conductivity,

$$\lambda_e = \tfrac{5}{2}(k^2 T_e n_e)/(m_e \nu_e) \tag{6.5.25}$$

and the thermoelectric coefficient

$$\beta_e = \tfrac{5}{2}(kT_e/|e|) \tag{6.5.26}$$

Equation (6.5.3) is thereby recovered.

The Earth's magnetic field makes the ionosphere an anisotropic medium since electrons and ions are subjected to the $\mathbf{v} \times \mathbf{B}$ Lorentz force. Transport processes involving electrons (and ions) are therefore not only a function of a collision frequency, ν_e, but also of the gyrofrequency

$$\omega_e = -eB/m_e c \tag{6.5.27}$$

the thermal conductivity coefficient is a tensor quantity,

$$\lambda_e = \begin{bmatrix} \lambda_\perp & \lambda_H & 0 \\ \lambda_H & \lambda_\perp & 0 \\ 0 & 0 & \lambda_{eo} \end{bmatrix} \tag{6.5.28}$$

where λ_{eo} is given by Equation (6.5.25) and

$$\lambda_\perp = \frac{\lambda_{eo}\nu_e^2}{\omega_e^2 + \nu_e^2} \tag{6.5.29}$$

is the thermal conductivity coefficient perpendicular to **B** and parallel to **E** while

$$\lambda_H = \frac{\lambda_{eo}\nu_e\omega_e}{\omega_e^2 + \nu_e^2} \tag{6.5.30}$$

is perpendicular to both **E** and **B**. The gyrofrequency for electrons is very large by comparison with the collision frequency, $\lambda_\perp$ and λ_H are small in the ionosphere, and $\lambda_{eo} \approx \lambda_e$.

Heat conduction in the electron gas is due primarily to Coulomb collisions with the ambient ions, but the ionosphere is only partially ionized and electron–neutral collisions contribute to heat transport throughout the ionosphere and are dominant in the E-region. The effective thermal conductivity coefficient is given by an expression of the form of the reduced mass and reduced temperature,

$$\lambda_e = \frac{\lambda_{ei}}{1 + \lambda_{ei}/\sum_n \lambda_{en}} \tag{6.5.31}$$

where λ_{ei} is the contribution from electron–ion collisions and λ_{en} from electron–neutral collisions.

Laboratory measurements of the thermal conductivity coefficient λ_{ei} of a fully ionized gas have been carried out at pressures that are far larger than are found in the ionosphere, and are not considered applicable to the natural medium. Instead, theoretical values are adopted; λ_{ei} has been evaluated for a Lorentz gas, i.e., a fully ionized gas in which electrons do not interact with each other and the positive ions are assumed to be at rest. The interaction potential between charged particles in a plasma has the form of a screened Coulomb potential (Equation (4.2.7)) and is identified as the Debye potential,

$$V(r) = -\frac{Ze^2}{r}\exp\left(-\frac{r}{\lambda_D}\right) \tag{6.5.32}$$

where Ze is the ionic charge. The Debye length (Equation (6.4.29)) is

$$\lambda_D = (kT_e/4\pi e^2 n_e)^{1/2} \tag{6.5.33}$$

Assuming that screening may be ignored for a collision impact parameter (Figure 4.4.3) $b < \lambda_D$ and that perfect screening prevails for $b > \lambda_D$, the collision cross section appropriate to electron thermal conduction is,

$$\phi_{e,i}^{(1)} = \frac{4\pi e^4}{m_e^2 v_e^4}\ln\Lambda \tag{6.5.34}$$

where

$$\Lambda = (3kT_e/e^2)\lambda_D \tag{6.5.35}$$

Based on the above definitions and approximations the collision integral is evaluated for a Maxwellian velocity distribution to yield the thermal conductivity coefficient for a Lorentz gas composed of electrons and singly charged ions,

$$\lambda'_{ei} = 20\left(\frac{2}{\pi}\right)^{3/2}\frac{k}{m_e^{\frac{1}{2}}e^4\ln\Lambda}(kT_e)^{5/2} \tag{6.5.36}$$

The reader may consult references to this chapter for the derivation of the quantities given above. The effective thermal conductivity coefficient is somewhat reduced by the thermoelectric effect. A current that flows as a result of a temperature gradient tends to set up an electric field. Since the current must be divergence free in a steady state, a secondary electric field is produced to cancel the flow of current. The secondary electric field has the effect of reducing the heat flow. The reduction factor for singly charged ions is 0.225, as given in one of the references.

$$\lambda_{ei} = 0.225\,\lambda'_{ei} \tag{6.5.37}$$

The electron–neutral component of thermal conduction is based on laboratory measurements of the momentum transfer collision cross section, $\phi_{en}^{(1)}(v_e)$. This parameter has been obtained for all major atmospheric species as a function of electron energy or velocity. A compilation of some experimental results is given in Appendix 4. The thermal conductivity coefficient then becomes

$$\lambda_{en} = \frac{4}{3}\frac{n_e}{n_n}\frac{k^2 T_e}{\pi m_e}\frac{1}{\Omega_{en}^{(1,1)}} \tag{6.5.38}$$

where $\Omega_{en}^{(1,1)}$ is the collision integral for momentum transfer defined by Equation (4.5.6).

The reader may recall that the collision integral for viscosity and heat conduction (Equation (4.5.15)) derived from the Chapman–Enskog solution to the Boltzmann equation differs from the diffusion or momentum transfer integral. The Chapman–Enskog formulation is most appropriate to gases that are composed primarily of neutral species and, indeed, we have used Equation (6.5.10) in deriving the thermal conductivity coefficient for the neutral atmospheric gases (Equation (6.5.13)). For inverse square law forces between particles the velocity distribution function is governed largely by many small angle scattering processes in long-range encounters. Neglecting the effects of any large deflection close encounters, Chandrasekhar advanced a formulation that uses a diffusion type equation for the velocity distribution function, an approach that he first adopted to describe the spatial distribution function in Brownian motion. This approach underlies our treatment of heat conduction in an ionized gas (Equation (6.5.34)) and accounts for the appearance of $\Omega_{en}^{(1,1)}$ in Equation (6.5.38).

Computations of the electron temperature profile in the ionosphere make use of the vertical one-dimensional component of the electron energy equation. The Earth's magnetic dipole field is inclined with respect to the vertical; the angle between the horizontal and the magnetic field vector is defined as the dip angle, I. The electron density and temperature are assumed to vary in the vertical z direction but electron drift, photoelectron and auroral electron fluxes, and heat flow all occur along the magnetic field. The electron drift and fluxes are written in terms of the current, following Equation (6.5.21), and the vertical component is $J \sin I$. Under these conditions Equation (6.5.19) assumes the form (Problem 33).

$$\frac{3}{2}\frac{\partial(n_e T_e)}{\partial t} = J \sin I \frac{kT_e}{|e|}\left(5\frac{\partial \ln J}{\partial z} + 4\frac{\partial \ln T_e}{\partial z} - \frac{\partial \ln n_e}{\partial z}\right) + \sin^2 I\left(\lambda_e \frac{\partial^2 T_e}{\partial z^2} + \frac{\partial \lambda_e}{\partial z}\frac{\partial T_e}{\partial z}\right) + Q_e - L_e \tag{6.5.39}$$

The electron temperature and density profiles in the ionosphere are strongly interdependent, as already noted in Chapter 5, demonstrating yet again the requirement for coupled solutions of the governing conservation equations in the ionosphere. Initial conditions and boundary conditions are usually derived from observational data.

The ion temperature in the ionosphere is governed by the ion energy balance equation. Several parameters in this equation also appear in the ion continuity and momentum equations requiring coupled solutions of the set. In addition, ionospheric ions are usually heated by collisions with electrons and cooled by ion–neutral collisions, coupling the energy balance of all three species. Finally, it has already been noted that different species of positive ions are heated and cooled by different collisional processes, resulting in somewhat different temperatures for various species. Taking account of all the complexities identified above in solving for the ion temperatures leads to a mathematical problem that has yet to be solved. In our exposition we are more concerned with the physical and chemical processes that operate to establish the ion temperature and we attempt to evaluate the relative importance of each at various altitude levels and under varying geophysical conditions.

For each ion species the energy balance equation, Equation (6.5.1), becomes

$$\tfrac{3}{2}k[\partial(n_i T_i)/\partial t + \mathbf{v}_i \cdot \nabla(n_i T_i)] = -\tfrac{5}{2}kn_i T_i \nabla \cdot \mathbf{v}_i - \nabla \cdot \mathbf{q}_i + Q_i - L_i \qquad (6.5.40)$$

but the equation has also been applied to the ion gas as a whole, characterized by a temperature T_i and density n_i. Local heating sources, Q_i, are discussed in Sections 6.2 and 6.3, while heat sinks, L_i, are given in Section 6.4. The ion gyrofrequency is smaller than the electron gyrofrequency by the inverse of their mass ratio, allowing ionospheric ions to drift orthogonal to the magnetic field more readily than electrons. Thus, while the field-aligned drift velocity and current are determined by the electron gas, orthogonal drift and currents are governed by ion motion.

The ion temperature has been measured extensively by instruments carried on board satellites. The observations show variability with latitude and longitude, due to solar effects and the influence of auroral energy input to the ionosphere, but by far the largest variation occurs with altitude (as is the case for the electron gas already discussed). This result is confirmed by rocket borne experiments. Energy transport in the ion gas is due to vertical temperature gradients and, to a much lesser extent, to ion diffusion,

$$\mathbf{q}_i = -\lambda_i \nabla T_i + \tfrac{5}{2}n_i k T_i \mathbf{v}_i \qquad (6.5.41)$$

where λ_i is the ion thermal conductivity coefficient. The second term on the right side of the equation represents diffusion. The ionospheric plasma carries no net charge and both electrons and positive ions contribute to heat conduction. The thermal conductivity coefficient for ions is considerably smaller than for electrons, however, approximately by the square root of their mass ratio, as may be inferred by inspection of Equation (6.5.36). In a reference given at the end of the chapter it is shown that the net thermal conductivity coefficient for a Lorentz gas can be expressed in terms of the electron thermal conductivity coefficient,

$$\lambda \approx \lambda_e \left[\frac{1}{(m_i/m_e)^{\frac{1}{2}} + 15/2\sqrt{2}} + \frac{1}{1 + 13/4\sqrt{2}} \right] \qquad (6.5.42)$$

where the first term in the square bracket represents the ion contribution, assuming singly ionized species only. When several kinds of ions are present, as is the case throughout much of the ionosphere, the overall contribution is the sum of the components. As with electron heat conduction (Equation (6.5.31)), collisions between ions and neutrals have the effect of reducing the ion thermal conductivity coefficient.

6.5.3 A special case: the stable auroral red arc

Airglow stations at middle latitudes routinely measure radiation from an O line at 6300 Å which is emitted in the transition $O(^3P_2)$–$O(^1D)$ of the ground configuration of O (Appendix 3). A chemical recombination process that is discussed in Chapter 7 is responsible for the excitation and emission in the airglow. Occasionally, following large solar flare activity and its geomagnetic response, the 6300 Å line is measured photometrically in a broad arc-like configuration, aligned along parallels of geomagnetic latitude with an emission rate well above the airglow level, but not visible to the naked eye. Excitation of the $O(^1D)$ state was discussed in Section 6.4 (Equation (6.4.26)) and it was noted that this low lying electronic state can be excited by a high temperature electron gas. Embedded electron heating sources such as photoelectrons and auroral electrons act primarily in the lower F- and E-regions of the ionosphere. The midlatitude subvisual red arc, however, originates at a mean altitude of about 400 km which is above the F-region electron density peak. Coordinated ground based and satellite borne experiments have identified the region of 6300 Å enhancement with a high electron temperature, upwards of 5000 K. Since there is no evidence of local heat sources the electron gas can only be heated by conduction from a high temperature electron gas above the atmosphere. The bibliography to this chapter lists articles that discuss possible mechanisms for producing a high temperature electron gas in the magnetosphere while here we investigate the consequences in the upper F-region.

The midlatitude subvisual red arc is a stable phenomenon compared to auroral arcs and the steady state electron energy equation adequately describes its behaviour. Without local heat sources and currents Equation (6.5.39) becomes

$$\sin^2 I \frac{\partial}{\partial z}\left(\lambda_e \frac{\partial T_e}{\partial z} \right) = L_e \qquad (6.5.43)$$

In the 300–500 km altitude interval the principal energy loss mechanisms are the excitation of the fine structure levels in O, $L_{FS}(e, O)$ (Equation (6.4.24)), vibrational excitation of N_2, $L_{vib}(e, N_2)$ (Appendix 6), elastic collisions with ambient O^+ ions (Equation (6.4.32)), and the excitation of $O(^1D)$, L_{O^1D} (e, O) (Equation (6.4.28)). The last energy loss reaction is, of course, the source of $O(^1D)$ atoms and the 6300 Å emission line. Although the electron equation for the special case of a red arc is not as complex as the full Equation (6.5.19), it is still a second order non-linear differential equation that is not trivial to solve. It is usually assumed that $T_e = T_i = T_n$ at the lower boundary, about 100 km altitude, and a heat flux, or a temperature gradient, or a temperature is specified at the upper boundary, about 1000 km. The loss rate, L_e, contains T_e, T_n, T_i, ion densities and neutral densities and the energy equations and continuity equations must be solved self-consistently. Numerical solutions are given in references listed with the chapter but all represent approximations of one sort or another. More complete observational data will require less approximate solutions of the governing equations (Problem 34).

Bibliography

Neutral gas heating by solar UV photons is discussed in a paper by

R.G. Roble and B.A. Emery, On the global mean temperature of the thermosphere, *Planet. Space Sci.*, **31**, 597–614, 1983.

which also lists previous papers on the subject. The global mean structure of the thermosphere and ionosphere is modelled in the paper by

R.G. Roble, E.C. Ridley and R.E. Dickinson, On the global mean structure of the thermosphere, *J. Geophys. Res.*, **92**, 8745–58, 1987.

Electron heating, including the origin of the stopping cross section (Equation (6.2.8)) is analyzed by

K. Stamnes and M.H. Rees, Heating of thermal ionospheric electrons by suprathermal electrons, *Geophys. Res. Lett.*, **10**, 309–12, 1983.

The contribution of auroral precipitation to neutral and ion heating is summarized separately for auroral electrons,

M.H. Rees, *et al.*, Neutral and ion gas heating by auroral electron precipitation, *J. Geophys. Res.*, **88**, 6289–300, 1983.

and for heavy ions, H^+ and O^+.

M.H. Rees, Modeling of the heating and ionizing of the polar thermosphere by magnetospheric electron and ion precipitation, *Physica Scripta*, **T18**, 249–55, 1987.

Ion and neutral heating by orthogonal electric fields and the ionospheric consequences have been explored by

M.H. Rees and J.C.G. Walker, Ion and electron heating by auroral electric fields, *Ann. Geophys.*, **24**, 193–9, 1968.

The basic concepts of frictional heating are discussed by

J.M. Burgers, *Flow Equations for Composite Gases*, (Chapter 6), Academic Press, New York, 1969.

The origin and consequences of a non-Maxwellian ion velocity distribution are discussed in detail by

J.P. St. Maurice and R.W. Schunk, Ion velocity distributions in the high latitude ionosphere, *Rev. Geophys. Space Phys.*, **17**, 99–134, 1979.

Various aspects of collisional energy transfer between charged and neutral species are summarized in

P.M. Banks and G. Kockarts, *Aeronomy*, Academic Press, New York, 1973.

who also give relevant tables and graphs.

Two review articles that focus on the electron and ion energy equations, sources and sinks of energy, heating and cooling rates and that include comprehensive lists of references are

M.H. Rees and R.G. Roble, Observations and theory of the formation of stable auroral arcs, *Rev. Geophys. Space Phys.*, **13**, 201–42, 1975.

R.W. Schunk and A.F. Nagy, Electron temperatures in the F-region of the ionosphere: Theory and observations, *Rev. Geophys. Space Phys.*, **16**, 355–99, 1978.

Derivation of the energy equation from a general formulation of the transport equations is given by

R.W. Schunk, Transport equations for aeronomy, *Planet. Space Sci.*, **23**, 437–485, 1975.

Theoretical and experimental work on rotational, vibrational and fine structure excitation by low energy electron impact is reviewed in the following articles:

K. Takayanagi and Y. Itikawa, The rotational excitation of molecules by slow electrons, *Adv. Atom. Mol. Phys.*, **6**, 105–53, 1970.

D.G. Thompson, The vibrational excitation of molecules by electron impact, *Adv. Atom. Mol. Phys.*, **19**, 309–44, 1983.

G.J. Schulz, Resonances in electron impact on diatomic molecules, *Rev. Mod. Phys.*, **45**, 423–86, 1973.

H.S.W. Massey, *Electronic and Ionic Impact Phenomena*, Vol. 2, Oxford University Press, Oxford, 1969.

G. Kockarts, Nitric oxide cooling in the terrestrial thermosphere, *Geophys. Res. Lett.*, **7**, 137, 1980.

G. Kockarts and W. Peetermans, Atomic oxygen infrared emission in the Earth's upper atmosphere, *Planet. Space Sci.*, **18**, 271, 1970.

B.F. Gordiets, M.N. Markov and A.L. Shelepin, I.R. radiation of the upper atmosphere, *Planet. Space Sci.*, **26**, 933, 1978.

Theoretical concepts that underly the transport phenomena of viscosity and thermal conductivity are presented by

L. Spitzer and R. Härm, Transport phenomena in a completely ionized gas, *Phys. Rev.*, **89**, 977, 1953.

S. Chapman, The viscosity and thermal conductivity of a completely ionized gas, *Astrophys. J.*, **120**, 151, 1954.

Mathematical techniques that have been found to be useful for solving the different equations encountered in this chapter are described by

J.T. Hastings and R.G. Roble, An automatic technique for solving coupled vector systems of non-linear parabolic partial differential equations in one space dimension, *Planet. Space Sci.*, **25**, 209–15, 1977.

The technique given in the preceding paper is applied by

R.G. Roble and J.T. Hastings, Thermal response properties of the Earth's ionospheric plasma, *Planet. Space Sci.*, **25**, 217–31, 1977.

Problems 23–34

Problem 23 Estimate the energy released at 115, 130 and 145 km by photodissociation of O_2 in the Schumann Runge continuum. Let the solar zenith angle be 45° and use the 23 April, 1974 reference solar spectrum.

Problem 24 Show that electron heating by an electric field is small in the F-region of the ionosphere by comparison with ion heating. At what altitude are ion and electron heating rates comparable? Use the collision frequencies given in Appendix 6.

Problem 25 Compute the cross section for rotational excitation in N_2 from the $J = 12$ to the $J = 14$ rotational level as a function of electron energy. Adopt a quadrupole moment $q = 1.04ea_0^2$.

Repeat the calculation neglecting polarization interaction.

Problem 26 Compute the energy loss rate coefficient ($\text{eV cm}^3\,\text{s}^{-1}$) for electrons in rotational excitation of N_2 by two methods, the exact equation (Equation (6.4.15)) and the approximate expression (Equation (6.4.17)). Assume a neutral temperature of 800 K and an electron temperature of 3000 K. The rotational constant for N_2 is $B = 2.5 \times 10^{-4}$ eV. Assume unit concentrations of N_2 and electrons.

Problem 27 Compute the energy loss rate coefficient ($\text{eV cm}^3\,\text{s}^{-1}$) for electrons in exciting O_2 by rotation, vibration and low lying electronic states for electron temperatures between 800 and 6000 K. Assume a neutral temperature of 800 K. Repeat for vibrational excitation of N_2 using the expression given in Appendix 6. Assume unit concentration of O_2, N_2 and electrons.

Problem 28 Compute the energy loss rate coefficient ($\text{cm}^3\,\text{s}^{-1}$) for electrons in exciting the fine structure levels of O per unit $n(\text{O})$ and unit $n(\text{e})$ as a function of $(T_e - T_n)$. Assume a neutral gas temperature of 800 K.

Problem 29 Compute the energy loss rate coefficient ($\text{cm}^3\,\text{s}^{-1}$) for electrons in exciting the $\text{O}(^1\text{D})$ electronic level per unit $n(\text{O})$ and unit $n(\text{e})$ as a function of $(T_e \;\; T_n)$, holding the neutral temperature fixed at 800 K.

Problem 30 Derive Equation (6.4.32) and verify its dimensionality. Assume that $\ln \Lambda$ varies very slowly with velocity.

Problem 31 Derive Equation (6.4.33) for the non-resonant ion–neutral energy exchange rate.

Problem 32 Evaluate the coefficients of viscosity for N_2, O_2 and O between 200 K and 2000 K using the empirical equation, Equation (6.5.12). The viscosity coefficient equation derived from the first approximation of the Chapman–Enskog method for hard elastic spheres (Equation (6.5.11)), contains the molecular (atomic) diameter. Estimate the value of σ that best fits the experimental data.

Problem 33 Starting with Equation (6.5.19) derive Equation (6.5.39) adopting the definitions and assumptions described in the text.

Problem 34 Thermal conduction above 300 km is dominated by electron–ion interaction and $\lambda_e \approx \lambda_{ei}$. Assume that $\ln\Lambda$ in Equation (6.5.36) is a slowly varying function of T_e and $n(\text{e})$, solve Equation (6.5.43) for T_e as a function of z. Assume a downward energy flux at 1000 km of $2.5 \times 10^{10}\,\text{eV cm}^{-2}\,\text{s}^{-1}$, $T_e = T_n$ at 200 km, a magnetic dip of 70° and the neutral atmospheric model listed in Appendix 1.

7

Spectroscopic emissions

7.1 Introduction

In our discussion of the structure and energetics of the upper atmosphere in the two preceding chapters the production of excited states of atoms, molecules, and ions was introduced in various terms of the continuity and energy equations that govern the ion and neutral composition and the temperature profiles. It was noted that a radiative transition between states can be an energy loss while producing atoms, molecules, or ions in lower energy states. In Section 3.3 a brief discussion of an aurorally associated optical emission, the N_2^+ 1NG band system, was undertaken, showing how radiation from an excited ion is related to energy deposition by a beam of energetic electrons. Spectroscopic emissions are the focus of this chapter but we also investigate the production of the excited states whence radiation is emitted.

It was stated in the overview presented in the first chapter that most of the atmospheric spectroscopic radiations measured by ground based and space borne instruments originate in the thermosphere. This follows from the expositions in Chapters 2 and 3 where we learned that solar UV photons are absorbed and energetic particles are stopped in the thermosphere. Below the mesopause, radiation is primarily thermal and in IR molecular bands, while above the thermosphere, H and He are the dominant gases and resonance scattering and fluorescence of solar photons produce the geocorona.

Radiation out of the thermosphere in the visible and UV regions of the spectrum accounts for only a small fraction of the energy that is absorbed but the spectral lines and bands are an invaluable means for remotely sensing several physical properties and processes in the emitting region. Some examples are introduced here and will be discussed further in subsequent sections. The absolute emission rate, combined with excitation cross sections can yield the column density of a species or the intensity of the source that produces the excited state. Doppler profiles and line shifts of certain radiations lead to temperatures and neutral winds or ion drift data. Vibrational and rotational distributions in molecular bands yield the respective temperatures or provide insight to possible excitation mechanisms. The interconnection between structure, composition, energetics and dynamics has repeatedly been stressed in preceding chapters. Such coupling should manifest itself in relationships between various emission features or other spectroscopic details. The technology required to measure weak spectral features at high resolution and simultaneously cover a large wavelength interval is rapidly developing, even as this chapter is being written. Many spectroscopic emission features can therefore be usefully employed provided that their origin is well understood and perturbing effects are correctly taken into account.

There is a profusion of emission lines and bands that originate in the thermosphere. The spectrum mirrors the neutral and ion composition of the region together with the excitation source. Solar UV photons and auroral particles have sufficient energy to excite, ionize, and dissociate all species into any accessible electronic state while the exothermicity of chemical reactions is frequently adequate to produce some excited states. The low density of the unconfined gas in the thermosphere is conducive to radiation of lines and bands from highly metastable states, features that are absent in laboratory spectra. Thus, the atomic O red lines at 6300 Å and 6364 Å are known as the 'nebular lines' while the atomic O green line at 5577 Å is called the 'auroral line' because both features were first detected in the extraterrestrial laboratory.

Radiation measurements yield the intensity along the line of sight of the receiver. Spectra, therefore, represent a superposition of all the emission and absorption features that fall in the bandpass of an instrument. Identification of lines, bands and continua becomes a necessary first task. The standard spectroscopic tools for identification are adopted, wavelengths, line shapes and band development. The fine structure of a line, the appearance of several members of an atomic multiplet and several bands of a molecular system, strengthens tentative identifications. The coarse wavelength resolution of the early instruments employed in thermospheric spectroscopy resulted in numerous incorrect identifications of emission lines and bands, and even with present day instruments it is not possible to resolve features such as the H-Ly-β line at 1025.72 Å and the OI triplet ($2p^4\,{}^3P_{2,1,0}$–$3d\,{}^3D^0$) at 1025.77, 1027.42 and 1028.16 Å. High spectral resolution is essential for identification of neutral and ionized atomic lines for comparison with standard wavelength tables, not forgetting the possibility of a Doppler shifted wavelength. Comparing synthetic spectra of molecular bands with an observed spectrum is frequently helpful for identifying weak spectral features. It is virtually impossible today to acquire sky spectra with ground based instruments that are free from man-made contaminating emissions such as Hg and Na lines. The possibility of their presence must not be forgotten. The problem is especially serious in the case of Na because the Na-D doublet occurs in the natural airglow spectrum.

The altitude region where an emission originates is a crucial parameter for relating the spectral feature to other thermospheric properties through its excitation mechanism. While measurements yield a column emission rate, interpretation relies on a volume emission rate height profile. Rocket borne instruments have been flown through emission layers and the data reduced to volume emission rates by differentiation. Tomographic techniques have recently been applied to satellite observations to obtain volume emission rate data. References given with this chapter describe this useful new technique.

The bright Rayleigh scattered continuum precludes acquiring spectra from the ground during daylight conditions. (It is, however, possible to observe individual lines at very high resolution using a multietalon interferometer.) Moreover, the wavelength region accessible from the ground is limited to the visible and near IR. Rocket and satellite borne instruments are required to observe in the UV and IR regions of the spectrum, and this has become routine within the last two decades. Observations have been carried out under day and night conditions and over a wide geographic area. The weakest spectrum is observed at middle and low latitudes at night; this is the night airglow. Under sunlit conditions the spectrum is both brighter and richer in variety and is named the dayglow. Twilight spectra show the transition resulting from a continuous change in the shadow height as a function of wavelength of solar photons and partial illumination of the thermosphere. In the auroral region, at high latitudes, the night and dayglow spectra are frequently supplemented by auroral emissions. While most, but not all, auroral spectral features are identical to features in the dayglow, the emission rate can be orders of magnitude larger than in the dayglow. The auroral spectrum is complex and rich in spectral features; it is also highly variable in time and space, straining the capability of the most advanced instruments currently available.

7.2 The night airglow spectrum

A composite of the night airglow spectrum assembled from several sources is shown in Figure 7.2.1. Data for the wavelength region between 750 and 3000 Å were acquired with satellite, rocket or balloon borne spectrometers. In the visible and beyond, into the near IR region, ground based instruments were used. Various spectral intervals were recorded at different locations, local times, and seasons using instruments with different spectral resolutions and sensitivities and viewing through varying layers of atmosphere from beneath or above the emitting region. It would be misleading, therefore, to attempt to establish a relationship between various emission features in the different spectral regions, based on the composition, thermal structure and dynamic behaviour of the thermosphere. Moreover, various lines and bands originate from different altitude layers by processes that are specific to the particular emission.

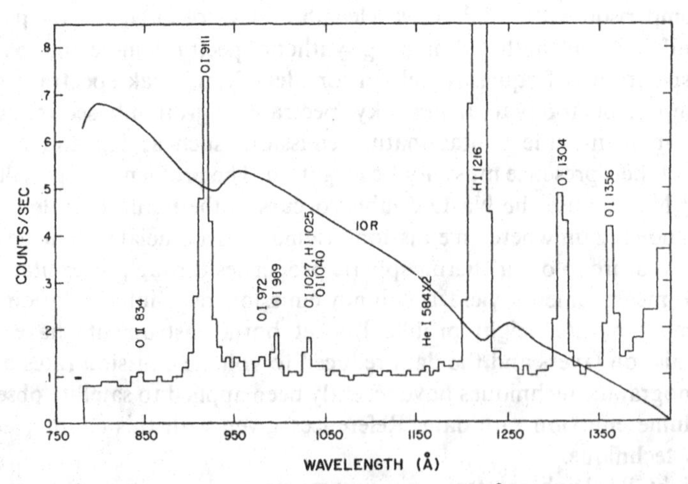

Fig. 7.2.1 (*a*) Nightglow spectrum between 750 and 1400 Å: average downlooking UV nightglow spectrum obtained from 600 km when the STP78-1 spacecraft was within ± 30° magnetic latitude range and outside the South Atlantic Anomaly region. (S. Chakrabarti, *Geophys. Res. Lett.*, **11**, 565, 1984.)

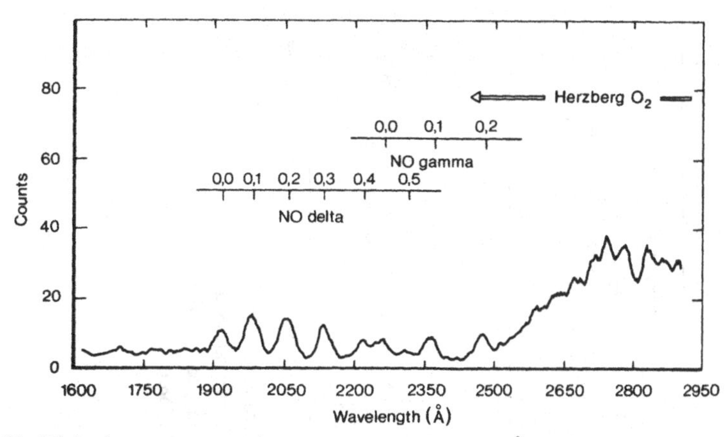

Fig. 7.2.1 (*b*) Nightglow spectrum between 1600 and 2950 Å: nadir viewed UV spectrum acquired by the S3-4 satellite in the equatorial region at a resolution of about 25 Å. (R.E. Huffman *et al.*, *J. Geophys. Res.*, **85**, 2201, 1980.)

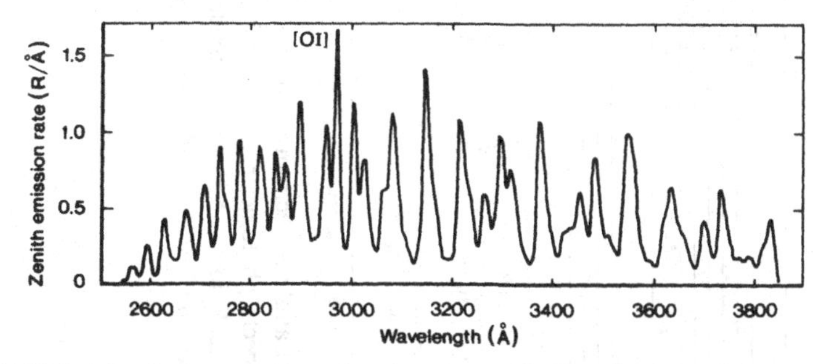

Fig. 7.2.1 (*c*) UV night airglow spectrum between 2500 and 3900 Å: acquired by a spectrograph carried on board a rocket launched from White Sands, New Mexico, USA. The resolution is 12 Å. All the spectral features are attributed to the Herzberg band system of O_2, except for the atomic O line identified at 2972 Å. (J.P. Hennes, *J. Geophys. Res.*, **71**, 763, 1966.)

The UV night airglow spectrum from 750 Å to about 1400 Å is characterized by lines of neutral O, OI, of singly ionized O^+, OII, by H and HeI lines with one exception, the feature at 911 Å. The reader should refer to the energy level diagrams collected in Appendix 3 for identification of the transitions and excited states.

At night, the ionospheric ions recombine, resulting in the decay of ionization from dusk to dawn. While the principal loss process for atomic ions is ion–atom interchange (see Section 4.4), another reaction that occurs in the F-region is two-body radiative recombination, for example,

$$O^+(^4S) + e \rightarrow O(2p^4\,^3P) + h\nu(<911\,\text{Å}) \tag{7.2.1}$$

The feature at 911 Å is the long wavelength cutoff of a continuum that is emitted when electrons are captured into the ground state of the atom. The energy level diagram (Appendix 3) shows the ionization potential of OI to be 13.62 eV. The process is the

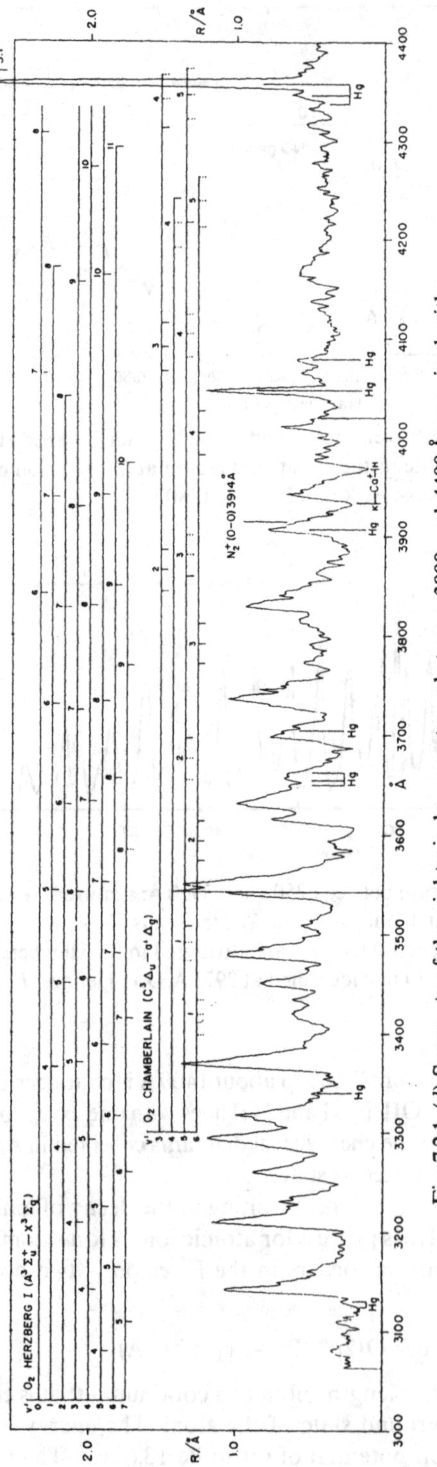

Fig. 7.2.1 (*d*) Segments of the night airglow spectrum between 3000 and 4400 Å: acquired with a Fastie–Ebert scanning spectrophotometer from the Kitt Peak, Arizona, observatory at 2080 m elevation. The resolution is 5 Å. Several unidentified features appear in the spectrum. (A.L. Broadfoot and K.R. Kendall, *J. Geophys. Res.*, 73, 426, 1968.)

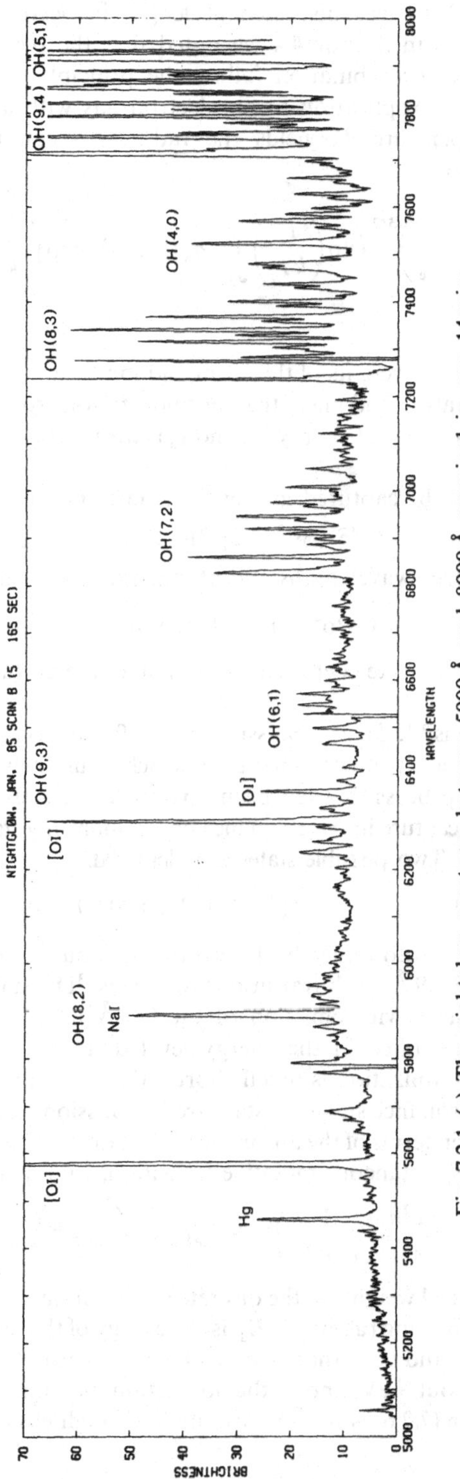

Fig. 7.2.1 (*e*) The nightglow spectrum between 5000 Å and 8000 Å acquired in one 14 min exposure at 4 AM LT on January 17, 1985 with an image intensified charge coupled detector on a spectrograph at the Catalina Observatory, Mt. Lennon, Arizona, USA. (Courtesy of L.A. Broadfoot, 1987.)

inverse of photoionization which was discussed at length in Section 2.2. In our discussion of reactive collisions in Section 4.4 we noted that the two processes of photoionization and radiative recombination follow the principle of microscopic reversibility (Equation (4.4.20)). Application to an electron gas with a Maxwellian velocity distribution at temperature T_e yields the radiative recombination rate coefficient by the Milne relation

$$\alpha_{i,a}(T_e) = \frac{\tilde{\omega}_a}{\tilde{\omega}_i}\left(\frac{2}{\pi}\right)^{1/2}\frac{1}{c^2}\left(\frac{1}{m_e k T_e}\right)^{3/2}\exp\left(\frac{I_a}{kT_e}\right)\int_{I_a}^{\infty}\sigma_{ai}(h\nu)(h\nu)^2\exp\left(\frac{-h\nu}{kT_e}\right)\mathrm{d}(h\nu) \tag{7.2.2}$$

where $\tilde{\omega}_a$ and $\tilde{\omega}_i$ are the statistical weights of the atom and ion states, respectively. T_e and m_e are the electron temperature and mass, the photoionization cross section from atomic state a to ionic state i is $\sigma_{ai}(h\nu)$ at energy $h\nu$, and I_a is the ionization potential of the atom (Problem 35).

The electron may, of course, be captured into excited states of the atom,

$$O^+(^4S) + e \rightarrow O^*(3s\,^3S^0, 3s\,^5S^0, 3p\,^5P, \text{etc.}) + h\nu \tag{7.2.3}$$

with emission continua at longer wavelengths and atomic lines, for example,

$$O^*(3s\,^5S^0) \rightarrow O(2p^4\,^3P) + h\nu(1356\,\text{Å}) \tag{7.2.4}$$

The total radiative recombination rate coefficient is the sum of the partial coefficients given by Equation (7.2.2).

A different process is responsible for the emission line at 989 Å because the upper state, $3s'\,^3D^0$, requires excitation of the atomic core which is unlikely to occur by radiative recombination. The process that has been invoked is dielectronic recombination which involves electron capture into a resonance or auto-ionizing state within the continuum, i.e., above 13.62 eV. Two possible states are identified,

$$O^+(^4S) + e \rightleftarrows O_d^*(3p'\,^3D \text{ or } 3p'\,^3F) \rightarrow O_b^*(3s'\,^3D^0) + h\nu \tag{7.2.5}$$

The resonance states can either auto-ionize back into the continuum or stabilize by emission of lines at 8227 and 7949 Å, both transitions to the $(3s'\,^3D^0)$ state. The latter then decays to the ground state with emission of the 989 Å line. The states and transitions cited above are identified in the energy level diagram in Appendix 3. Although the lifetime for auto-ionization is much shorter than for radiative stabilization, a fraction of atoms in resonance states do stabilize by emission and the reaction rate coefficient may be written in terms of the absorption oscillator strength f_{bd} (related to the Einstein coefficient for spontaneous radiative transition, see Equation (7.3.17))

$$\alpha_{diel}(T_e) = \frac{\tilde{\omega}_b}{\tilde{\omega}_i}(2\pi)^{1/2}\frac{e^2 h}{m_e c^3}\left(\frac{1}{m_e k T_e}\right)^{3/2}(h\nu_{bd})^2\exp\left(\frac{-E_{di}}{kT_e}\right)f_{bd} \tag{7.2.6}$$

where $\tilde{\omega}_b$ and $\tilde{\omega}_i$ are the statistical weights of the discrete atomic state and the ion, $h\nu_{bd}$ is the energy of the emitted stabilizing radiation, E_{di} is the energy of the resonance state above the ionization potential and the other symbols have their usual meaning. The resonance states, O_d^*, are about $\frac{1}{2}$eV above the ionization potential so that the exponential factor in Equation (7.2.6) is not unduly small for high electron temperatures (Problem 36).

The H-Ly-α and H-Ly-β lines, the HeI 584 Å line and the OII 834 Å line are due to resonance scattering of the solar UV radiation (Figures 2.2.1 and 2.2.2) by thermospheric and geocoronal H, He and O^+.

The night airglow in the wavelength region between 1600 Å and 2900 Å has emissions of vibrational bands of two systems of NO, the $v' = 0$ progressions of $NO(C^2\Pi, v' = 0) \rightarrow NO(X\,^2\Pi, v'')$ and $NO(A\,^2\Sigma^+, v' = 0) \rightarrow NO(X\,^2\Pi, v'')$ transitions. The relevant electronic states may be identified in the potential diagram of NO shown in Appendix 3 and in the partial schematic diagram shown in Figure 7.2.2. The chemiluminescent process believed to be responsible is preassociation, also known as inverse predissociation. The predissociation mechanism was discussed in Section 4.3 in connection with N_2 (Figure 4.3.3). With reference to Figure 7.2.2 the N and O atoms approach along the a $^4\Pi$ potential curve, hence to the C $^2\Pi$ state by a radiationless transition. The C $^2\Pi$ state decays by radiation, either directly to the ground X $^2\Pi$ state with emission of the δ-bands or by way of the A $^2\Sigma^+$ state to the ground state with emission of the γ-bands. The reaction sequence may be written as

$$N(^4S) + O(^3P) \rightleftarrows NO(a\,^4\Pi) \rightleftarrows NO(C\,^2\Pi) \rightarrow NO(X\,^2\Pi) + h\nu(\delta - \text{bands})$$
$$\rightarrow NO(A\,^2\Sigma^+) + h\nu(1.224\,\mu\text{m}) \quad (7.2.7)$$

$$NO(A\,^2\Sigma^+) \rightarrow NO(X\,^2\Pi) + h\nu\ (\gamma\text{-bands}) \quad (7.2.8)$$

where a $^4\Pi$ is the preassociation state. The reaction rate coefficient for emission of the bands, summed over v'', has been determined experimentally at a value $1.5 \times 10^{-17}\,\text{cm}^3\,\text{s}^{-1}$ at 300 K. Curve crossings in the potential diagram suggest that only the $v' = 0$ vibrational level of C $^2\Pi$ is populated and that radiative transition to A $^2\Sigma^+$ likewise leads to $v' = 0$. The minima of the above two states, however, do not occur at

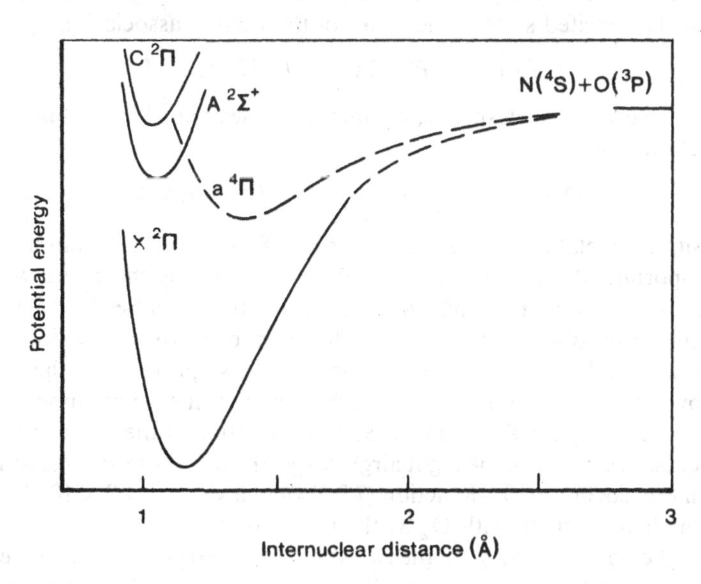

Fig. 7.2.2 Partial potential curve diagram for NO showing the electronic states that contribute to the airglow emission bands.

the same internuclear separation as the ground $X\,^2\Pi$ state. This leads to emission of several bands in the $v' = 0$ progression, as is observed in the night airglow spectrum.

Even a casual perusal of airglow (and auroral) spectra shows that there is overlap of vibrational bands from band systems, sometimes from more than one molecular species, with atomic lines interspersed here and there. Identifying individual vibrational bands and sorting them into the various systems needs to be approached systematically. The method adopted in spectral analysis is to construct synthetic spectra, i.e., compute the intensity of the vibrational bands under excitation conditions believed to occur in the atmosphere, and using molecular constants derived from laboratory measurements. The distribution of vibrational bands depends on the mechanism by which the excited state is produced. The complexity of the problem increases if more than one reaction contributes to excitation since various bands may be populated at different rates.

Contributions from individual bands to the emission rate are added and adjusted to attain a best match with the airglow spectrum after convolving the synthetic spectra with the instrumental resolution. Identification of atomic lines is facilitated by the presence in a spectrum of several lines in a multiplet, by high resolution measurements showing fine structure and by the realization that an appreciable population of highly metastable atoms can survive in the low density thermosphere. Several works devoted to the identification and interpretation of atomic and molecular spectra are listed in the bibliography to this chapter and only a scant outline is provided here as part of our discussion of specific emissions in airglow and auroral spectra.

Beyond 2500 Å, bands of the Herzberg I system of O_2 dominate the night airglow spectrum in the wavelength region not accessible to ground based measurements. The Herzberg I system extends to longer wavelengths but well-identifiable bands appear only to about 4000 Å as shown in Figure 7.2.1. The Herzberg I bands originate from transitions between the $A\,^3\Sigma_u^+$ electronic state and the $X\,^3\Sigma_g^-$ ground state of O_2 (see Appendix 3). The excited state is the result of three-body association

$$O(^3P) + O(^3P) + M \rightarrow O_2(A\,^3\Sigma_u{}^+) + M \tag{7.2.9}$$

where M can be any third species, atomic or molecular. This is followed by the forbidden electronic transition

$$O_2(A\,^3\Sigma_u^+) \rightarrow O_2(X\,^3\Sigma_g{}^-) + h\nu(\text{Herzberg I}) \tag{7.2.10}$$

The transition violates the selection rule $\Sigma^+ \not\leftrightarrow \Sigma^-$. Association of O is the only important three-body collision process in the thermosphere where a low density favours two-body reactions. As already mentioned in Sections 2.4 and 5.4, O atoms produced by photodissociation in the thermosphere diffuse downward to a region where three-body association becomes the loss process for the atoms. This occurs below about 100 km. Downward diffusion continues after sunset providing a steady source of atoms for Reaction (7.2.9) and excitation of the Herzberg I bands, as well as other band systems in the night airglow spectrum. Laboratory measurements of the reaction rate coefficient for Reaction (7.2.9) yield a value of $1.2 \times 10^{-34}\,\text{cm}^6\,\text{s}^{-1}$ at 300 K for the band system, with O_2 as the third body.

Details of the potential curves of the two states involved in emission of the Herzberg I bands are shown in Figure 7.2.3. According to the Frank–Condon principle, the most likely transitions are those for which the overlap integral $\int \Psi_{v'} \Psi_{v''}\,dr$ is large

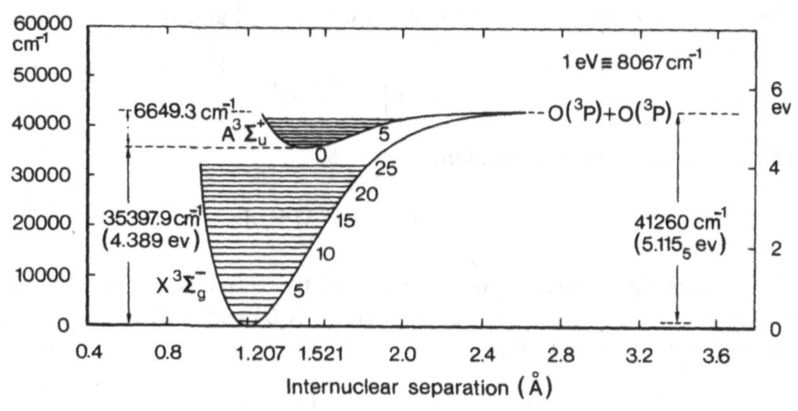

Fig. 7.2.3 Partial potential curve diagram for O_2 showing the electronic states and vibrational levels involved in the Herzberg I airglow bands.

(Equation (7.2.17)). This occurs when the vibrational wave functions $\Psi_{v'}$ and $\Psi_{v''}$ of the initial and final states have maxima at about the same value of the internuclear distance r. Accordingly, some of the stronger bands in the system are the (6, 3), (5, 3), (4, 3) and (4, 4), in agreement with airglow measurements. There are others, of course.

The emission rate of a band, integrated over the rotational lines, is given by

$$I_{v'v''} = N_{v'} A_{v'v''} h\nu_{v'v''} \qquad (\text{erg cm}^{-3}\,\text{s}^{-1}) \tag{7.2.11}$$

where $N_{v'}$ is the population density (cm^{-3}) of the upper vibrational level, $A_{v'v''}$ is the Einstein coefficient for spontaneous transition (s^{-1}), and $h\nu_{v'v''}$ is the energy of the transition (erg).

The nascent vibrational population of the A $^3\Sigma_u^+$ state resulting from the three-body association Reaction (7.2.9) is not expected to follow a Boltzmann distribution since the excited molecular state is formed as the two O(^{3}P) atoms approach along the potential curve shown in Figure 7.2.3. The third body serves to stabilize formation of the molecule. The Herzberg I bands are an example of the complementary aspects of theory and observations toward enhancing our understanding of atmospheric phenomena. The equilibrium population of the vibrational levels, $N_{v''}$, cannot be derived theoretically but must be inferred by fitting synthetic spectra to measured airglow spectra. The relative population favours levels $v' = 4$, 5 and 6. Vibrational decay by collisional deactivation (especially with atomic O) probably depopulates the higher levels. There are several other emission bands in the airglow and the aurora that must be similarly analysed.

The emission rate specified by Equation (7.2.11) contains an important parameter, the transition probability, A. The transition probability for electric and magnetic dipole radiation is given by

$$A_{v'v''} = \frac{64\pi^4}{3h}\left(\frac{\nu_{v'v''}}{c}\right)^3 \frac{S_{v'v''}}{\tilde{\omega}_{v'}} \tag{7.2.12}$$

where $\tilde{\omega}_{v'}$ is the statistical weight of the upper level and $S_{v'v''}$ is the band strength,

$$S_{v'v''} = \left|\int \Psi_{v'} R_e(r) \Psi_{v''} \, dr\right|^2 \tag{7.2.13}$$

The electronic transition moment is

$$R_e(r) = \int \Psi_e' M \Psi_e \, d\tau \tag{7.2.14}$$

where M is the electric dipole moment of the transition and $d\tau$ is an element in configuration space. For most band systems $R_e(r)$ varies slowly with r and the band strength may be rewritten as

$$S_{v'v''} = R_e^2(\bar{r}_{v'v''}) \left|\int \Psi_{v'} \Psi_{v''} \, dr\right|^2 \tag{7.2.15}$$

with $\bar{r}_{v'v''}$, the r-centroid of the band,

$$\bar{r}_{v'v''} = \frac{\int \Psi_{v'} r \Psi_{v''} \, dr}{\int \Psi_{v'} \Psi_{v''} \, dr} \tag{7.2.16}$$

The overlap integral squared,

$$q_{v'v''} = \left|\int \Psi_{v'} \Psi_{v''} \, dr\right|^2 \tag{7.2.17}$$

has been called the Frank–Condon factor of the band. These factors give the relative band strengths and are tabulated in references listed with this chapter. Solutions of the Schrödinger equation yield the wave function, Ψ, and their accuracy depends on how well the molecular potential is known and represented analytically. Transition probabilities may therefore be determined by applying quantum mechanical principles and adopting the relevant molecular parameters. If the matrix elements of the electric dipole moment are zero then magnetic dipole or electric quadrupole moments may have non-zero matrix elements and the transitions are forbidden. The transition probabilities are then orders of magnitude smaller. The Einstein coefficient for electric quadrupole radiation is

$$A_{v'v''} = \frac{32\pi^6}{5h} \left(\frac{v_{v'v''}}{c}\right)^5 \frac{S_{v'v''}}{\tilde{\omega}_{v'}} \tag{7.2.18}$$

In the preceding discussion of molecular band intensities, integration over the rotational structure was assumed. With the development of instruments of higher sensitivity and spectral resolution measurements of individual rotational lines can be applied to investigating the dynamical and thermal properties of the atmosphere. Thus, in place of the band strength $S_{v'v''}$ (Equation (7.2.15)) interest focuses on the rotational line strength

$$S_{v'v''J'J''} = R_e^2(\bar{r}) q_{v'v''} \frac{\zeta_{J'J''}}{\tilde{\omega}_e(2J+1)} \tag{7.2.19}$$

where $\zeta_{J'J''}$ is related to the so-called Hönl–London factors that determine the relative intensity of various branches within a band and are a function of the type of coupling of the orbital, spin and nuclear angular momenta in the molecule (Hund's cases); $\bar{\omega}_e$ is the electronic statistical weight which depends on the type of state, $2S + 1$ for Σ states and $2(2S + 1)$ for all others, and $(2J + 1)$ is the rotational statistical weight. The vibrational statistical weight is unity. Hönl–London factors are derived from the quantum mechanical description of molecular rotation and are a function of the selection rules for transitions between rotational lines. Tabulations may be found in the research literature for several molecules of atmospheric interest.

The preceding discussion of transition probabilities focused on the theoretical concepts and inputs required to obtain this important parameter for constructing synthetic spectra. The Herzberg I bands have also been extensively studied in the laboratory, leading to values for the band strengths, $S_{v'v''}$. The transition probabilities are computed with Equations (7.2.12) or (7.2.18) and synthetic spectra are constructed by adjusting the relative population of vibrational levels. Figure 7.2.4 shows the synthetic spectrum of the Herzberg I system in the wavelength region and at a resolution that matches the night airglow spectrum obtained by a rocket borne spectrometer, Figure 7.2.1(c). Sources of reading on this topic are listed in the bibliography to this chapter.

The energy level diagram for O_2 (Appendix 3) identifies several bound electronic states, in addition to the A $^3\Sigma_u^+$, that could be formed by the three-body association of O. One such band system that is present in the spectrum shown in Figure 7.2.1 occurs between about 3400 and 4000 Å. The few weak bands are attributed to the Chamberlain system, corresponding to the transition A′ $^3\Delta_u \rightarrow$ a $^1\Delta_g$. Additional bands occur in different wavelength intervals and will be discussed subsequently.

The wavelength region between 4200 and 5200 Å consists of unresolved bands, continua, and several lines of Hg due to anthropogenic contamination.

There is a faint feature at 5200 Å attributed to a forbidden transition in the ground configuration of N, $^4S^0_{3/2} - {}^2D^0_{5/2,3/2}$. The energy level diagram (Appendix 3) shows that the upper state is highly metastable with a radiative lifetime of about 26 h. Any measurable emission must therefore originate at high altitude where collisional

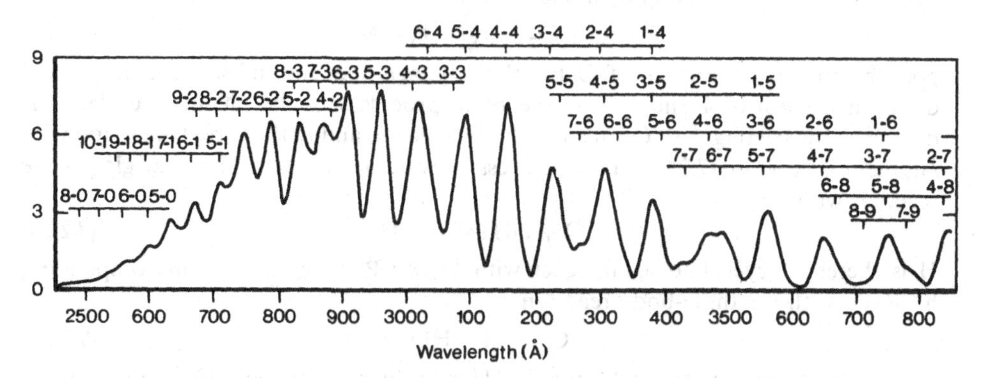

Fig. 7.2.4 Synthetic spectrum of the Herzberg I bands that contribute to the night airglow between 2500 Å and 3850 Å. (Adapted from V. Degen, *J. Geophys. Res.*, **74**, 5145, 1969.)

deactivation is minimal. The ionospheric F-region slowly decays at night by ion–atom exchange between O^+ and molecular species. The reaction

$$O^+ + N_2 \rightarrow NO^+ + N \tag{7.2.20}$$

followed by dissociative recombination

$$NO^+ + e \rightarrow O + N(^2D, ^4S) \tag{7.2.21}$$

most likely accounts for production of $N(^2D)$ atoms since, according to laboratory experiments, about 80% of the N atoms are produced in the excited state. The reaction rate coefficient is listed in Appendix 5.

Molecular bands of the Meinel system of the OH radical appear beyond about 5200 Å, dominating the night airglow spectrum into the IR ($\sim 4\mu m$). The emission is due to fundamental ($\Delta v = 1$) and overtone ($\Delta v > 1$) transitions from vibrationally excited radicals in the ground electronic state, OH ($X\,^2\Pi_{\frac{3}{2},\frac{1}{2}}$).

While there may be small contributions from other reactions, the principal source of vibrationally excited radicals is the displacement reaction between H and O_3,

$$H + O_3 \rightarrow OH(v' \leqslant 9) + O_2 + 3.3\,\text{eV} \tag{7.2.22}$$

The exothermicity is sufficient to excite vibrational levels up to $v' = 9$. Bands originating from this level, and no higher, are observed in the airglow.

Altitude profiles of the Meinel bands obtained by rocket borne photometers and satellite borne instruments place the peak emission between 85 and 90 km, with some variability. Although the Meinel bands originate at the lower boundary of the atmospheric region described in this text their pervasiveness and prominence requires the inclusion of OH emissions in assembling synthetic spectra that are used for analysis and interpretation of airglow and auroral spectra. Temporal variations in the emission rate of the Meinel bands have been cited as evidence of gravity waves propagating upward through the mesosphere. If this interpretation is correct such waves could transport energy into the thermosphere where dissipation would occur. This possibility provides an additional justification for including OH radiation in the physics and chemistry of the thermosphere.

The reactants H and O_3 in Reaction (7.2.22) are minor constituents in the lower thermosphere and in the mesosphere and their production and loss needs to be understood in order to appreciate the importance of the OH emissions.

O_3 is formed by three-body associations,

$$O + O_2 + M \rightarrow O_3 + M \tag{7.2.23}$$

pointing to the production of O by photodissociation and subsequent downward diffusion (Sections 2.4 and 5.4) as the basic processes in the chemical cycle. O is destroyed by reacting with O_3 in a two-body collision and in three-body reaction with another O, Reaction (7.2.9), but the process relevant to the Meinel band airglow is the reaction with OH,

$$O + OH \rightarrow O_2 + H \tag{7.2.24}$$

H is thereby recycled to again react with O_3 by Reaction (7.2.22) in competition, however, with the three-body reaction

$$H + O_2 + M \rightarrow HO_2 + M \tag{7.2.25}$$

to form the hydroperoxy (HO_2) radical. This, in turn, reacts with O forming OH,

$$HO_2 + O \rightarrow O_2 + OH + 2.3\,\text{eV} \tag{7.2.26}$$

The 2.3 eV exothermicity would be adequate to excite vibrational levels of OH up to $v' \leqslant 6$ but is more likely to be taken up by the O_2.

Photodissociation of water vapour in the thermosphere,

$$H_2O + h\nu(<2000\,\text{Å}) \rightarrow H + OH \tag{7.2.27}$$

is a daytime source of odd H while photodissociation of O_3

$$O_3 + h\nu(<3100\,\text{Å}) \rightarrow O_2 + O \tag{7.2.28}$$

is a sink of O_3. The complex chemistry of odd O (O, O_3) and odd H (H, OH, HO_2) is described in references listed with this chapter. Typical concentrations of the participating species are shown in Appendix 1 but these can vary widely, depending on the time of day, location and season. The densities are strongly influenced by transport, driven by diurnal and latitudinal inhomogeneities in energy and momentum sources, ultimately of solar origin. The physical and chemical processes that underlie the Meinel band emissions in the airglow are an important aspect of coupling between the thermosphere and the mesosphere.

Rotational relaxation is sufficiently rapid so that the distribution of rotational lines within a band represents the kinetic temperature of the gas. The intensity of individual rotational lines can be measured with medium spectral resolution instruments since the line spacing is appreciable. From the airglow survey spectrum assembled in Figure 7.2.1 a limited wavelength region (7260–7380 Å) is extracted and shown in Figure 7.2.5 to illustrate the rotational structure of the OH (8, 3) band. The photon emission rate of rotational lines within a band is given by

$$I_{v'J',v''J''} = N_{v'J'} A_{v'J',v''J''} \tag{7.2.29}$$

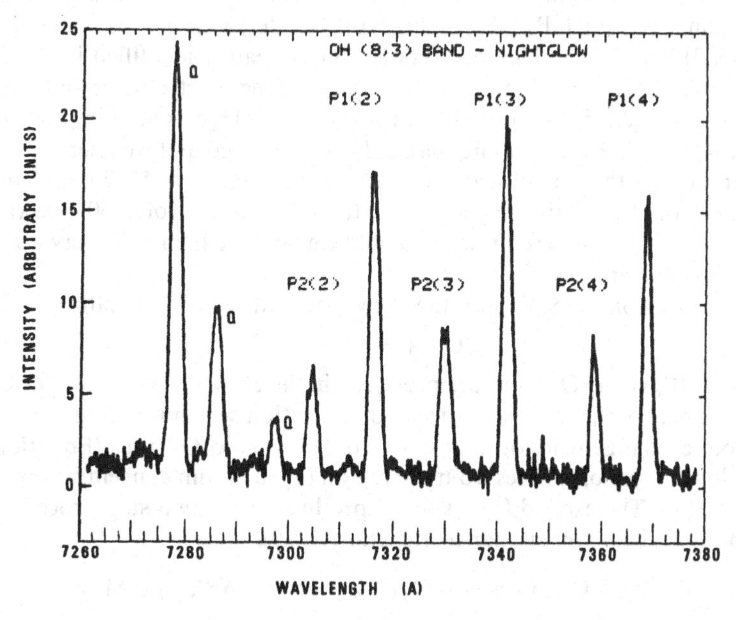

Fig. 7.2.5 Spectrum of the OH (8, 3) band observed in the nightglow between 7260 and 7380 Å.

where $A_{v'J',v''J''}$ is the transition probability between rotational lines and $N_{v'J'}$ is the population of the Jth state in the v' level given by,

$$N_{v'J'} = N_{v'},\tilde{\omega}_{J'} \exp(-E'_{v'J'}/kT)/Q_r \tag{7.2.30}$$

where $N_{v'}$ is the total number of radicals in the v' level, the rotational statistical weight of each state is

$$\tilde{\omega}_{J'} = 2J' + 1 \tag{7.2.31}$$

and the rotational partition function is

$$Q_r = \sum_{J'}^{\infty}(2J' + 1)\exp(-E_{v'J'}/kT) \tag{7.2.32}$$

The energy of a rotational state is

$$E_{v'J'} = hcBJ'(J' + 1) \tag{7.2.33}$$

where B is the rotational constant specific to the molecule. OH is a heteronuclear species and each Meinel band therefore has three branches, P, Q and R, corresponding to the selection rules for rotational transitions $J' - J'' = -1,\ 0,\ +1$ respectively. Adopting the molecular constants and Einstein coefficients that have been obtained through laboratory experiments and theory, a rotational temperature may be inferred from the relative photon emission rates of various lines in a band by constructing synthetic spectra that best match the observations. Prediction of the observed emission rates requires knowledge of reaction rate coefficients and concentrations of reactants as well as of horizontal and vertical transport. The interested reader may wish to consult the references given with this chapter for more detailed expositions of this important topic.

The brightest emission line in the visible region of the night airglow is the ($^1D_2$–$^1S_0$) forbidden transition in the ground configuration of O at 5577 Å. The weak companion transition ($^3P_1$–$^1S_0$) at 2972 Å is blended with the rotational development in the (5, 3) band of the Herzberg system but is readily identified in the 20 year old photographic spectrum obtained with a rocket borne spectrometer and reproduced as part of Figure 7.2.1. Suggestions for the production of the excited $^1S_0$ state in the night airglow were already made more than half a century ago and we refer the reader to the bibliography for the convoluted history of the problem. The 5577 Å emission actually originates from two distinct layers, one in the F-region at about 250–300 km, the other in the lower thermosphere around 95–100 km, and each emission layer is associated with a different process.

In the F-region O($^1S_0$) is produced by dissociative recombination of O_2^+ ions,

$$O_2^+ + e \rightarrow O + O(^1S_0) \tag{7.2.34}$$

Less than 10% of the O atoms are produced in the excited ($^1S_0$) state. The O_2^+ ions are formed as part of the night time decay of ionization which is controlled by the initial ion–atom exchange reaction that converts O^+ ions to O_2^+ ions (Equation (5.4.10)).

The lower layer originates from the region of maximum concentration of atomic O at 95–100 km. The excited ($^1S_0$) state is produced by a two-step process. The first is three-body association, similar to Reaction (7.2.9),

$$O + O + M \rightarrow O_2(c^1\Sigma_u^-, A'^3\Delta_u, A^3\Sigma_u^+) + M$$

The excited O_2 is formed not only in the $A^3\Sigma_u^+$ state with emission of the Herzberg bands, but also in two additional electronic states, the $c\,^1\Sigma_u^-$ which is the upper state of the very faint Herzberg II bands, and the $A'\,^3\Delta_u$ which is the upper state of the weak Chamberlain bands identified in Figure 7.2.1. Collisions with atomic O may cause the O_2 molecules to switch between excited electronic states but the important second step for production of $O(^1S_0)$ is

$$O_2(c\,^1\Sigma_u^-) + O \rightarrow O_2(X\,^3\Sigma_g^-) + O(^1S_0) \tag{7.2.36}$$

followed by

$$O(^1S_0) \rightarrow O(^1D_2) + h\nu(5577\,\text{Å}) \tag{7.2.37}$$

or

$$O(^1S_0) \rightarrow O(^3P_1) + h\nu(2972\,\text{Å}) \tag{7.2.38}$$

The two-body reaction (Reaction (7.2.36)) competes with the three-body reaction (Reaction (7.2.23)), discussed earlier, for the formation of O_3.

Although the emission rate of 5577 Å is not large compared to other spectral features in the night airglow it has probably received more attention by experimenters than any other. There are at least two reasons. A line is usually more amenable to identification and recording than a band and the green wavelength region offers an optimal combination of good instrument sensitivity and small atmospheric extinction.

Proceeding to longer wavelengths in the nightglow spectrum we find the (unresolved) Na doublet, Na-D at 5896–5890 Å, resulting from the $3s\,^2S_{\frac{1}{2}}$–$3p\,^2P^0_{\frac{3}{2},\frac{1}{2}}$ transition in NaI. Metallic atoms in the upper atmosphere originate from meteor ablation. Loss of metallic atoms is by downward transport. Observational and theoretical evidence convincingly favours production of the excited state through the oxidation–reduction cycle

$$Na + O_3 \rightarrow NaO + O_2 \tag{7.2.39}$$

$$NaO + O \rightarrow Na(^2P, ^2S) + O_2 \tag{7.2.40}$$

followed by the transition

$$Na(^2P_{\frac{3}{2},\frac{1}{2}}) \rightarrow Na(^2S_{\frac{1}{2}}) + h\nu(5896, 5890\,\text{Å}) \tag{7.2.41}$$

The symmetry properties of the correlated molecular states that characterize the electron jump type processes invoked for Reactions (7.2.39) and (7.2.40) allow for an appreciable fraction of Na atom production in the excited ^2P state. The reactions are sufficiently rapid to maintain chemical equilibrium between free Na and NaO and the photon emission rate is therefore given by

$$\eta(\text{Na-D}) = f_{2P}k_{39}n(O_3)n(\text{Na}) \qquad (\text{photons cm}^{-3}\,\text{s}^{-1}) \tag{7.2.42}$$

where f_{2P} is the fraction of Na atoms produced in the ^2P excited state and k_{39} is the reaction rate coefficient for the oxidation reaction which has a value $3.4 \times 10^{-10}\,\text{cm}^{-3}\,\text{s}^{-1}$ at 200 K. The emission rate of the Na-D lines is dependent upon the abundance of odd O in the lower thermosphere and becomes a measure of the rate of destruction of odd O by Na atoms. Rocket borne photometer measurements locate the Na nightglow emission layer in the altitude range 83–95 km. With the development of sensitive lidar systems the Na atom abundance–height profile may be measured with good accuracy and altitude resolution providing data for investigating some aspects of the complex chemistry and dynamics of the lower thermosphere (Problem 37). The Na-D lines occur in the wavelength region of the Meinel OH (8, 2) band requiring application of the synthetic spectral technique to separate the blended features correctly.

Additional Meinel bands appear as we progress to longer wavelengths, and in the midst of the P-branch of the OH (9, 3) band there appears a doublet at 6300 Å and 6364 Å due to the forbidden transition in atomic O ($^3P_{2,1}$–$^1D_2$). The $^1D_2$ state is only 1.97 eV above the ground state of the atom and is produced by a variety of inelastic collisions. It is also highly metastable (see Appendix 3), with a radiative lifetime of the order of 100 s, and therefore suffers severe collisional deactivation below the F-region. In the night time F-region dissociative recombination of O_2^+ ions,

$$O_2^+ + e \rightarrow O + O(^1D_2) \tag{7.2.43}$$

is the source of excitation and about one half of the O atoms are produced in the $^1D_2$ state. As already discussed in connection with the F-region source of 5577 Å emission (Equation (7.2.34)), the O_2^+ ions result from the night time decay of ionization, i.e., the conversion of O^+ ions to molecular ions. The emission of every 5577 Å photon by Reaction (7.2.37) leaves the O atom in the $^1D_2$ state yielding an additional small source, cascading, for red line doublet emission,

$$O(^1D_2) \rightarrow O(^3P_{2,1}) + h\nu(6300, 6364\ \text{Å}) \tag{7.2.44}$$

in competition with collisional deactivation

$$O(^1D_2) + X(N_2, O, e) \rightarrow O(^3P) + X \tag{7.2.45}$$

In addition to several other bright Meinel OH bands, there are two band systems from transitions in O_2 that appear in the nightglow spectrum beyond the red line doublet. A partial potential energy curve diagram, excerpted from Appendix 3, is reproduced in Figure 7.2.6 and the transitions between the bound states are identified. The excited states are produced by the same three-body association responsible for the higher lying states (Reactions (7.2.9)) and (7.2.35),

$$O + O + M \rightarrow O_2(b\,^1\Sigma_g^+, a\,^1\Delta_g) + M \tag{7.2.46}$$

followed by

$$O_2(b\,^1\Sigma_g^+) \rightarrow O_2(X\,^3\Sigma_g^-) + h\nu\ \text{(Atmospheric)} \tag{7.2.47}$$

and

$$O_2(a\,^1\Delta_g) \rightarrow O_2(X\,^3\Sigma_g^-) + h\nu\ \text{(IR Atmospheric)} \tag{7.2.48}$$

The upper states are both metastable, with radiative lifetimes of about 12 s for the b state and 3.9×10^3 s for the a state. The electronic state deactivation coefficients for collisions with the major atmospheric species are sufficiently small, however, that, even at the density of the lower thermosphere, quenching has little effect on the emission rate. Energy considerations favour excitation of low lying electronic states by several reactions. It has been suggested, and there is some supporting experimental evidence, that the O_2 states are also excited by Reaction (7.2.24),

$$OH(v' \geqslant 4) + O \rightarrow O_2(b\,^1\Sigma_g^+) + H \tag{7.2.49}$$

and

$$OH(v' \geqslant 1) + O \rightarrow O_2(a\,^1\Delta_g) + H \tag{7.2.50}$$

Electronic state quenching of excited atomic O,

$$O(^1D_2) + O_2(X\,^3\Sigma_g^-) \rightarrow O_2(b\,^1\Sigma_g^+, v' \leqslant 2) + O \qquad (7.2.51)$$

is another possible source of the Atmospheric bands, especially at altitudes where a two-body reaction becomes more effective than the three-body Reaction (7.2.46).

Alignment of the potential curves in Figure 7.2.6 suggests that excitation of the lower vibrational levels of the a and b states should be favoured. This is borne out by the observed spectrum shown in Figure 7.2.1. The (0, 1) band of the Atmospheric system appears at 8645 Å but the (0, 0)band at 7619 Å, with about 20 times the emission rate, cannot be observed from the ground because of absorption by the intervening

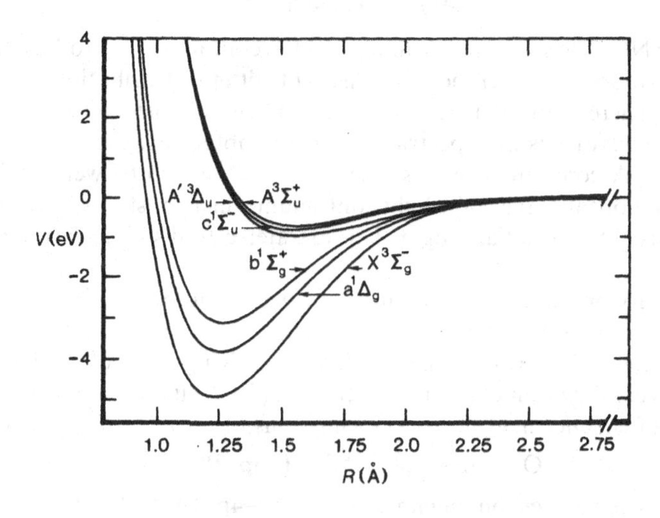

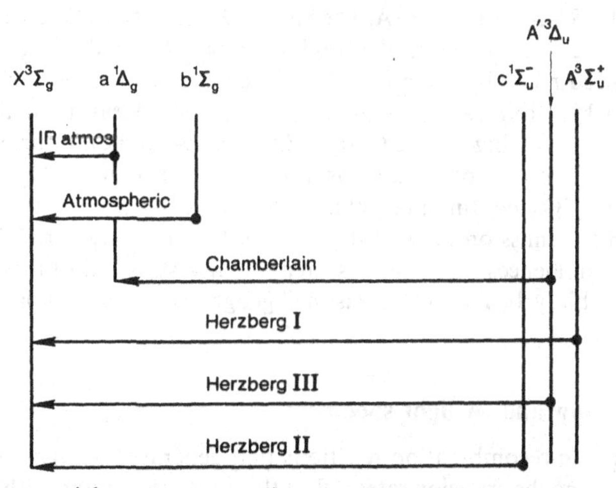

Fig. 7.2.6 Partial potential energy curve diagram of O_2 and band systems arising from transitions between various states. (Adapted from D.R. Bates, Airglow and Auroras, Chapter 6 in *Applied Atomic Collision Physics*, Vol. 1, Ed. H.S.W. Massey, *et al.*, Academic Press, New York, 1982.)

column of O_2. The (0, 0) band has been measured with rocket borne spectrometers that place the peak emission altitude at about 95 km, consistent with the principal excitation source (Reaction (7.2.46)). The (0, 0) band of the IR Atmospheric system appears at 1.27 μm. Due to absorption, only a very small fraction of the emission reaches the ground, but it is detectable.

Careful absolute photometry provides good evidence for the occurrence of a weak continuum in the night airglow throughout the visible wavelength region. The emission height is at about 100 km, according to rocket measurements. The reaction that has been proposed as being responsible for the continuum is

$$NO + O \rightarrow NO_2 + h\nu(>4000\,\text{Å}) \tag{7.2.52}$$

followed by

$$NO_2 + O \rightarrow NO + O_2 \tag{7.2.53}$$

to recover the NO. The spectral distribution of the continuum due to Reaction (7.2.52) obtained in a laboratory experiment is consistent with the distribution observed in the night airglow. There is considerable uncertainty inherent in subtracting all the bands of O_2 and OH from the measured spectrum in order to obtain an absolute emission rate of the residual weak continuum but its occurrence appears to be well established. An additional component of the white light continuum which must be subtracted is due to integrated starlight and zodiacal light. This extraterrestrial source varies in space and with time.

The night airglow spectrum of Figure 7.2.1 was acquired at midlatitude sites and does not show a very few faint emission lines that appear in spectra acquired at equatorial latitudes. These are transitions from high lying states of OI that are excited by radiative recombination of O^+ ions, Reaction (7.2.3), and their detection coincides with regions of enhancement of the UV emissions of OI at 1302 Å and 1356 Å,

$$O^+ + e \rightarrow O(4p\,^3P, 3p\,^5P, 3p\,^3P) + h\nu \tag{7.2.54}$$

The transitions and corresponding lines are (3s $^3S^0$–4p ^3P) at 4368 Å, (3s $^5S^0$–3p ^5P) at 7774 Å and (3s $^3S^0$–3p ^3P) at 8446 Å. The higher O^+ density and higher electron density in the upper F-region at equatorial latitudes compared to midlatitude regions account for the enhanced recombination rate and concomitant spectroscopic emission rates.

It is worthwhile noting at this point how many spectral features in the nightglow are due to reactions involving atomic O: the NO δ- and γ-bands, the various band systems of O_2, the OH vibration–rotation bands through the formation of O_3, the 5577 Å line of [OI], and the NO_2 continuum. While other factors besides the O concentration contribute to the emission rates of the spectral features, any variability in atomic O concentration influences them all. A systematic investigation of the consequences of atomic O variability, be it diurnal, seasonal, geographic, or with solar activity, has not yet been made.

7.3 The dayglow and twilight spectrum

The chemical and recombination reactions that generate the nightglow also proceed during the day and the emission rates reflect the diurnal variation in the concentrations of some of the reactants. Several lines and bands are substantially enhanced by higher production rates and by additional processes. Many new features also appear in the dayglow spectrum. Twilight spectra reflect the transition between the dark and fully

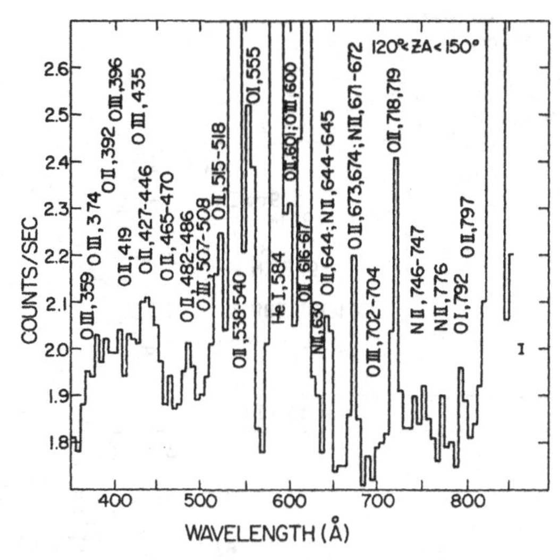

Fig. 7.3.1 (*a*) Nadir viewed UV dayglow spectrum between 350 and 900 Å acquired by the STP78-1 satellite from 600 km between 50° N and 50° S latitudes. (S. Chakrabarti *et al., J. Geophys. Res.*, **88**, 4898, 1983.)

illuminated atmosphere as the shadow height or screening height changes in various wavelength bands of the solar spectrum. The dayglow and twilight emission features result from direct excitation by solar UV photons, from excitation by photoelectron impact and by resonance scattering and fluorescence of solar radiation. The production of photoelectrons has been discussed in Chapter 2 while inelastic electron impact processes were described in a general way in Chapter 4.

A composite dayglow spectrum has been assembled in Figure 7.3.1 from several sources, representing the highest available spectral resolution and sensitivity at the time of writing this chapter. A few observations, over very limited wavelength intervals, have been acquired at higher resolution. The UV spectra are, of course, acquired with rocket or satellite borne instruments. The high level of white light continuum vitiates acquiring dayglow spectra with ground based instruments. However, as already mentioned in the introduction to this chapter, very high resolution interferometers are capable of isolating single spectral lines. These provide a valuable tool for atmospheric dynamics investigations (see Chapter 8).

In twilight, direct solar illumination does not extend to the lower levels of the atmosphere where most of the scattering occurs and the level of sky brightness is substantially reduced. The thermosphere, however, remains fully illuminated although the column density of gas that solar photons must traverse to reach local zenith is very much larger than for an overhead sun and considerable attenuation may occur. The mathematical expression of the phenomenon was given in Chapter 2.

The short-wavelength UV spectrum, starting at 400 Å in Figure 7.3.1(*a*) is dominated by singly ionized and neutral atomic lines of O and N. Even the highest available spectral resolution, indicated in some figure captions, is inadequate to resolve several

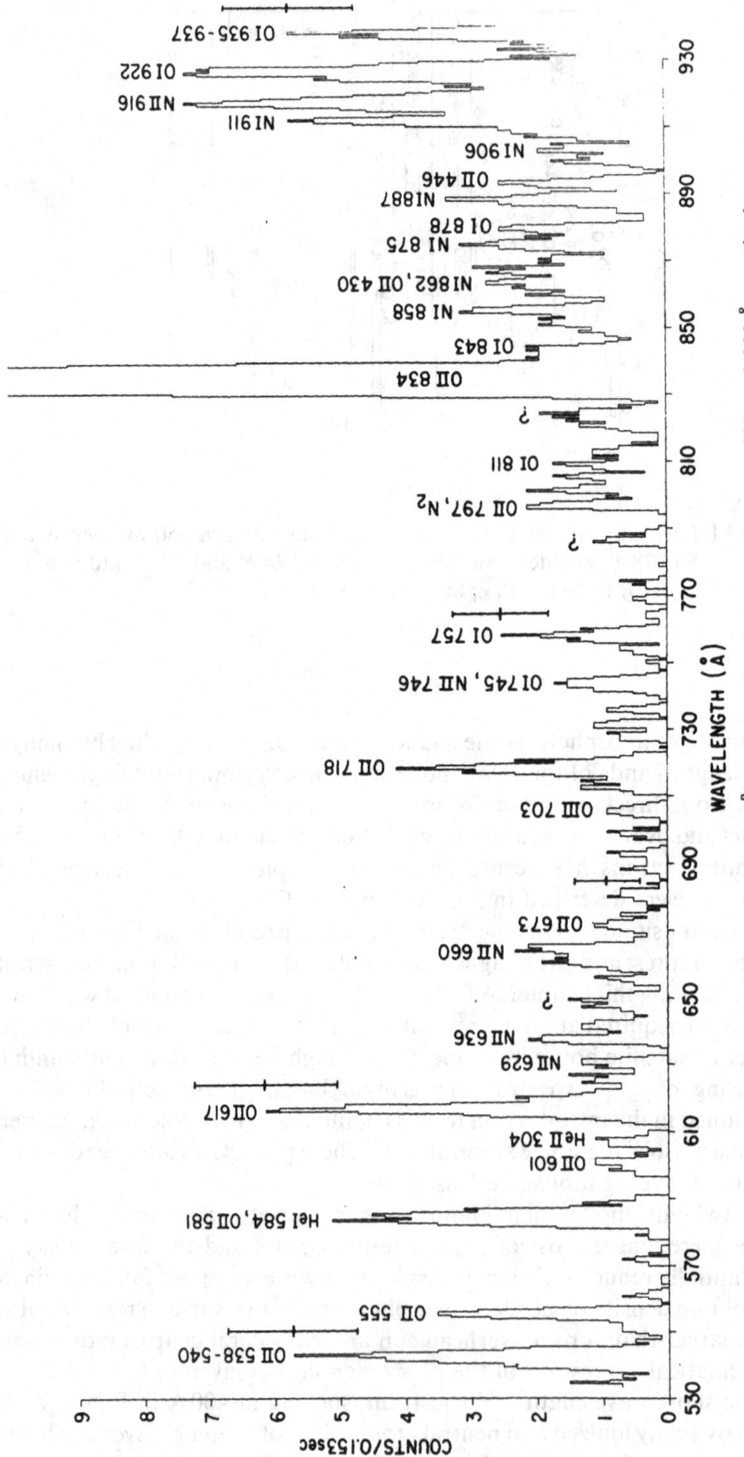

Fig. 7.3.1 (*b*) High resolution (3.5 Å) UV dayglow spectrum between 530 and 930 Å acquired with a scanning spectrometer carried by a rocket above White Sands, New Mexico. (E.P. Gentieu, *et al.*, *J. Geophys. Res.*, **89**, 11053, 1984)

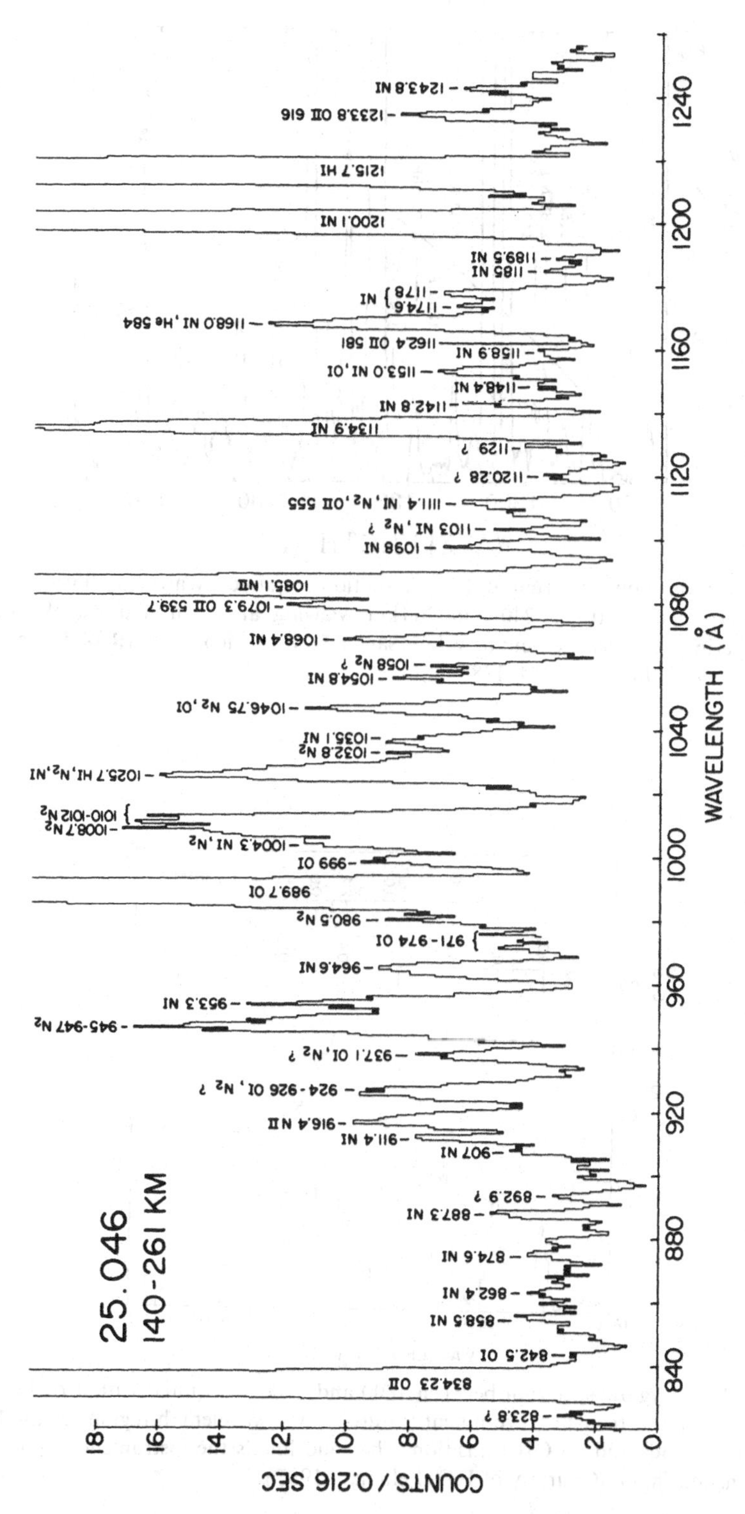

Fig. 7.3.1 (*c*) High resolution (3.5 Å) UV dayglow spectrum between 820 and 1250 Å acquired with a scanning spectrometer carried by a rocket above White Sands, New Mexico. (E.P. Gentieu, *et al.*, *Geophys. Res. Lett.*, **8**, 1242, 1981.)

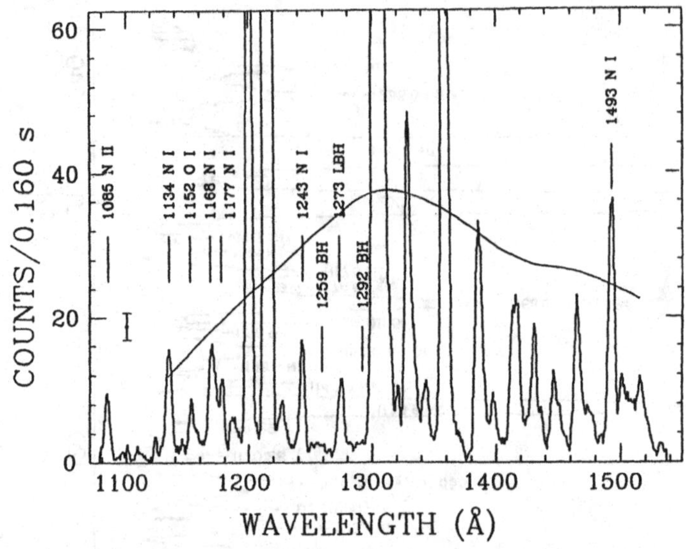

Fig. 7.3.1 (*d*) UV dayglow spectrum at 4.4 Å resolution acquired by a rocket borne scanning spectrometer between 240 and 260 km, viewing at a zenith angle of 98°. The experiment was carried out at White Sands, New Mexico, USA. (R.W. Eastes, *et al.*, *J. Geophys. Res.*, **90**, 6594, 1985.)

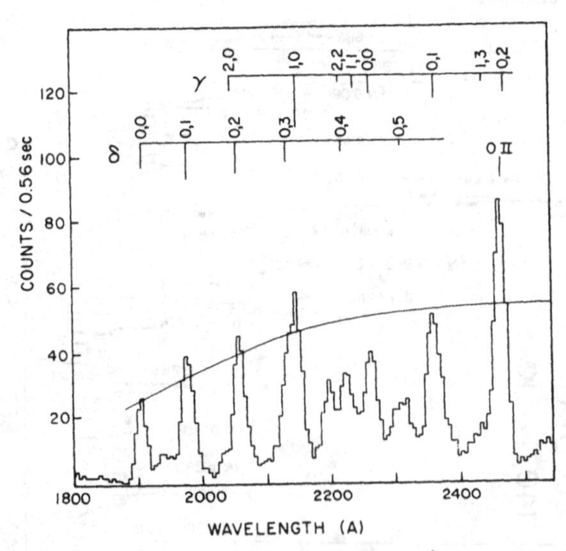

Fig. 7.3.1 (*e*) UV dayglow spectrum between 1800 and 2500 Å acquired with a rocket borne scanning spectrometer. Prominent features in the wavelength region are the NO δ.- and γ-bands and an OII transition. The solid line is the instrument response to a constant input. (Courtesy of P.D. Feldman, 1987.)

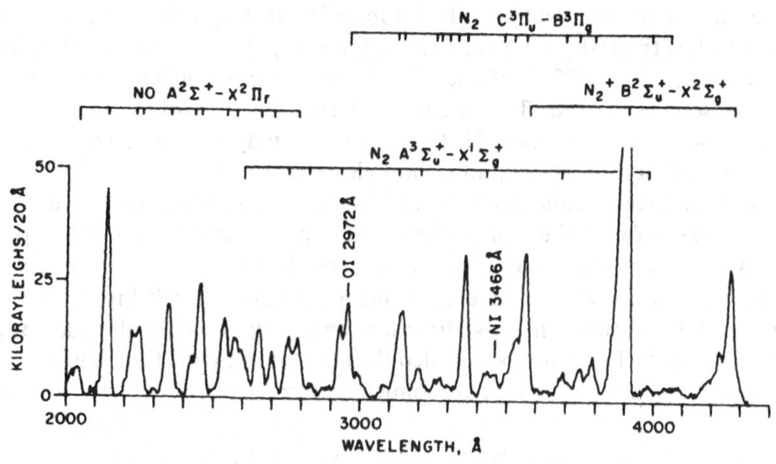

Fig. 7.3.1 (*f*) Spectrum of the Earth's dayglow between 2000 and 4300 Å at a resolution of 20 Å, obtained from an altitude of 150 km above White Sands, New Mexico. The scanning spectrometer viewed horizontally toward the north at a local time of approximately noon. (C.A. Barth, *et al.*, *J. Geophys. Res.*, **76**, 2213, 1971.)

individual features, leading to uncertainty in some identifications and in evaluating the emission rates of individual lines. There is little doubt, however, that the excited states of OI, NI, OII and NII are produced by energetic photon and photoelectron impact. The energy required to produce the upper states is too large for chemical or recombination reactions to have any part in exciting the observed lines. The absorption cross sections of O, N_2 and O_2 in the wavelength region under consideration (400–820 Å) are large and the solar photons do not penetrate to the lower E-region, as shown in Figure 6.2.1. The observed lines therefore originate in the F-region. The oxygen lines are produced mostly by ionization and excitation of atomic O with a small contribution from dissociative ionization and excitation of O_2,

$$O + h\nu \rightarrow O^* \tag{7.3.1}$$

$$O + h\nu \rightarrow O^{+*} + e_{p,e} \tag{7.3.2}$$

$$e_{p,e} + O \rightarrow e + O^* \tag{7.3.3}$$

$$O_2 + h\nu \rightarrow O^* + O \tag{7.3.4}$$

$$O_2 + h\nu \rightarrow O^{+*} + O + e_{p,e} \tag{7.3.5}$$

Due to the low abundance of N the nitrogen lines result principally from dissociation and dissociative ionization of N_2,

$$N_2 + h\nu \rightarrow N^* + N \tag{7.3.6}$$

$$N_2 + h\nu \rightarrow N^{+*} + N + e_{p,e} \tag{7.3.7}$$

but at high altitudes photoelectron impact on N also contributes to excitation of some observed NI lines,

$$e_{p,e} + N \rightarrow e + N^* \tag{7.3.8}$$

The reader has probably already noticed that several atomic and ionic lines in the dayglow spectrum are also lines that appear prominently in the solar spectrum that irradiates the Earth's atmosphere. The list includes the brightest single feature in both spectra, which is the H-Ly-α line at 1216 Å, the H-Ly-β line at 1025 Å, the HeI line at 584 Å, a triplet of O at 1302, 1305 and 1306 Å, the OII triplet around 834 Å and others. Resonance scattering and fluorescent scattering of the solar emission lines by thermospheric atoms and ions, H, O, He or O^+ may therefore contribute to the dayglow, in addition to photon and photoelectron excitation.

Atomic H and He are minor constituents in the thermosphere but become the major species in the exosphere. The prominence of their spectral emissions in the dayglow and nightglow justifies a brief discussion. The lines of H and He are brighter viewed away from the Earth than toward it even from an altitude of 800 km. The radiations are observed by viewing in any direction and over a large altitude range. The glow enveloping the Earth has been called the H geocorona and the He geocorona. The geocoronal H-Ly-α is excited by resonance scattering of the intense solar chromospheric line,

$$\mathrm{H(1s\,^2S)} + h\nu(1215.7\,\text{Å}) \rightleftarrows \mathrm{H(2p\,^2P)} \tag{7.3.9}$$

The geocoronal H-Ly-β line is produced similarly. However, a fraction of the excited atoms (about 1:7) decays to a different level by fluorescence with emission of the H-Balmer-α line,

$$\mathrm{H(1s\,^2S)} + h\nu(1025\,\text{Å}) \rightleftarrows \mathrm{H(3p\,^2P^0)} \rightarrow \mathrm{H(2s\,^2S)} + h\nu(6563\,\text{Å}) \tag{7.3.10}$$

H-Balmer-α is observed in the dayglow and the nightglow. Multiple scattering carries photons from the dayside to the non-illuminated region of the exosphere.

Transport of photons is described by the radiation transfer equation,

$$\mu\, \mathrm{d}I/\mathrm{d}\tau_\nu = S(\nu, \mu, \tau) - I(\nu, \mu, \tau) \tag{7.3.11}$$

where $I(\nu, \mu, \tau)$ is the photon intensity (photons $\mathrm{cm^{-2}\,s^{-1}\,sr^{-1}\,Hz^{-1}}$) in the direction $\mu = \cos\theta$, $S(\nu, \mu, \tau)$ is the source function, and τ_ν is the optical depth at frequency ν. The Lambert–Beer absorption relationship introduced in Chapter 2 (Equation (2.1.1)) is applicable in the absence of the source function, but the occurrence of multiple scattering and other sources requires the adoption of the complete equation. The mathematical form of the radiative transfer equation is very similar to the electron transport equation introduced in Chapter 3 (Equation (3.2.9)) but the physical parameters are, of course, different.

The two components of the source function, S_ν, are solar radiation incident at the top of the atmosphere and photons produced within the atmosphere by impact excitation, chemical excitation, and multiple scattering. The source function is specified by ε_ν, the photon emission rate per unit volume, time, frequency, and solid angle, and by α_ν the absorption coefficient per unit length.

$$S_\nu = \varepsilon_\nu/\alpha_\nu \tag{7.3.12}$$

The optical depth in a single species gas is

$$\tau_\nu(z) = \int_z^\infty \alpha_\nu(z)\,\mathrm{d}z = \sigma_\nu^{\mathrm{a}} \int_z^\infty n(z)\,\mathrm{d}z \tag{7.3.13}$$

where σ_ν^{a} is the absorption cross section and the integral represents the vertical column density of absorbing gas. The optical depth for a non-vertical column is given by Equation (2.2.5). The cross section for self-absorption of line radiation resulting from transitions between two levels of an atom is

$$\sigma_\nu^{\mathrm{a}} = \frac{c^2}{8\pi}\frac{\tilde{\omega}_m}{\tilde{\omega}_n}\frac{1}{\nu^2}A_{mn}\phi(\nu) \tag{7.3.14}$$

for a gas in which the number of atoms in the upper state m is very small compared to those in the lower state n. The $\tilde{\omega}$s are the statistical weights, A_{mn} is the Einstein coefficient for spontaneous radiative transitions and $\phi(\nu)$ is the normalized spectral intensity distribution over the line,

$$\int_{-\infty}^{\infty} \phi(\nu)\,\mathrm{d}\nu = 1 \tag{7.3.15}$$

which is discussed below.

The absorption cross section varies over the frequency range of the spectral line shape as determined by various broadening mechanisms. Since the integral over frequency depends on the total number of photons absorbed in the transition between two states,

$$\int_0^{\infty} \sigma_\nu^{\mathrm{a}}\,\mathrm{d}\nu = (\pi e^2/m_{\mathrm{e}}c) f_{nm} \tag{7.3.16}$$

where f_{nm} is the absorption oscillator strength for the transition. This is related to the Einstein coefficient A_{mn} by

$$A_{mn} = \frac{\tilde{\omega}_n}{\tilde{\omega}_m}\frac{8\pi e^2}{m_{\mathrm{e}}c}\left(\frac{\nu}{c}\right)^2 f_{nm} \tag{7.3.17}$$

As a consequence of the Heisenberg uncertainty principle, spectral lines exhibit natural broadening which has a Lorentzian spectral intensity distribution,

$$\phi_{\mathrm{L}}(\nu) = \frac{\Delta\nu_{\mathrm{L}}/2\pi}{(\nu-\nu_0)^2 + (\Delta\nu_{\mathrm{L}}/2)^2} \tag{7.3.18}$$

The frequency at the line centre is ν_0 and the line width at half maximum is

$$\Delta\nu_{\mathrm{L}} = \Gamma_{\mathrm{R}}/2\pi \tag{7.3.19}$$

The radiative damping constant, Γ_{R}, is given by

$$\Gamma_{\mathrm{R}} = \Gamma_m + \Gamma_n = 1/t_m + 1/t_n \tag{7.3.20}$$

where t_m and t_n are the mean radiative lifetimes of the atom in levels m and n respectively. Applied to atmospheric conditions,

$$\Gamma_m = \sum_{n<m} A_{mn} \tag{7.3.21}$$

Spectral lines are also subject to collisional broadening. This mechanism likewise results in a Lorentz intensity distribution. It yields an expression identical to Equation (7.3.18), except for the parameter Γ, which becomes a collisional damping constant, Γ_{coll}, or the mean collision frequency.

The kinetic temperature T in the thermosphere and exosphere is sufficiently high to produce Doppler broadening of spectral lines. The intensity distribution is given by a Gaussian profile

$$\phi_{\mathrm{G}}(\nu) = \frac{c}{\nu_0}\left(\frac{M}{2\pi kT}\right)^{\frac{1}{2}} \exp\left[-(\nu-\nu_0)^2 \Big/ \frac{\nu_0^2}{c}\frac{2kT}{M}\right] \tag{7.3.22}$$

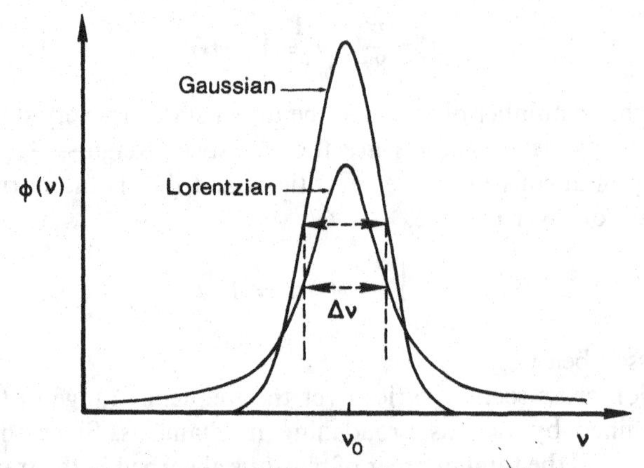

Fig. 7.3.2 Comparison of Lorentzian and Doppler line profiles. The two curves have the same total area (i.e., the same equivalent width) and the same half value width $\Delta\nu$. (Adapted from A.P. Thorne, *Spectrophysics*, Chapman and Hall, London, 1974.)

where ν_0 is the centre frequency and M is the mass of the atom. The line width at half maximum is

$$\Delta\nu_D = \frac{\nu_0}{c}\left(8\ln 2\frac{kT}{M}\right)^{\frac{1}{2}} \tag{7.3.23}$$

The Lorentzian and Gaussian profiles are both normalized to unity (Equation (7.3.15)). The profiles are contrasted in Figure 7.3.2.

The effective line profile is the superposition of the two independent distributions in which each element of the Lorentzian profile is broadened by a Gaussian function. The convolution of the two profiles is

$$\phi(\nu) = \int_{-\infty}^{\infty} \phi_G(\nu')\phi_L(\nu-\nu')\,d\nu' \tag{7.3.24}$$

Substitution of Equations (7.3.18) and (7.3.22) yields

$$\phi_V(\nu) = \frac{(\ln 2)^{\frac{1}{2}}}{\pi^{3/2}}\frac{\Delta\nu_L}{\Delta\nu_D}\int_{-\infty}^{\infty}\frac{\exp\{-\nu'^2/[\Delta\nu_D/2(\ln 2)^{\frac{1}{2}}]^2\}}{[(\nu-\nu_D-\nu')^2+(\Delta\nu_L/2)^2]}\,d\nu' \tag{7.3.25}$$

The resulting profile is called the Voigt profile and is illustrated in Figure 7.3.3. The folding of two Voigt profiles yields another Voigt profile, a useful property since natural and collisional broadening yield Lorentzian profiles, and instrumental broadening is frequently represented likewise. The Voigt profile does not reduce to an analytic expression, but may be rewritten as follows

$$\phi_V = \frac{2(\ln 2)^{\frac{1}{2}}a}{\pi^{3/2}\Delta\nu_D}\int_{-\infty}^{\infty}\frac{\exp(-y^2)}{(b-y)^2+a^2}\,dy \tag{7.3.26}$$

with

$$a = \frac{\Delta\nu_L}{\Delta\nu_D}(\ln 2)^{\frac{1}{2}}, \quad b = 2(\ln 2)^{\frac{1}{2}}\frac{\nu-\nu_0}{\Delta\nu_D}, \quad y = 2(\ln 2)^{\frac{1}{2}}\frac{\nu'}{\Delta\nu_D} \tag{7.3.27}$$

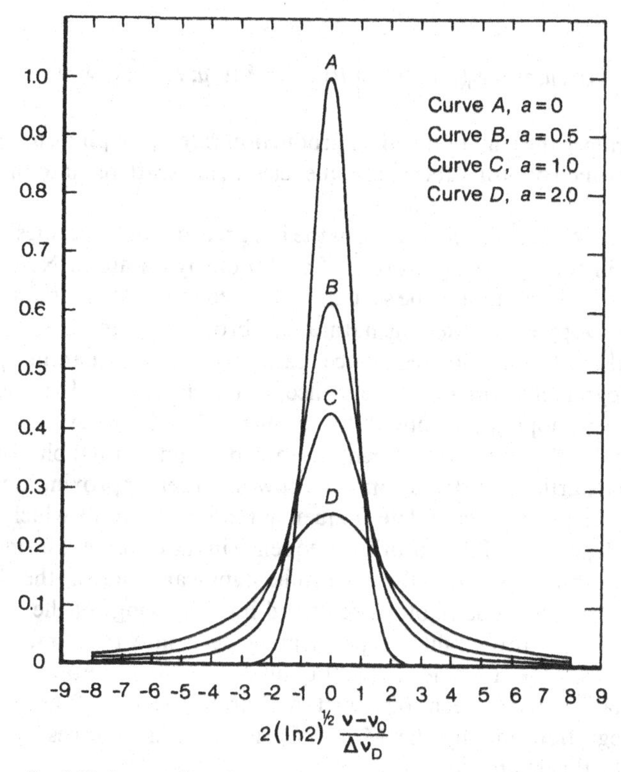

Fig. 7.3.3 Normalized Voigt profiles for a range of ratios of the Lorentzian to Doppler half widths, $a = (\ln 2)^{1/2}\, \Delta\nu_L/\Delta\nu_D$. (Adapted from A.C.G. Mitchell and M.W. Zemansky, *Resonance Radiation and Excited Atoms*, Cambridge University Press, Cambridge, 1961.)

Tables of the Voigt integral may be found in several references (Problem 38).

Combining Equations (7.3.13), (7.3.14) and (7.3.26) we obtain, for the optical depth,

$$\tau_\nu(z) = \frac{e^2}{m_e c} f_{nm} \phi_V(\nu) \int_z^\infty n(z)\,dz \tag{7.3.28}$$

A medium is said to be optically thin to radiation of frequency ν when $\tau_\nu(z) < 1$ and optically thick when $\tau_\nu(z) \gtrsim 1$. Since the absorption profile varies widely over the frequency spread of the line the atmosphere may be optically thick to resonance radiation in the centre of the line and optically thin in the wings (Problem 39).

The formal solution to the radiative transfer equation (Equation (7.3.11)) is

$$I(\nu,\mu,\tau) = I_0(\nu,\mu)\exp(-\tau_\nu) + \int S(\nu,\mu,\tau')\exp[-(\tau-\tau')/\mu]\,d\tau'/\mu \tag{7.3.29}$$

The first term on the right side of the equation takes account of the attenuation of the intensity, I_0, incident at the top of the atmosphere, while the integral includes local sources and photons scattered from levels at optical depths τ' into the level at optical depth τ. The source function $S(\nu,\mu,\tau)$ is written in terms of the photon emission rate

(Equation (7.3.12)),

$$\varepsilon(\nu, \mu, z) = \varepsilon_0(\nu, \mu, z) + \int I(\nu', \mu', z) K(\nu, \mu; \nu', \mu'; z)\, d\nu'\, d\Omega \qquad (7.3.30)$$

The emission rate is the sum of the local production rate, ε_{0_ν} of photons at frequency ν and the rate at which photons at other frequencies, ν', are scattered into the interval ν to $\nu + d\nu$.

The scattering kernel, $K(\nu, \mu; \nu', \mu', z)$, in the integral equation specifies the frequency redistribution in the scattering process of a photon by an atom. Non-coherence in frequency is due to the motion of the scatterers. The redistribution in frequency therefore depends on the Doppler broadening, natural line broadening and collisional broadening. The kernel appropriate for resonance scattering, a single quantum process, is an intractable integral (given in a basic reference to this chapter) and various approximations have been adopted for solving the problem. The least severe approximation retains the angular dependence between the absorbed and emitted photon. Next is the angle averaged partial redistribution. The lowest order approximation applied to dayglow analyses assumes complete frequency redistribution in which the emission frequency is independent of the absorbed frequency in the observer's frame of reference. Redistribution of photons within the line profile significantly affects the altitude profile of the emission rate because photons scattered into the wings of the absorption line profile have a higher probability of penetrating deep into the atmosphere, i.e., the optical depth varies over the line profile (Equation (7.3.28)). Solutions to the radiative transport equation have been obtained at various levels of approximation by numerical integration and by the Monte Carlo method. Sources are listed in the bibliography to this chapter.

One of the most prominent UV dayglow spectral features in Fig. 7.3.1 is the (unresolved) triplet of atomic O at 1304 Å. The relevant electronic states are shown in the energy level diagram given in Appendix 3. Radiative transfer theory has been particularly valuable in analyzing this spectral feature leading to prediction of its intensity, and the problem is still a current research topic. There are two excitation sources of comparable importance, absorption of solar photons incident at the top of the atmosphere and photoelectron impact excitation,

$$e^* + O(2p^4\,{}^3P) \rightarrow e + O(3s'\,{}^3S) \qquad (7.3.31)$$

The line profiles for the two sources are very different. A high resolution photographic spectrum of the solar OI triplet emission is shown in Figure 7.3.4. The observation was made when the rocket was at an altitude of about 180 km and all components of the triplet show strong central absorption due to the O atoms in the terrestrial thermosphere. At the top of the atmosphere (an expression that refers to the level above which solar photons have not been scattered) the line profile is free of the terrestrial absorption core but retains the broad absorption that is characteristic of the solar Fraunhofer spectrum. Figure 7.3.5 shows an assumed profile of the solar OI lines that fits the observations reasonably well. It is so broad as to be almost a continuum over the line width of terrestrial O atoms. By contrast, the photoelectron excitation source produces a narrow frequency profile characteristic of the thermospheric atomic O (Figure 7.3.3).

Resonance scattering produces multiple absorption and emission within the line profile,

$$O({}^3P_{2,1,0}) + h\nu(1302.17, 1304.96, 1306.03\ \text{Å}) \rightleftarrows O({}^3S_1) \qquad (7.3.32)$$

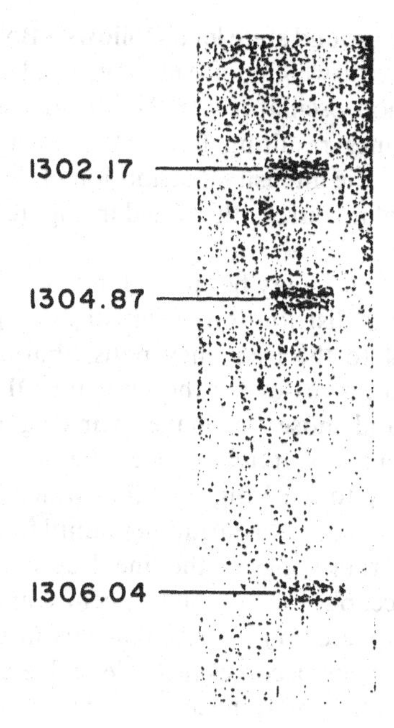

Fig. 7.3.4 Photograph of the rocket film showing the OI triplet at 1302–1306 Å. The central absorption is primarily due to absorption by O in the Earth's atmosphere. (E.C. Bruner and W.A. Rense, *Astrophys. J.*, **157**, 417, 1969.)

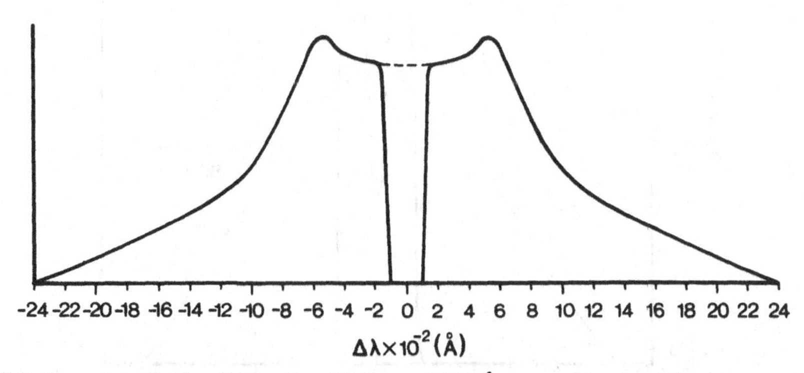

Fig. 7.3.5 Assumed profile of the solar OI line at 1304 Å at an altitude of 184 km. Dotted line indicates the continuation of the solar profile in the spectral region where the (predicted) absorption of atomic O in the Earth's atmosphere takes place. When instrumental broadening effects are applied to this assumed profile, the best fit with observational data is obtained. (E.C. Bruner and W.A. Rense, *Astrophys. J.*, **157**, 417, 1969.)

The population of the three ground state levels follows a Boltzmann distribution which is a function of the statistical weight of each level and the altitude-dependent gas temperature. The absorption oscillator strength is the same for all three lines. The intensity distribution of lines resulting from photoelectron excitation follows the Voigt profile (Figure 7.3.3). The intensity distribution of the solar photons within each line is determined by the solar source, illustrated in Figure 7.3.5, as is the partitioning amongst the three lines.

In a scattering event a photon may be emitted in any of the three lines of the triplet (the probability following the distribution of statistical weights). It may acquire a new direction, and it is subject to the frequency redistribution discussed earlier in this section. While the physical processes are the same for OI triplet photons created by photoelectron excitation and those that have their origin in the solar flux, the line profiles are rather different. As noted above, the internal photoelectron source produces photons according to the Voigt function which has a maximum at the line centre. This results in strong radiation entrapment until frequency redistribution shifts photons into the optically thin wings of the line, leading to upward and downward escape. For the solar source, photons in the line centre are scattered out of the beam near the top of the atmosphere. With most photons in the wings of the line, deep penetration into the atmosphere becomes possible and less multiple scattering occurs. In addition to the escape of photons at the upper and lower boundaries of the scattering

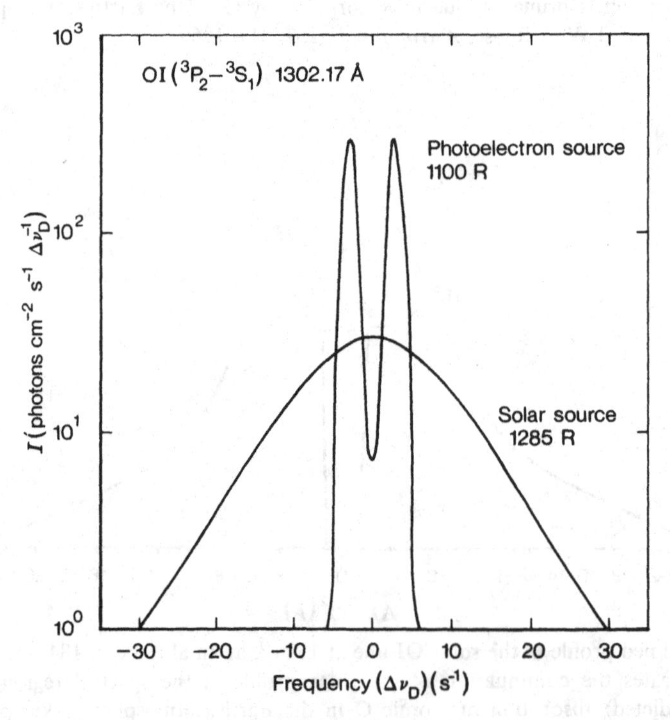

Fig. 7.3.6 Profile of the OI (1302.17 Å) line predicted to emerge from the atmosphere at a zenith angle of 140° (looking downward) from an altitude of 600 km. The intensity is plotted as a function of the Doppler width, ν_D, defined by Equation (7.3.23). (Adapted from R.R. Meier and J.-S. Lee, *Planet. Space Sci.*, **30**, 439, 1982.)

medium, they also are subject to true absorption by O_2 molecules. A result of the physical processes discussed above is illustrated in Figure 7.3.6 by the line profile of OI(1302.17 Å) predicted to emerge from the atmosphere, as viewed in the nadir from an altitude of 600 km. Experimental measurements at such a high resolution have not yet been achieved.

Another prominent emission line in the dayglow spectrum is the OI doublet at 1356 Å resulting from the intercombination transition ($2p^4\,{}^3P_{2,1}$–$3s\,{}^5S_2$). The forbidden character of the transition eliminates resonant absorption of solar photons as a source, and the line is due solely to photoelectron impact excitation,

$$e^* + O({}^3P_{2,1}) \rightarrow e + O({}^5S_2) \tag{7.3.33}$$

$$O({}^5S_2) \rightarrow O({}^3P_{2,1}) + h\nu(1355.5,\ 1358.5\ \text{Å}) \tag{7.3.34}$$

The vertical optical depth at the lower boundary of the thermosphere slightly exceeds unity and multiple scattering and radiation transport of 1356 Å photons should be included in an accurate analysis. The photoelectron production rate, ε_0 in Equation (7.3.30), exceeds the multiple scattering contribution for the 1356 Å line. By contrast, for the 1304 Å triplet discussed previously, the scattering term is several powers of ten larger than the local production term in the optically thick region of the thermosphere.

The general expression for the excitation rate by photoelectron impact at altitude z is

$$\eta_j^l(z) = n_j(z) \int_{E_{th}}^{E_{max}} \sigma_j^l(E) I(z, E)\,dE \quad (\text{cm}^{-3}\,\text{s}^{-1}) \tag{7.3.35}$$

where $n_j(z)$ is the number density of species j, $\sigma_j^l(E)$ is the cross section for excitation of state l of species j, E_{th} is the excitation threshold energy, and E_{max} is an energy above which the omnidirectional photoelectron intensity $I(z, E)$ is too insignificant to contribute to the process. The reader may note here that the symbol I is used to denote both the photoelectron intensity and the photon intensity since the two quantities are obtained by similar methods: electron transport and radiation transport, respectively. The appropriate identification should emerge in context.

We have already noted in Chapter 6 that excited metastable states do not decay solely by radiation but are also deactivated by collisions, transferring energy to the collision partners. The volume emission rate of species j in an excited state l at altitude z is given by

$$\varepsilon_j^l(z) = \eta_j^l(z) \frac{A_{lm}}{\sum_{m<l} A_{lm} + \sum_i n_i(z) k_i} \tag{7.3.36}$$

where A_{lm} is the radiative transition probability between states l and m, the summation is over all possible radiative transitions with $m < l$, and the second term in the denominator gives the collisional deactivation. The rate coefficient, k_i, may be a function of temperature (neutral kinetic, vibrational, electron or ion).

The volume emission rate is not a measurable quantity. Observations measure the photon flux at the detector. This is generally less than the emitted flux due to attenuation of the photons by absorption and scattering. A special unit has been adopted in airglow and auroral photometry. It is called the rayleigh (R), defined as a measure of the omnidirectional emission rate in a column of unit cross section along the line of sight

$$1\,\mathrm{R} = 10^6\,\text{photons}\,\mathrm{cm}^{-2}\,\mathrm{s}^{-1} \tag{7.3.37}$$

A derivation of this unit in terms of the specific photon intensity $I(\nu, \mu, \tau)$, Equation (7.3.11), is given in a reference listed with this chapter. An implicit assumption is that the field of view of the detector is uniformly filled. This is usually the case for airglow radiation. The rayleigh is, therefore, a unit of surface brightness. It is not a unit of energy since photons at different wavelengths have different energies.

Emission lines of OI, OII, NI and NII that appear in the UV dayglow spectrum originate from one or more of the following processes,

$h\nu + \mathrm{A} \rightarrow \mathrm{A}^* + \mathrm{k.e.}$	Photon impact excitation
$h\nu + \mathrm{A} \rightarrow \mathrm{A}^{+*} + \mathrm{e}$	Photon impact ionization excitation
$h\nu + \mathrm{AB} \rightarrow \mathrm{A}^* + \mathrm{B}^* + \mathrm{k.e.}$	Photodissociative excitation
$\mathrm{e} + \mathrm{A} \rightarrow \mathrm{e} + \mathrm{A}^*$	Photoelectron excitation
$\mathrm{e} + \mathrm{A} \rightarrow \mathrm{e} + \mathrm{A}^{+*} + \mathrm{e}$	Photoelectron ionization excitation
$\mathrm{e} + \mathrm{AB} \rightarrow \mathrm{e} + \mathrm{A}^* + \mathrm{B}$	Photoelectron impact dissociative excitation
$\mathrm{e} + \mathrm{AB} \rightarrow \mathrm{e} + \mathrm{A}^{+*} + \mathrm{B}^* + \mathrm{e}$	Dissociative ionization excitation

A, A^+, and AB signify atoms, atomic ions and molecules and the asterisk signifies the formation of an excited state. Resonant scattering or fluorescence are additional sources. Several dayglow emissions are due to cascading from higher levels but these lines are in the visible and IR regions of the spectrum, not the UV. The energy level diagrams in Appendix 3 should be helpful in sorting out the multitude of emission features that appear in Figure 7.3.1 (Problem 40). Photons resulting from transitions to the ground state of an atom have the potential for multiple scattering. The importance of this additional source is determined by the optical depth which is a function of the absorption oscillator strength, the column density and the line profile (Equation (7.3.28)). The viewing direction needs to be specified since the column density of scatterers is smallest in the vertical and largest for observations near the limb. For example, multiple scattering contributes to the bright OII (834 Å) and NI (1200 Å) features in Figure 7.3.1, the spectra in this wavelength region having been acquired by side viewing rocket borne spectrometers. Most, but not all, the identifications shown in the composite dayglow spectrum are probably correct. It should be noted that some spectral features are identified as lines appearing in the second order of diffraction.

The H lines H-Ly-α (1215.67 Å) and H-Ly-β (1025.72 Å) are excited solely by resonant scattering of the solar emissions. Multiple scattering strongly enhances the observed dayglow emissions and also transports radiation to the dark hemisphere. Analysis of altitude profiles of the H-Ly-α emission rate using a global radiative transfer model can yield the distribution of atomic H in the thermosphere. At the resolution currently available (Figure 7.3.1) the H-Ly-β line is blended with OI and N_2 features, introducing substantial difficulties in the analysis.

Several features in the UV dayglow are identified as electronic transitions in N_2 and the appearance of several bands in a system reinforces the identification. The occurrence of these bands is not surprising because laboratory experiments on electron scattering by N_2 show that the high lying states $\mathrm{b}'^1\Sigma_u^+$, $\mathrm{b}^1\Pi_u$, and $\mathrm{c}_4'\,{}^1\Sigma_u^+$ of the molecule are readily excited. These are the predissociation states discussed in

Section 4.3. Evidently, a small fraction of the excited N_2 molecules undergo radiative transitions instead of dissociation.

In the wavelength interval from 1273 Å to about 1700 Å, over a dozen vibrational bands of the Lyman–Birge–Hopfield (LBH) system of N_2 appear in the dayglow spectrum. The band system originates in the $a\,^1\Pi_g$–$X\,^1\Sigma_g^+$ transition which is electric dipole forbidden. The (3, 0) vibrational band of the LBH system lies at about 1354 Å and cannot be resolved from the OI line at 1356 Å discussed earlier. The band intensity is estimated from the emission rates of other bands of the system and the Frank–Condon factors. Non-uniform absorption by O_2 in the wavelength region of the LBH bands further complicates the analysis. The most likely process for exciting the upper state of the transition is electron impact which also accounts for the excitation of various other electronic states,

$$e^* + N_2(X^1\Sigma) \rightarrow e + N_2^*(a^1\Pi, b^1\Pi, b'^1\Sigma, c_4'^1\Sigma, \text{etc.}) \tag{7.3.38}$$

Excitation of the $a\,^1\Pi_g$ state proceeds from the ground state of the molecule; however, the absorption oscillator strength is small and resonant scattering is negligible.

Proceeding to longer wavelengths, a prominent band system appears between 2000 and 2800 Å that originates from a minor thermospheric species. This is the γ-band of NO. The excitation mechanism is resonant scattering,

$$NO(X^2\Pi) + h\nu \rightleftarrows NO(A^2\Sigma^+) \tag{7.3.39}$$

and the radiation has been used extensively to investigate the abundance of NO in the thermosphere. The observed column brightness is inverted to yield the volume emission rate as a function of altitude, a procedure that justifiably assumes that the thermosphere is optically thin to the radiation. The volume emission rate is the product of the concentration and an emission rate factor g,

$$\eta(z) = n_z(NO)_{v''}g(v', v'') \tag{7.3.40}$$

where

$$g(v', v'') = \sum_{v''} \pi F(\nu) \frac{\pi e^2}{m_e c} f_{v''v'} \frac{A_{v'v''}}{\sum_{v'} A_{v'v''}} \tag{7.3.41}$$

is the emission rate factor for the (v', v'') band of the system, $\pi F(\nu)$ is the solar photon flux, $f_{v''v'}$ is the absorption oscillator strength and each $A_{v'v''}$ is a transition probability for spontaneous emission. The molecular parameters are derived from laboratory data and the solar flux from observational data. The NO concentration may therefore be obtained. Sources and sinks of NO are listed in Appendix 5. The source reactions depend indirectly on the solar UV flux, while the brightness of the NO γ-bands in the dayglow and twilight is a direct function of the solar photon flux. However, in the wavelength region responsible for resonant scattering, 2000–2800 Å, the solar flux is less variable than it is in the wavelength region ($\lambda < \sim 1000$ Å) where ionization and dissociation of the major atmospheric species initiates the formation of NO molecules in the thermosphere. Variation of NO during a solar cycle is therefore expected. The dayglow and twilight source of the NO γ-bands, resonant scattering, is substantially larger than the chemical reaction (Reactions (7.2.7) and (7.2.8)) which is the sole source in the non-sunlit atmosphere. In the same wavelength region NO ε-bands ($D\,^2\Sigma^+$–$X\,^2\Pi$) have also been identified in the dayglow by comparing observations with synthetic spectra.

Two band systems of N_2 appear between 2600 and 4100 Å, ($A\,^3\Sigma_u^+ - X\,^1\Sigma_g^+$) called the Vegard–Kaplan (VK) bands, and $C^3\Pi_u - B^3\Pi_g$) known as the Second Positive (2P) bands. Both upper states are excited solely by electron impact; the VK bands represent a singlet–triplet forbidden transition and the 2 P bands do not originate from the ground state of the molecule,

$$e^* + N_2(X^1\Sigma_g^+) \rightarrow e + N_2(A^3\Sigma_u^+, B^3\Pi_g, C^3\Pi_u) \qquad (7.3.42)$$

With reference to the potential energy curves for the N_2 molecule (Appendix 3) an additional state may be excited by electron impact, and the ($B\,^3\Pi_g - A\,^3\Sigma_u^+$) transition is the source of the First Positive (1P) bands in the dayglow beyond about 6000 Å.

The N_2^+ First Negative (1NG) bands are bright dayglow features in the near UV and visible wavelength regions of the spectrum. While a small fraction of the $N_2^+(B\,^2\Sigma_u^+)$ molecular ions are excited by photoelectron impact and photon impact ionization of N_2, the principal mechanism for exciting these bands is resonant scattering of sunlight by the ambient N_2^+ ground state ions,

$$N_2^+(X^2\Sigma_g^+) + h\nu(1N) \rightleftarrows N_2^+(B^2\Sigma_u^+) \qquad (7.3.43)$$

N_2^+ is not the most abundant ion species in the thermosphere but the emission rate factor, g, (Equation (7.3.41)) is large and the resulting band emission dominates the blue spectral region.

Two atomic lines are identified in the low resolution spectrum of Figure 7.3.1 in the near UV. The line at 2972 Å is attributed to a transition in OI and the line at 3466 Å to NI. The reader may refer to the energy level diagrams in Appendix 3 for a more complete characterization of these lines; they will be discussed in greater detail in the next section, together with many other spectral features that are common to the dayglow and the aurora.

7.4 The spectrum of the aurora

Somewhat arbitrarily, auroral radiation is attributed to the consequences of energetic electron and proton bombardment of the atmosphere usually, but not always, confined to relatively narrow, high latitude, circumpolar bands in each hemisphere, called the auroral ovals. Energetic H^+, H and O^+ also precipitate at middle and low latitudes but their magnitude is only a very small fraction of typical auroral precipitation and concomitant spectroscopic emissions are so weak as to require specialized instrumentation for detection. None appear in the survey nightglow and dayglow spectra in Figures 7.2.1 and 7.3.1.

In Chapter 3 we discussed transport of auroral electrons into the thermosphere, and the resulting energy deposition. It was noted that many inelastic collisions lead to the production of excited states that may decay by emission of lines and bands. Auroral proton (and O^+) precipitation likewise becomes a source of spectral emissions. Auroral particle precipitation occurs day and night; auroral spectroscopic emissions are therefore superposed on dayglow and nightglow emissions. Ground based measurements yield spectra when the sun is below the local horizon and only in the accessible visible and infrared regions. These restrictions are lifted for auroral observations acquired from satellites and rockets. The auroral spectra shown in Figure 7.4.1 are daytime observations in the UV acquired from space borne platforms. Non-sunlit observations are reproduced in Figures 7.4.2–7.4.8 and these include ground based

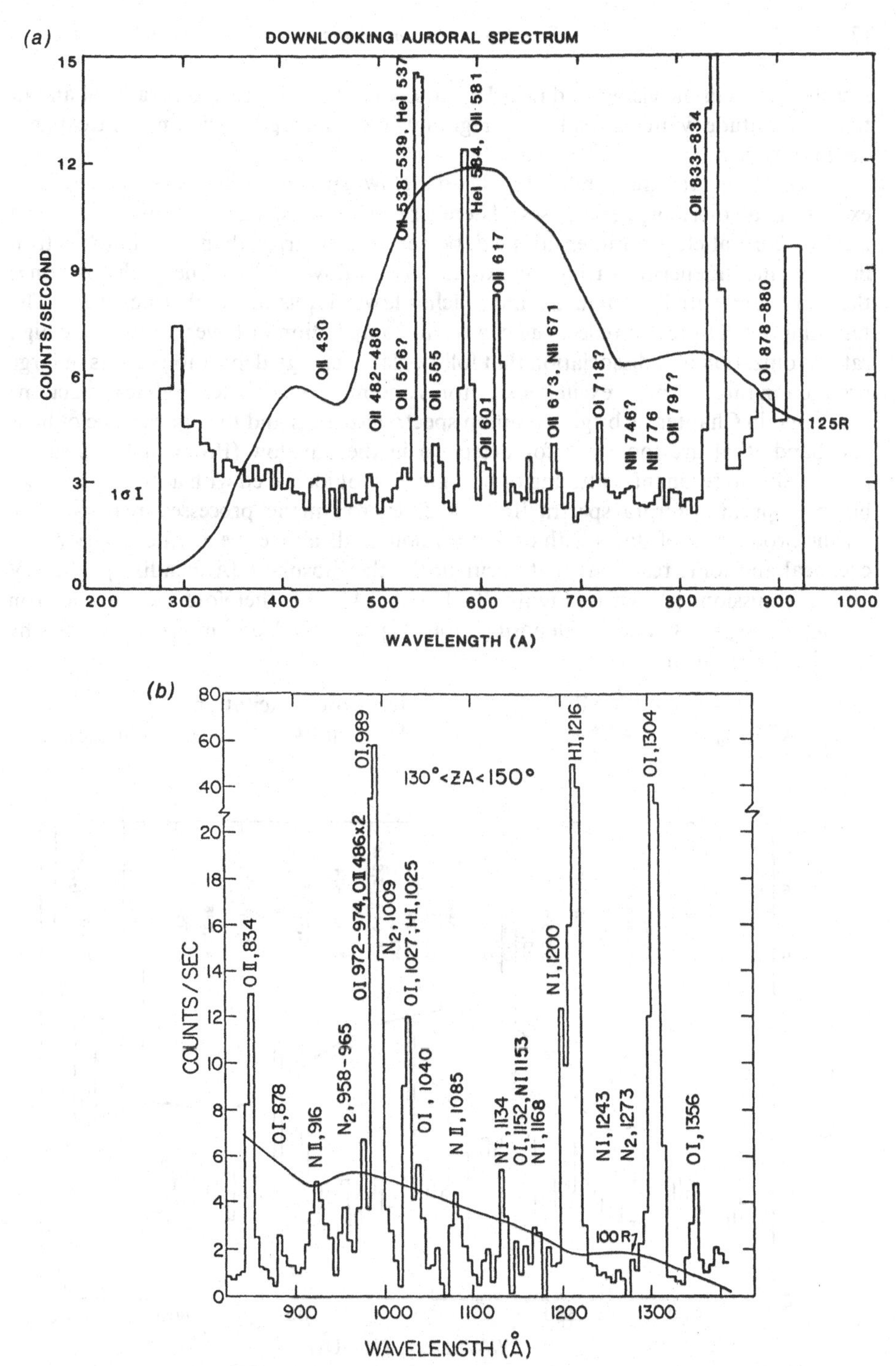

Fig. 7.4.1 Nadir viewed dayside auroral spectra acquired by the STP78-1 satellite from 600 km near local noon: (*a*) between 300 and 900 Å (S. Chakrabarti, *J. Geophys. Res.*, **90**, 4421, 1985); (*b*) between 800 and 1400 Å (F. Paresce, *et al.*, *J. Geophys. Res.*, **88**, 4905, 1983). In each case the solid line is the instrument's response to constant input.

measurements in the visible and near IR acquired in the nightside auroral oval and in the high latitude winter dayside oval region where the troposphere and stratosphere are in darkness.

Auroral spectra are similar to the dayglow spectra since both derive from excitation of thermospheric gases. There are differences, however, due to several factors. Auroral electron intensities can be very much larger than the photoelectron intensity and the energy distribution extends to the kilovolt range. Energy deposition is therefore substantially enhanced and reaches lower levels in the thermosphere. The resulting optical spectra reflect the change in composition at lower altitudes. A high rate of ionization and dissociation that follows large energy deposition results in large production rates of many excited ions, atoms and molecules, faster chemical reactions (discussed in Chapter 5), bright common spectral features and the appearance of lines and bands that are too weak for detection in the dayglow (Problem 41). Spectral features due to resonant scattering of sunlight are not identified with auroral processes but may appear in auroral spectra. In aurorae, electron impact processes are responsible for the production of states with excitation potentials above the exothermicity of the chemical and ionic reactions that occur in the thermosphere (Appendix 5). The UV auroral emissions shown in Figures 7.4.1 and 7.4.2 are, therefore, due to electron impact on N_2, O_2, O and N. Denoting atomic species by A and molecular species by AB, the processes are

$e_p + A, AB \rightarrow e_p + A^{+*}, AB^{+*} + e_s$ Ionization excitation
$e_p + AB \rightarrow e_p + A^{+*} + B^* + e_s$ Dissociative ionization excitation

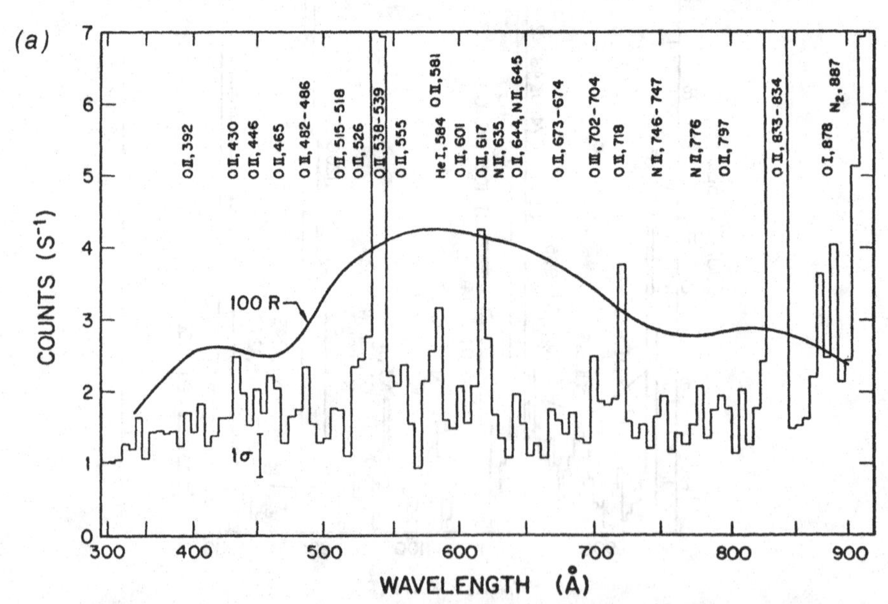

Fig. 7.4.2 (*a*) Nightside auroral spectrum between 300 and 900 Å acquired by the STP78-1 satellite from 600 km altitude viewing in the nadir direction near local midnight in the northern hemisphere. The solid line is the instrument response to a constant input. (F. Paresce, *et al.*, *J. Geophys. Res.*, **88**, 10247, 1983.)

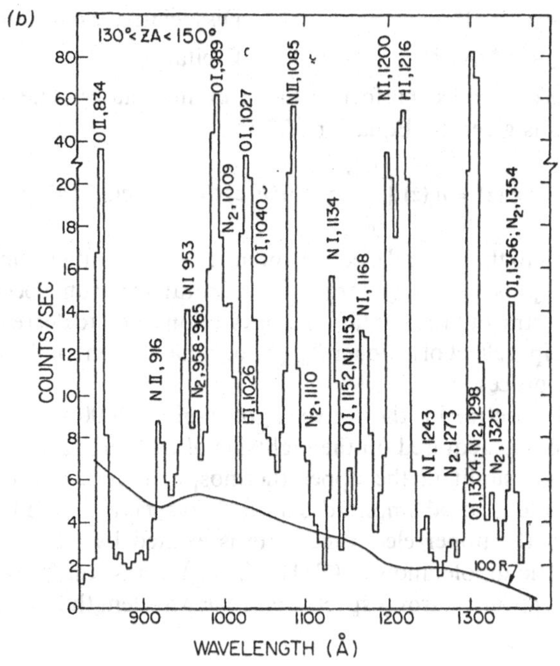

Fig. 7.4.2 (*b*) UV auroral spectrum acquired during night time over the southern polar region by the STP78-1 satellite from 600 km altitude viewing in the nadir direction. The solid line is the instrument response to a constant input. (F. Paresce, *et al.*, *J. Geophys. Res.*, **88**, 4905, 1983.)

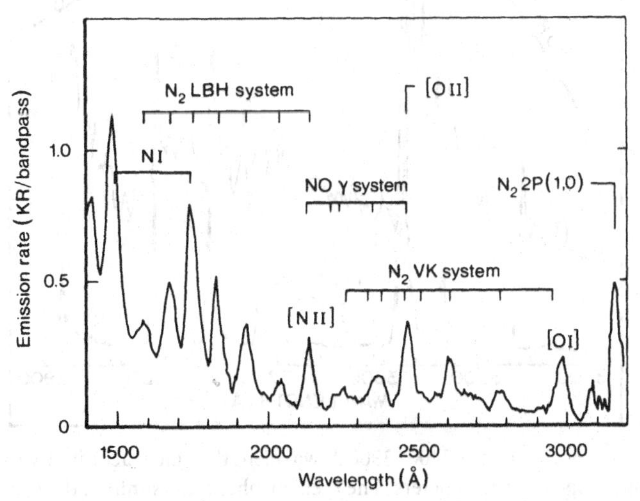

Fig. 7.4.2 (*c*) UV auroral spectrum acquired by a nadir viewing spectrometer on board the OGO-4 satellite. (Courtesy of C.A. Barth, 1987.)

$e_p + AB \rightarrow e_p + A^* + B^*$ Dissociative excitation

$e_s + A, AB \rightarrow e_s + A^*, AB^*$ Excitation

where the asterisk denotes the formation of excited states. The excitation rate by auroral electrons is given by Equation (7.3.35),

$$\eta_j^l(z) = n_j(z) \int_{E_{th}}^{E_{max}} \sigma_j^l(E) I(E, z)\, dE \qquad (\text{cm}^{-3}\,\text{s}^{-1}) \qquad (7.3.35)$$

where $I(E, z)$ refers to the auroral electron intensity and the other symbols have already been defined. E_{max} is generally larger for auroral electron spectra compared to photoelectron spectra, but the excitation cross sections tend to decrease with increasing energy. The principal effect of a higher E_{max} is an enhancement of the various ionization and dissociation processes.

Two auroral spectra in the wavelength region 3400–3900 Å are shown in Figure 7.4.3. One was acquired on the nightside of the auroral oval, the other on the dayside which was sunlit in the upper thermosphere. The vertical scale has been adjusted to give about equal amplitudes in the two spectra to the bands of the N_2 2P system for which the upper electronic state is excited by electron impact and the transition is electric dipole allowed ($C\,^3\Pi_u$–$B\,^3\Pi_g$). Emission from this band system dominates the night time auroral spectrum in the wavelength interval of Figure 7.4.3.

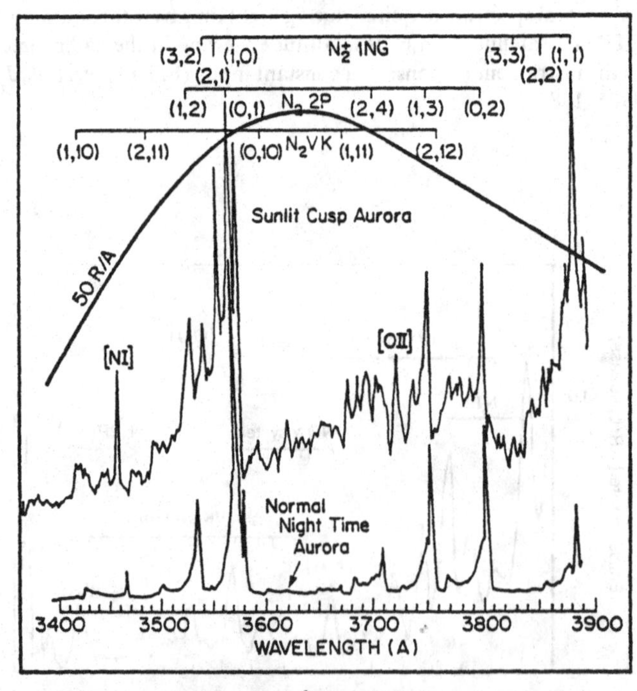

Fig. 7.4.3 Auroral spectra in the 3400–3900 Å wavelength region acquired with ground based scanning spectrophotometers. The thermosphere was sunlit in the top spectrum but in darkness in the bottom spectrum. Differences are discussed in the text. (C.S. Deehr, *et al.*, *J. Geophys. Res.*, **85**, 2185, 1980.)

The N_2 2P bands are also present in the dayside spectrum but additional emissions become prominent. There are several bands of the N_2^+ 1 NG system ($B\,^2\Sigma_u^+ - X\,^2\Sigma_g^+$) due in large measure to resonant scattering and fluorescence (Equation (7.3.43)) by N_2^+ ions in the $v' = 1$ vibrational level. Substantial enhancement of several bands of the N_2 VK system ($A\,^3\Sigma_u^+ - X\,^1\Sigma_g^+$), and of forbidden transitions in [OII] and [NI], respectively, show that the electron spectrum must be softer in the dayside auroral oval compared to the night-side region of the oval. The long radiative lifetimes of the upper states of the forbidden transitions that appear in the spectrum require excitation at high altitudes produced by low energy electrons to offset collisional deactivation at low altitudes. The [NI] line at 3466 Å and the [OII] line at 3726 Å are transitions in the ground configuration of the respective atom and ion (Appendix 3). Partial energy level diagrams for O, O^+, N and N^+ are presented in Figure 7.4.4 to show details that are not included in the overall diagrams in Appendix 3. Forbidden transitions are prominent in auroral spectra. They include not only lines in the near UV but also the bright 'auroral green line' at 5577 Å ($^1S_0 - {}^1D_2$) and the 'red doublet' at 6300 Å and 6364 Å ($^1D_2 - {}^3P_{2,1}$), both transitions of OI. The transition at 2972 Å ($^1S_0 - {}^3P_1$) has already been mentioned in connection with the dayglow spectrum; it is also a perennial feature of the auroral spectrum.

Prominence of the metastable ion lines is due to favourable branching in electron

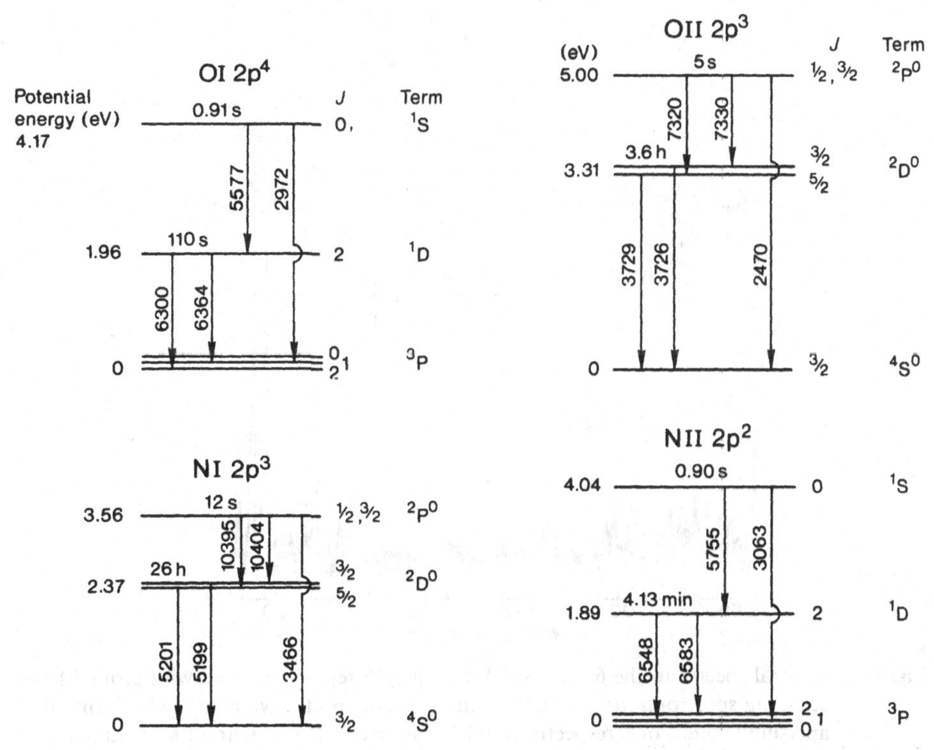

Fig. 7.4.4 Partial energy level diagrams for O, O^+, N and N^+ showing states in the ground configuration of the atoms and ions.

impact ionization of atomic O,

$$e_p + O \rightarrow e_p + O^+(^4S, ^2D, ^2P) + e_s \tag{7.4.1}$$

whereby about 40% of the product ions are produced in the excited ^2D state and about 20% in the excited ^2P state. In spite of the long radiative lifetime of the ^2P state, about 5 s, the 7320–7330 Å doublet appears in auroral spectra excited primarily by low energy electrons, as shown in Figure 7.4.5. However, in low altitude aurora produced by kilovolt electron precipitation, the [OII] lines are masked by strong band emission of the N_2 1P system. The radiative lifetime of the ^2D state is 3.6 h limiting the emitting region for the 3726–3728 Å doublet to very high altitudes only. Instead of radiating, a large fraction of the $O^+(^2D)$ ions produced in auroral ionization take part in ion chemical reactions, listed in Appendix 5, and eventually dissipate energy as heat.

The [NI] and [OI] lines are prominent for different reasons. The excitation potentials of the upper electronic states are relatively low, between 1.97 and 4.17 eV as

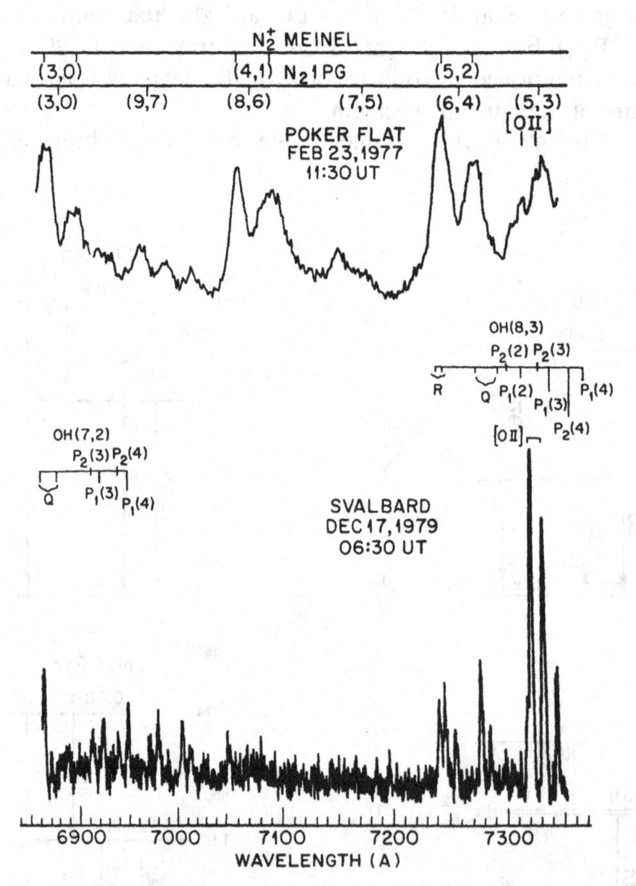

Fig. 7.4.5 Auroral spectra in the 6850–7360 Å wavelength region acquired with ground based scanning spectrophotometers. Top and bottom spectra were acquired during dark and sunlit conditions, respectively. (G.G. Sivjee and C.S. Deehr, in *Exploration of the Polar Upper Atmosphere*, Eds. C.S. Deehr and J.A. Holtet, Reidel Publ. Co., Dordrecht, 1981.)

shown in Figure 7.4.4. Several chemical reactions and ion–electron recombination processes are sufficiently exothermic to excite the metastable states involved in the processes (see Appendix 5). In addition, the intensity of low energy secondary electrons is large and excitation by electron impact contributes to the production rate of $O(^1S)$ and $O(^1D)$. On the other hand, impact excitation of the ground state of N is due primarily to dissociation of N_2 (see Section 4.3) because the concentration of N in the thermosphere is relatively low.

While electron impact contributes to excitation of the low lying states in O the dominant source is a chemical reaction. Several excitation sources for the $O(^1S_0)$ electronic state are possible. There is deactivation of the triplet metastable state of N_2,

$$N_2(A^3\Sigma_u^+) + O \rightarrow N_2(X^1\Sigma_g^+) + O(^1S_0) \tag{7.4.2}$$

Laboratory measurements of this energy transfer reaction support the importance of this mechanism which is currently (1987) favoured to be the principal source of the auroral green line. About 5–10% of the dissociative recombination reaction

$$O_2^+ + e \rightarrow 2O(^3P, {}^1D, {}^1S) \tag{7.4.3}$$

yields $O(^1S)$ atoms. Electron impact on O,

$$e_s + O(^3P) \rightarrow e + O(^1S) \tag{7.4.4}$$

has already been discussed. It has been suggested that the ion–atom interchange reactions,

$$N^+ + O_2 \rightarrow NO^+ + O(^1S) \tag{7.4.5}$$

$$O_2^+ + N \rightarrow NO^+ + O(^1S) \tag{7.4.6}$$

may be sources of $O(^1S)$ but current laboratory data do not provide any support. Finally, impact dissociation of O_2 is a small source of excited O,

$$e_p + O_2 \rightarrow e + O + O(^1S) \tag{7.4.7}$$

The relative importance of the various processes is investigated by model computations which attempt to reproduce measured altitude profiles of the 5577 Å volume emission rate. Collisional deactivation needs to be taken into account in evaluating the emission rate from a metastable state. $O(^1S)$ atoms are deactivated by collisions with $O(^3P)$ atoms and the effect becomes important below about 100 km.

The major source of $O(^1D)$ atoms is currently thought to be the atom interchange reaction

$$N(^2D) + O_2 \rightarrow NO + O(^1D) \tag{7.4.8}$$

for which a laboratory experiment has provided indirect support. Direct observation of the 6300 Å emission line in a confined volume is difficult because the long radiative lifetime of the excited $O(^1D)$ state, about 100 s, favours collisional deactivation. Other sources of $O(^1D)$ atoms are the dissociative recombination of O_2^+ (Reaction (7.4.3) with about half the product atoms formed in the ^1D state), electron impact excitation of O,

$$e_s + O(^3P) \rightarrow e + O(^1D), \tag{7.4.9}$$

dissociation of O_2, (Reaction (7.4.7)), and cascading from $O(^1S)$ and several other higher excitation states identified in the energy level diagram of Appendix 3,

$$O(^1S_0) \rightarrow O(^1D_2) + h\nu(5577\,\text{Å}) \tag{7.4.10}$$

If the electron temperature exceeds about 5000 K then the population of electrons with

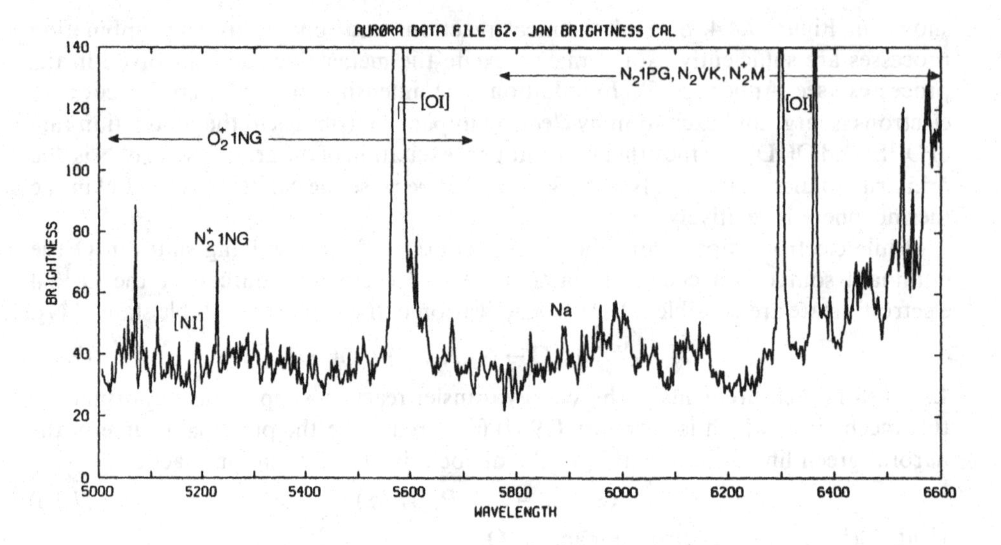

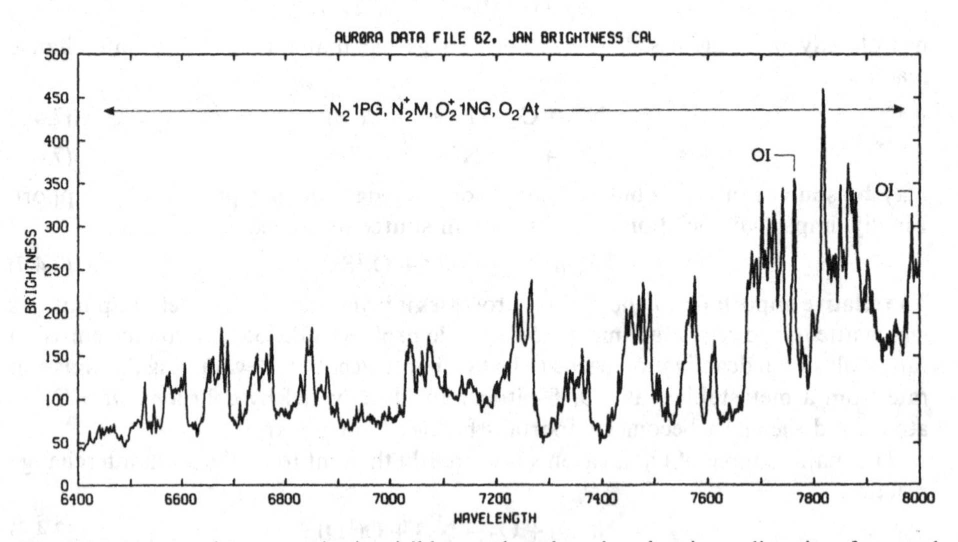

Fig. 7.4.6 (*a*)Auroral spectrum in the visible wavelength region showing a diversity of spectral emission features. (Courtesy of L.A. Broadfoot, 1987.)

energy in excess of 2 eV in the high energy tail of the Maxwellian distribution may be sufficient to contribute to electron impact excitation of O(^{1}D), (Reaction (7.4.9)). This mechanism is responsible for the midlatitude subvisual red arc discussed in Section 6.5.3.

Figure 7.4.6 covers the auroral spectrum in the visible wavelength region at low resolution. Many atomic lines and molecular bands are identified. The [NI] 5200 Å doublet, shown in Figure 7.4.7 at higher resolution, is a forbidden emission in the ground configuration of NI corresponding to the ($^2D_{\frac{5}{2},\frac{3}{2}} - {}^4S_{\frac{3}{2}}$) transition. The two

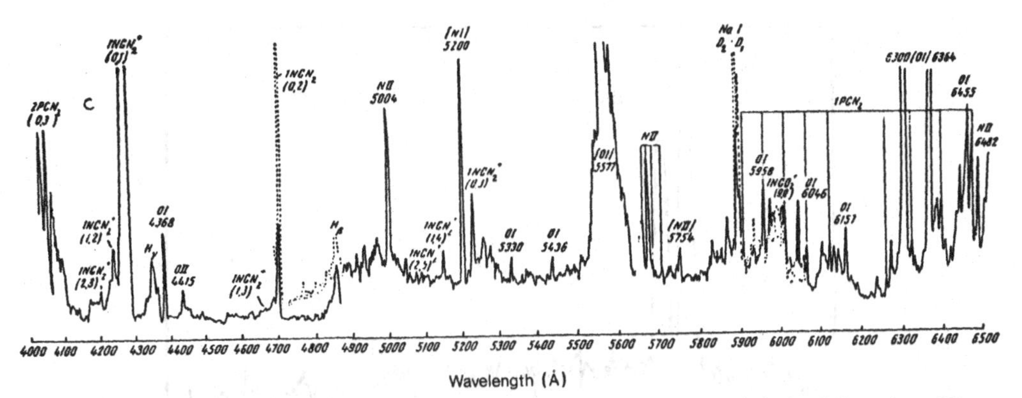

Fig. 7.4.6 (*b*) Composite spectrum of aurorae acquired at a geomagnetic latitude less than 55°. The predominance of atomic emissions over molecular bands is an indication of high altitude emission sources. (I.A. Yashchenko, Results of researches of the program of the I.G.Y., *Aurora and Airglow*, No. 11, Pub. House NAUKA, Moscow, 1965, p. 70.)

principal excitation sources are dissociative recombination,

$$NO^+ + e \rightarrow N(^2D) + O \tag{7.4.11}$$

$$N_2^+ + e \rightarrow N(^2D) + N(^2D) \tag{7.4.12}$$

and electron impact dissociation

$$e_p + N_2 \rightarrow e + N(^2D) + N(^4S) \tag{7.4.13}$$

A doublet emission of atomic sodium that appears in the spectrum near 5893 Å is an airglow feature originating at the 85–90 km level and is unrelated to the aurora but illustrates a statement made earlier that auroral emissions are superposed on the omnipresent airglow emissions, day or night.

Features identified as H_β and H_γ in Figure 7.4.6(b) are members of the Balmer series of H. These are excited as a result of proton precipitation into the atmosphere and are discussed in Section 3.4. The expected Doppler shift and broadening are not evident in this low resolution spectrum.

Auroral spectra in the wavelength region 6850–7360 Å acquired in the nightside and dayside segments of the oval are contrasted in Figure 7.4.5. The difference between the two spectra is even stronger than differences shown in Figure 7.4.3 in the near UV. Kilovolt electrons on the nightside excite molecules and molecular ions at low altitude and the band emissions are so intense as to mask all other features. The Meinel bands ($A\,^2\Pi_u$–$X\,^2\Sigma_g^+$) are produced by electron impact ionization excitation of N_2 while the $N_2$1 P bands ($B\,^3\Pi_g$–$A\,^3\Sigma_u^+$) are excited by electron impact as well as cascading from higher levels. The auroral emissions that appear in the dayside oval spectrum are the forbidden [OII] lines ($^2P_{\frac{3}{2},\frac{1}{2}}$–$^2D_{\frac{5}{2},\frac{3}{2}}$) at 7320–7330 Å. The molecular features are the OH bands of the airglow. The latter originate at about 80–90 km altitude (Section 7.2) while the atomic features are excited in the upper F-region. Auroral spectra are not yet available over a large wavelength interval at a resolution higher than 1 Å and with a wide dynamic range to record simultaneously both weak and strong emission features. Until better spectra become available we resort to the piecemeal presentation in Figures 7.4.1–7.4.3 and 7.4.5–7.4.8. The last figure covers the wave-

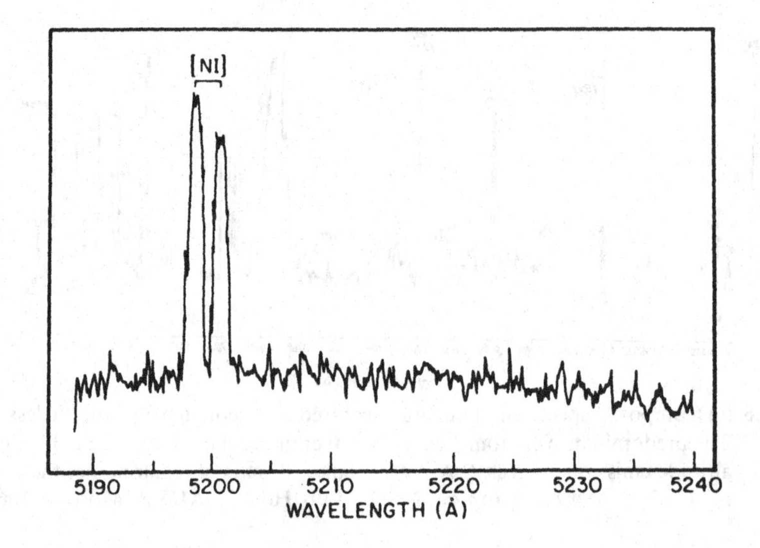

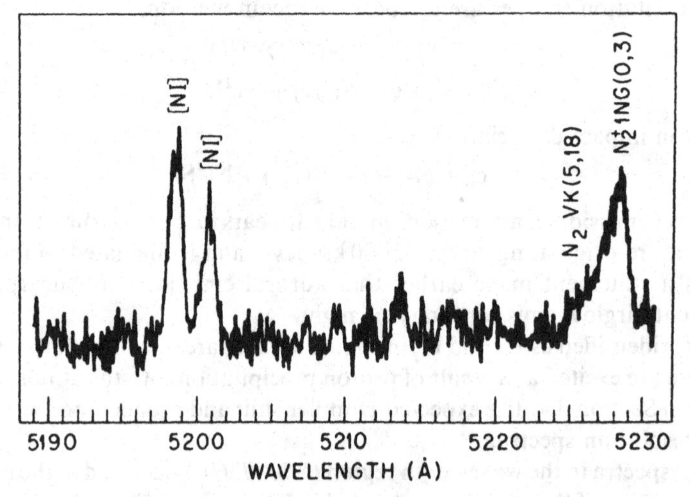

Fig. 7.4.7 Small wavelength range of an auroral spectrum at a resolution sufficiently high to resolve the two lines of the [NI] doublet at 5200 Å. Spectra acquired under sunlit (top) and dark (bottom) conditions in the thermosphere are shown. (G.G. Sivjee and C.S. Deehr, in *Exploration of the Polar Upper Atmosphere*, Eds C.S. Deehr and J.A. Holtet, Reidel (Publ. Co., Dordrecht, 1981.)

length region 7700–8600 Å which includes lines of OI and NI and bands of the N_2 1 P system, N_2^+ Meinel system, the forbidden O_2 Atmospheric bands ($b^1\Sigma_g^+ - X^3\Sigma_g^-$) and some OH airglow bands. At the low resolution used to acquire this spectrum, labelling of the O_2 bands and the OH bands refers to where these should occur rather than demonstrating their presence in the spectrum. The OI lines at 7774 Å, 7990 Å and 8446 Å are identifiable. These are excited by electron impact on O, with a small contribution to the 7774 Å feature from dissociative excitation of O_2. Considerable effort has been expended to obtain excitation and emission cross sections of auroral

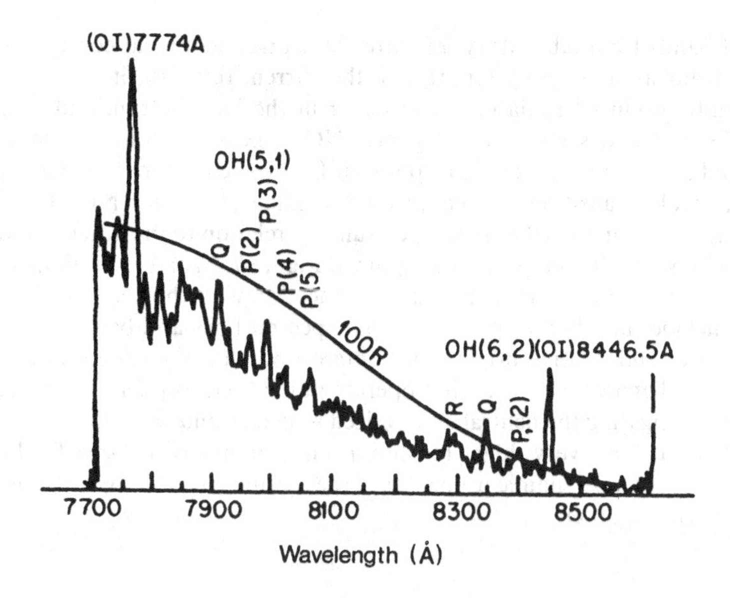

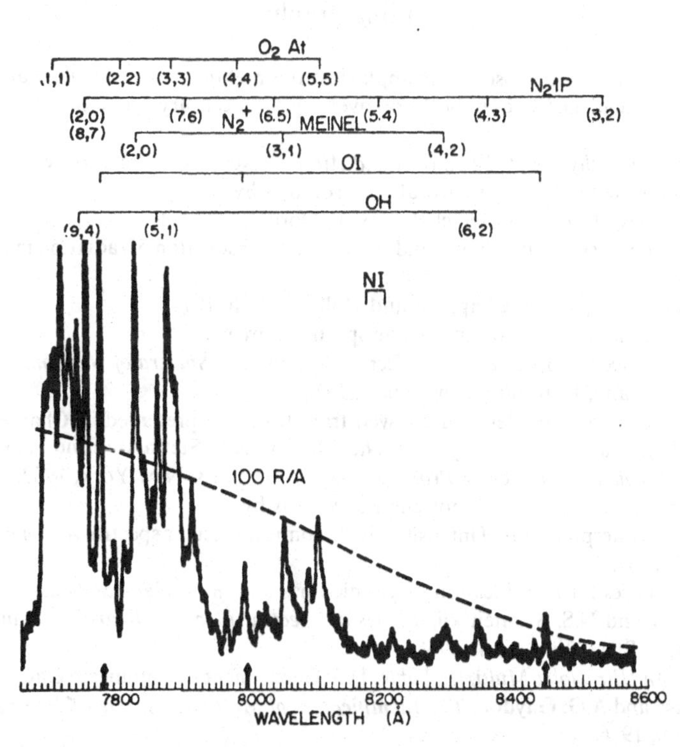

Fig. 7.4.8 Auroral spectra in the wavelength region 7700–8600 Å acquired with ground based scanning spectrophotometers. Spectra acquired under sunlit (top) and dark (bottom) conditions in the thermosphere are shown. (G.G. Sivjee and C.S. Deehr, in *Exploration of the Polar Upper Atmosphere*, Eds C.S. Deehr and J.A. Holtet, Reidel Publ. Co., Dordrecht, 1981.)

lines and bands from laboratory measurements and theoretical computations. New measurements are frequently reported in the current research literature.

Aurorally produced radiations also occur in the IR spectrum and these include bands of O_2 (the IR system), N_2, N_2^+, NO, NO^+ and several atomic lines. Electron impact and chemical reactions are responsible for these emissions in the thermosphere. The metastable transition in the ground configuration of [NI]($^2P_{\frac{3}{2},\frac{1}{2}}$–$^2D_{\frac{5}{2},\frac{3}{2}}$) at 10 400 Å is present in the IR auroral spectrum. Its relation to the other forbidden lines of [NI] is shown in the energy level diagram of Figure 7.4.4 and excitation of the upper state is by the same process discussed in connection with the [NI] 3466 Å emission.

We conclude this chapter by noting that spectral lines and bands, their absolute emission rates and relative brightness, contain a wealth of evidence bearing on the physical and chemical processes that operate in the thermosphere. They are of great value for investigating the neutral and ionized structure and the energy balance of the region. In addition, very high resolution measurements of specific lines yield parameters about the dynamic properties of the thermosphere, a topic that is taken up in the following chapter.

Bibliography

The basic concepts of spectroscopy as applied to the nightglow, the dayglow and the aurora have remained unchanged since publication over a quarter century ago of the excellent textbook by

J.W. Chamberlain, *Physics of the Aurora and Airglow*, Academic Press, New York, 1961.

and of the more recent book on auroral spectroscopy by

A. Vallance Jones, *Aurora*, D. Reidel Publ. Co., Dordrecht, 1974.

A lucid account of spectral line shapes and emission and absorption of radiation may be found in the book by

A.P. Thorne, *Spectrophysics*, Chapman and Hall, London, 1974.

An authoritative monograph on molecular spectroscopy is by

G. Herzberg, *Molecular Spectra and Molecular Structure*, I. *Spectra of Diatomic Molecules*, D. Van Nostrand, Princeton, New Jersey, 1950.

Useful summaries on forbidden and allowed transitions are presented in Chapter 1 (by R.H. Garstang) and Chapter 2 (by R.W. Nicholls and A.L. Stewart) of the book

D.R. Bates, *Atomic and Molecular Processes*, Academic Press, New York, 1962.

A well-written, concise review of molecular spectra is by

J.B. Tatum, The interpretation of intensities in diatomic molecular spectra, *Astroph. J. Suppl.* **14**, 21–56, 1967.

Standard references for identification of atomic lines and molecular bands are

A.R. Striganov and N.S. Sventeitskii, *Tables of Spectral Lines of Neutral and Ionized Atoms*, Plennum Press, New York, 1968.

C.E. Moore, *An Ultraviolet Multiplet Table*, U.S. Department of Commerce Circular 488, 1962.

R.W.B. Pearse and A.G. Gaydon, *The Identification of Molecular Spectra*, Chapman and Hall, London, 1976.

A derivation of the photometric unit of surface brightness, the rayleigh, is given by

J.W. Chamberlain, *Theory of Planetary Atmospheres* (Section 6.1.1), Academic Press, New York, 1978.

The photon emission rate factor, Equation (7.3.41), is also derived in the above reference and Table 6.1 lists g-values for several atomic and molecular transitions. The book also discusses the chemistry of odd O, odd H and odd N.

The mathematical procedure used to derive the volume emission rates of the airglow and the aurora from line-of-sight measurements is described by

S.C. Solomon, P.B. Hays and V.J. Abreu, Tomographic inversion of satellite photometry, *Appl. Optics*, **19**, 3409, 1984; **23**, 4134, 1985.

Resonant scattering of atomic H and O lines has been extensively discussed in the literature. The theory is developed by

S. Chandrasekhar, *Radiative Transfer*, Dover Publ. Inc., New York, 1952.

and application to planetary atmospheres is discussed by

D.E. Anderson and C.W. Hord, Multidimensional radiative transfer: Application to planetary coronae, *Planet. Space Sci.*, **25**, 563, 1977.

The theory is applied to line emissions in the terrestrial atmosphere by

D.E. Anderson, R.R. Meier, R.R. Hodges and B.A. Tinsley, H-Balmer alpha intensity distributions and line profiles from multiple scattering theory using realistic geocoronal models, *J. Geophys. Res.*, **92**, 7619, 1987.

and by

R.R. Meier and J.S. Lee, An analysis of the OI 1304 Å dayglow using a Monte Carlo resonant scattering model with partial frequency redistribution, *Planet. Space Sci.*, **30**, 439, 1982.

Problems 35–41

Problem 35 Using the photoionization cross section for atomic O given in Appendix 2, compute the radiative recombination rate coefficient for electron capture into the ground state of the atom. Compare the rate of loss of O^+ ions by this process with ion–atom interchange with N_2 and O_2 at 250 km. Assume an electron temperature of 1800 K.

Problem 36 At what value of electron temperature are the radiative recombination and dielectronic recombination rate coefficients for O^+ ions equal? Use a value of 15 eV for the energy E_{di} of the resonant state in dielectronic recombination.

Problem 37 Using the 'typical' night time concentrations given in Appendix 1 of various species, compute the height profile of the photon emission rate of the Na-D lines. Assume that the fraction of Na atoms produced in the excited state is determined by the relative statistical weights of various levels. What would be the column emission rate measured by a zenith viewing photometer?

Problem 38 Verify the equation for the Voigt profile, Equation (7.3.26), starting with the intensity distributions for the Gaussian and Lorentzian profiles.

Problem 39 Evaluate the optical depth of the atmosphere to the OI ($2p^4\,{}^3P_2$–$3d\,{}^3D_1^0$) 1027 Å line at 200, 150 and 120 km. Use typical values for the atmospheric density, composition and temperature. The absorption oscillator strength of the transition is 0.01. Assume that the instrumental resolution is 0.1 Å and that the response may be represented by a Lorentzian profile.

Problem 40 Using the energy level diagrams for various atomic and ionic species, identify the transitions corresponding to the emission features that appear in the dayglow spectrum. Group these by multiplets and attempt to rationalize the relative brightness of various lines. The count rates given in Figure 7.3.1 are not equivalent to the relative brightness because the detector sensitivity changes as a function of wavelength.

Problem 41 Discuss the differences between the day and night auroral spectra shown in Figures 7.4.1–7.4.8 and the dayglow spectra shown in Figure 7.3.1 in terms of the processes responsible for individual emissions.

8

Dynamics of the thermosphere and ionosphere

8.1 Introduction

Discussion of the dynamic behaviour of the thermosphere and ionosphere has been left for the last chapter for several reasons. The phenomenology is complex. The global scale wind pattern shows little geographic symmetry, being influenced by the seasonal variability over the globe of the input of solar energy and by geomagnetically controlled ion drift. Since momentum and energy sources vary non-uniformly with altitude, the dynamic response likewise has height-dependent structure. Superimposed on the quasi-steady winds and drifts there are waves that may be generated within the thermosphere or propagate from below. The perturbations may be systematic, such as tidal effects, or impulsive, such as auroral storms. The goal is to understand the large scale or global dynamics as well as the small scale or localized perturbations. The dynamic behaviour of the thermosphere and ionosphere cannot be fully understood without knowledge of the structure and energetics of the region. These properties, presented in the preceding chapters, are therefore incorporated as required.

The system is three-dimensional and time-dependent and there exists a wide range of scale lengths and time constants. Predicting the temporal and spatial morphology of the observable parameters that characterize the thermosphere and ionosphere provides a continuing challenge. As this chapter is being written, a fully self-consistent analysis of the coupled ionosphere and thermosphere is a topic that is at the cutting edge of upper atmosphere research. An effort of equal importance is the acquisition of measurements of the parameters that quantify the dynamics and thermodynamics of the system. Without such observations it would not be possible to identify the relative importance of various physical processes included in the models, any incorrect assumptions or unidentified contributions.

8.2 Observations

Many observational techniques are used to investigate the dynamic and thermodynamic properties of the ionosphere and thermosphere. Each technique has a particular advantage over the others; all have serious limitations. Some methods yield wind, drift, or temperature fields directly; others give related measurements that, combined with certain assumptions, lead to the desired parameters. Various measurements complement each other, yielding bits of information used to build up the global dynamic picture. The overall data base that is currently available is patchy and sparse. Parameters obtainable by each method are listed in this section but the reader is directed to the bibliography of this chapter for extended descriptions of the various experimental techniques.

Neutral and ion winds can be measured by recording optically the motion of luminous trails or clouds that are produced by chemicals released from rockets in the thermosphere during twilight. The release of a Ba compound results in ionization of the Ba atoms by solar UV radiation, followed by resonant scattering by the Ba^+ ions of visible radiation to produce a luminous signature. Motion of the luminous cloud represents the ion drift velocity vector. The Ba compound in these chemical releases contains a small amount of Sr that is not ionized, but the atoms resonantly scatter solar radiation at 4607 Å and their motion gives the velocity vector of the prevailing neutral wind. Release of the chemical trimethyl aluminium (TMA) also produces a luminous trail and this method has proven particularly effective in the E-region and lower F-region of the thermosphere.

An example of neutral wind data derived from the drift of a TMA trail is shown in Figure 8.2.1. Altitude profiles of the meridional and zonal wind obtained at a midlatitude station are shown for consecutive evening and morning twilights. It is assumed that the vertical neutral wind is small by comparison with the horizontal wind.

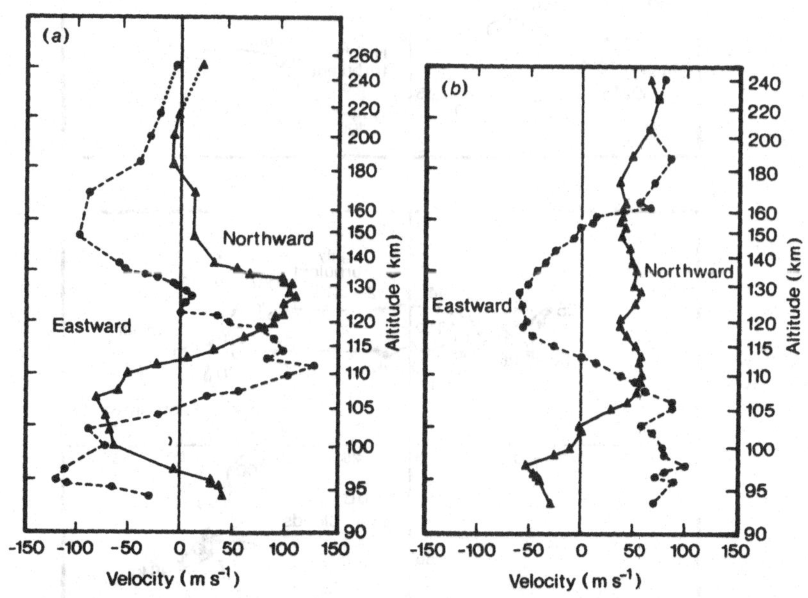

Fig. 8.2.1 Altitude profiles of the meridional and zonal winds acquired during evening and morning twilights. (*a*) Wind profile 0446 CST, Woomera, 16 October, 1969, (*b*) wind profile 1915 CST, Woomera, 17 October, 1969. (K.H. Lloyd, *et al.*, *Planet. Space Sci.*, **20**, 761, 1972.)

The validity of this assumption is borne out by experimental results of another observing technique. Averaged over time and space, vertical stability, or hydrostatic equilibrium, prevails in the neutral atmosphere (Equation (5.1.5)), but localized, short duration, vertical velocities do occur, particularly at high latitudes. Large altitude gradients in the zonal and meridional wind vectors are evident in the example shown in Figure 8.2.1 except for the meridional wind above 105 km at evening twilight. A different presentation of the same data is the hodograph shown in Figure 8.2.2 which combines the zonal and meridional components into a single vector. Circles on the curve are loci of the velocity vectors at the marked altitude. The hodographic presentation brings out significant differences between morning and evening neutral wind patterns.

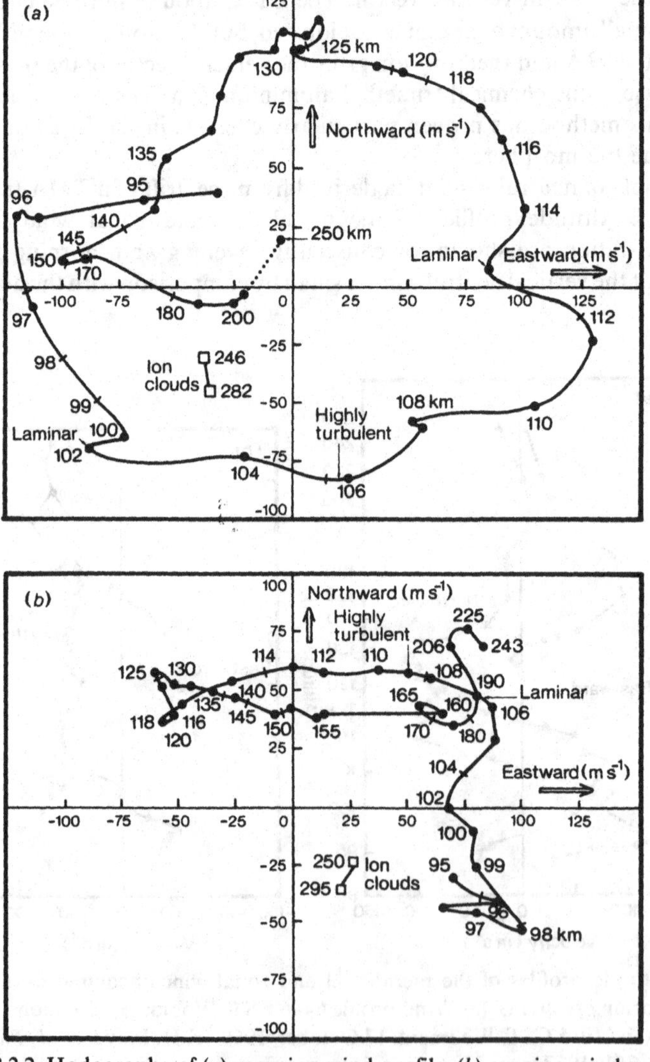

Fig. 8.2.2 Hodographs of (*a*) morning wind profile, (*b*) evening wind profile.

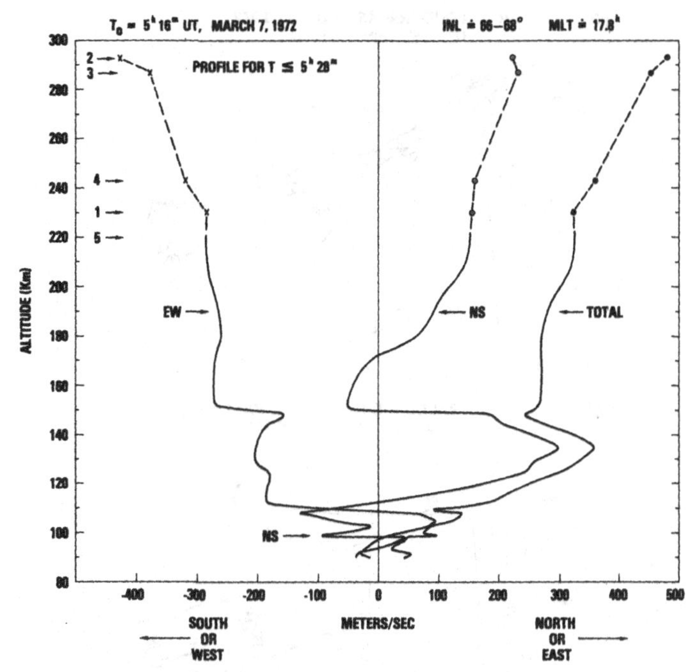

Fig. 8.2.3 Evening wind velocity profile in geographic coordinates. (J.P. Heppner and M.L. Miller, *J. Geophys. Res.*, **87**, 1633–47, 1982.)

Results of a combined TMA and Ba chemical release at Poker Flat, Alaska (65° N, 149° W) are shown in Figure 8.2.3. Altitude profiles of the meridional, zonal and total horizontal neutral wind were obtained from the TMA trail between 95 and 220 km and from the neutral Sr component of four successive Ba releases identified on the figure. The E-region is characterized by large wind shear while the variation of wind speed with altitude is less in the F-region. Figure 8.2.4 shows the trajectory of the Sr cloud from each release on a geographic and geomagnetic latitude–longitude grid projected to an arbitrary altitude (110 km). Time is marked in minutes along the trajectories showing that the westward drift of the clouds can be followed for many minutes after release. The drift of the luminous Ba^+ ion cloud is similarly displayed in the figure but it can no longer be assumed that the vertical velocity component is small. The feature of note is that, during the time interval when both the ion and neutral drifts could be followed, the horizontal components of the velocity vectors were in approximately the same direction but the ion speed exceeded the neutral speed. The significance of these observations in terms of physical mechanisms is discussed in Section 8.3. The examples of chemical release experiments illustrate the principal contribution of this observational technique: obtaining simultaneous direct measurements of ion drift and horizontal neutral wind and their variation with altitude over a period of tens of minutes. The few available rocket launch facilities limit the geographic coverage that may be obtained with chemical release experiments while the task of organizing rocket campaigns precludes frequent launchings. Moreover, chemical release experiments can be carried out only in twilight when the thermosphere is sunlit and the lower atmosphere is in darkness, as required for recording faint luminous trails.

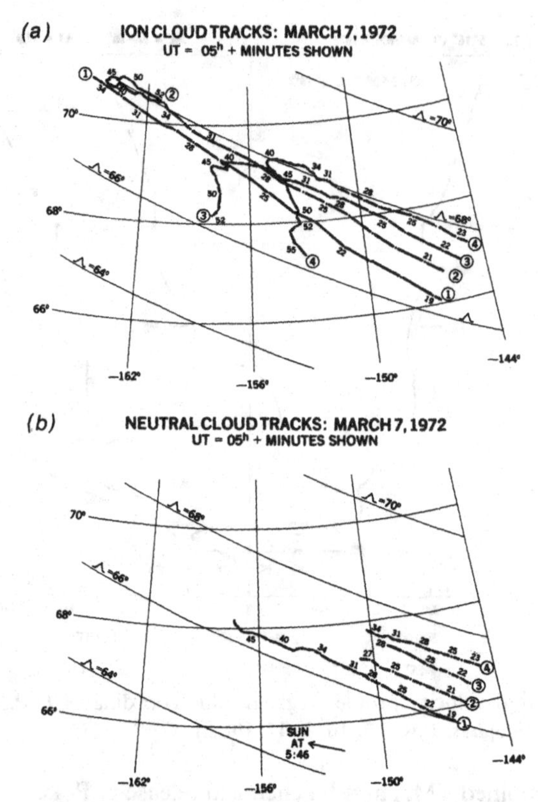

Fig. 8.2.4 (*a*) Motions of four Ba^+ ion clouds projected to the 110 km level, and (*b*) the simultaneous motions of the associated Sr neutral clouds. (J.P. Heppner and M.L. Miller, *J. Geophys. Res.*, **87**, 1633, 1982.)

In our discussion of airglow and auroral emission line profiles in Section 7.3 we noted that line broadening is due to natural or radiative damping, Γ_R, pressure effects, Γ_{coll}, and temperature effects, or Doppler broadening. Pressure broadening is negligibly small in the thermosphere and radiative damping is proportional to the spontaneous emission coefficient, A_{mn}. A forbidden transition has a small A value and natural broadening of the emission line is correspondingly small. The line profile is therefore determined solely by temperature effects; the line profile is Gaussian, and the emission temperature is equal to the kinetic temperature of the ambient gas, provided the excited atom has undergone a sufficient number of collisions to become thermalized before radiating. Highly metastable species are usually thermalized before radiating. Measurements of emission line profiles of forbidden transitions have been carried out for many years using high resolution Fabry–Perot and Michelson interferometers. While the shape of an emission line yields a temperature, any shift of the centre frequency from a reference value of an emitting atom at rest represents the velocity of the emitter along the line of sight of the measurement, the Doppler shift. High resolution optical instruments have been used to infer neutral winds and ion drifts in the thermosphere. The principal advantage of ground based optical observations is the capability of making continuous measurements from any location on the globe. However, most

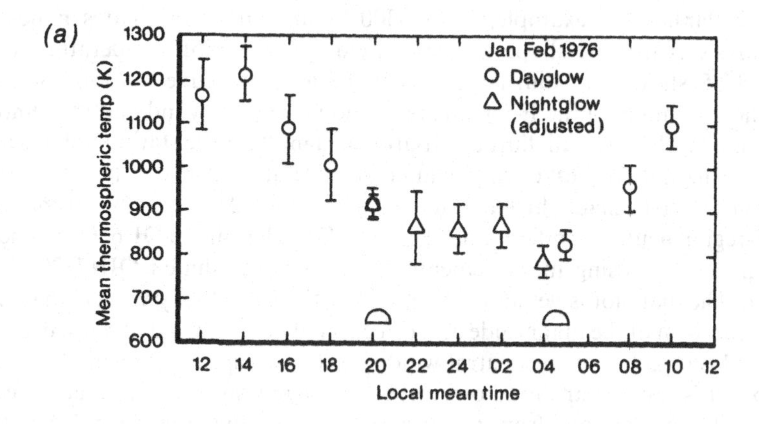

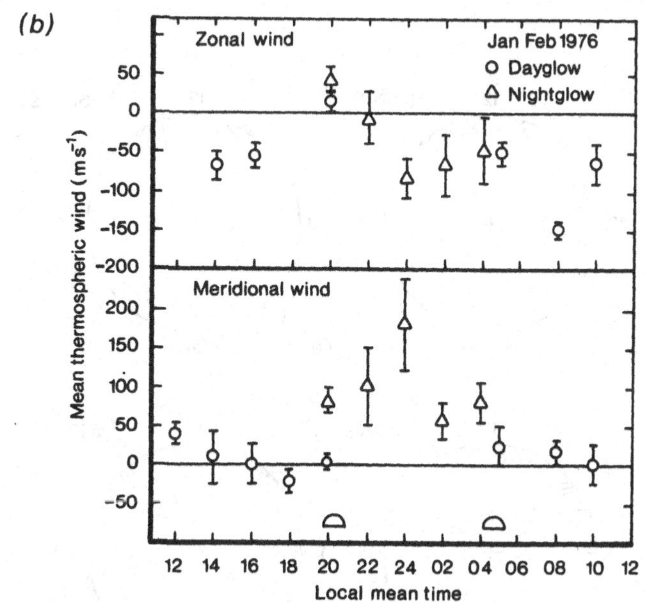

Fig. 8.2.5 (*a*) Mean diurnal variation of thermospheric temperature. Circles represent dayglow data; triangles represent 1972–3 nightglow data, adjusted for K_p and 10.7 cm solar flux. Half circles indicate times of twilight. (*b*) Mean diurnal variations of thermospheric zonal wind reckoned positive eastward and meridional wind reckoned positive northward. Circles represent dayglow data; triangles represent 1972–3 nightglow data. (T.D. Cocks, D.F. Creighton and F. Jacka, *J. Atmosph. Terrest. Phys.*, **42**, 499–511, 1980.)

instruments currently operate only during periods of darkness. Winds and temperatures measured optically are averages over the height interval of the emitting region of any particular line. For example, the OI 6300 Å emission line originates in the F-region, and analysis of the line shape and shift yields the F-region temperature and wind. Figure 8.2.5 shows the diurnal variation at a midlatitude station for (southern hemisphere) summer of the F-region zonal and meridional wind and the temperature. This data set is the only full diurnal variation at a midlatitude station published to date (1987). At high latitudes during winter solstice it is sufficiently dark to permit measurements to be made throughout the day. Figure 8.2.6 shows the diurnal variation of the F-region neutral wind obtained from the Doppler shifted OI (6300 Å) line and the F-region ion drift using measurements of the Doppler shifted OII (7320 Å) ion line emission. The dial plot is yet another representation of velocity vectors that allows an easy comparison of the magnitude and direction of neutral wind and ion drift. The ion drift speed is larger than the neutral wind and the two species flow in about the same direction. It is not a snapshot, however, since the measurements are acquired over a 24 h period. Observations from a station are usually made in several directions. At auroral sites geomagnetic coordinates are adopted. Measurements obtained in each

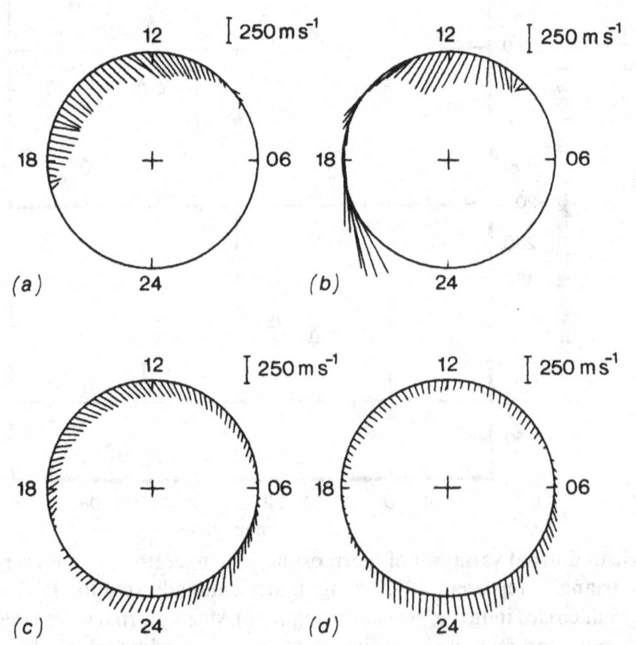

Fig. 8.2.6 Velocity vectors plotted at 15 min intervals on a circle at 75° geomagnetic latitude which traces the position of the observing station on Svalbard. The cross at the centre of the circle marks magnetic north and magnetic local time is indicated on the circumference. The magnitude and direction of the velocities are shown: (*a*) and (*b*) ion drift; (*c*) and (*d*) neutral wind. The observations shown in (*a*) and (*c*) were obtained when the interplanetary magnetic field B_y component was positive and in (*b*) and (*d*) the IMF B_y component was negative. (F.G. McCormack and R.W. Smith, *Geophys. Res. Letters.*, **11**, 935–8, 1984.)

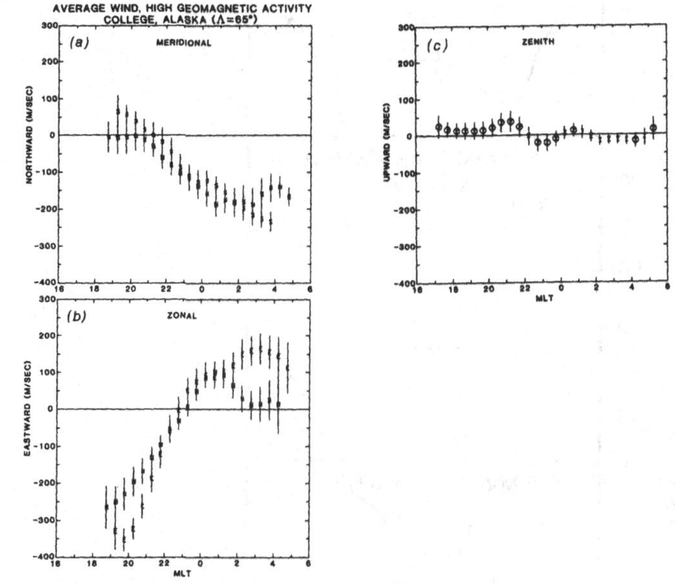

Fig. 8.2.7 Average thermospheric wind components during periods of high geomagnetic activity: (*a*) meridional components, (*b*) zonal components and (*c*) vertical component. The meridional components measured viewing N and S are shown separately and the zonal components measured toward the E and W are also plotted separately. The abscissa is magnetic local time. (R.J. Sica, *et al.*, *J. Geophys. Res.*, **91**, 3231, 1986.)

direction, N, S, E, W and zenith are shown separately in Figure 8.2.7 for observations made at an auroral station. The data are averages of several nights during periods of high geomagnetic activity. At times, different wind speeds are measured of both the meridional and the zonal components viewing N and S or E and W, respectively. These observations show that, at times, gradients occur in the wind that could be interpreted as regions of convergence or divergence. Vertical winds appear during periods of high geomagnetic activity in response to large energy input rates.

During geomagnetic quiet periods at midlatitudes there is an annual variation of the prevailing zonal and meridional thermospheric wind, illustrated in Figure 8.2.8. A 6 yr long experiment measuring the night time thermospheric temperature at a midlatitude station reveals not only an annual variation, maximum in summer and minimum in winter, but also a systematic change with the solar activity cycle, shown in Figure 8.2.9. These conclusions are based on about 15 000 individual measurements. An empirical formula

$$T(\mathrm{K}) = 195(\pm 3)F_{-1}^{0.34(\pm 0.005)} + 3.3(\pm 0.15)Ap + 67(\pm 4.0)\delta_{\mathrm{N}} + 30(\pm 4) \times \cos\{4\pi[d - 110(\pm 3)]/365.25\} \quad (8.2.1)$$

that represents the data well includes the variables that would be expected to influence the temperature: F_{-1} is the solar radio noise flux (in units of 10^4 Jansky, 1Jansky = $1\ \mathrm{W\,m^{-2}\,Hz^{-1}}$) for the day previous to the optical measurement of temperature, Ap is the daily index of magnetic activity, δ_{N} is the normalized solar declination $\delta_{\mathrm{N}} = \delta_{\theta}/23.44$,

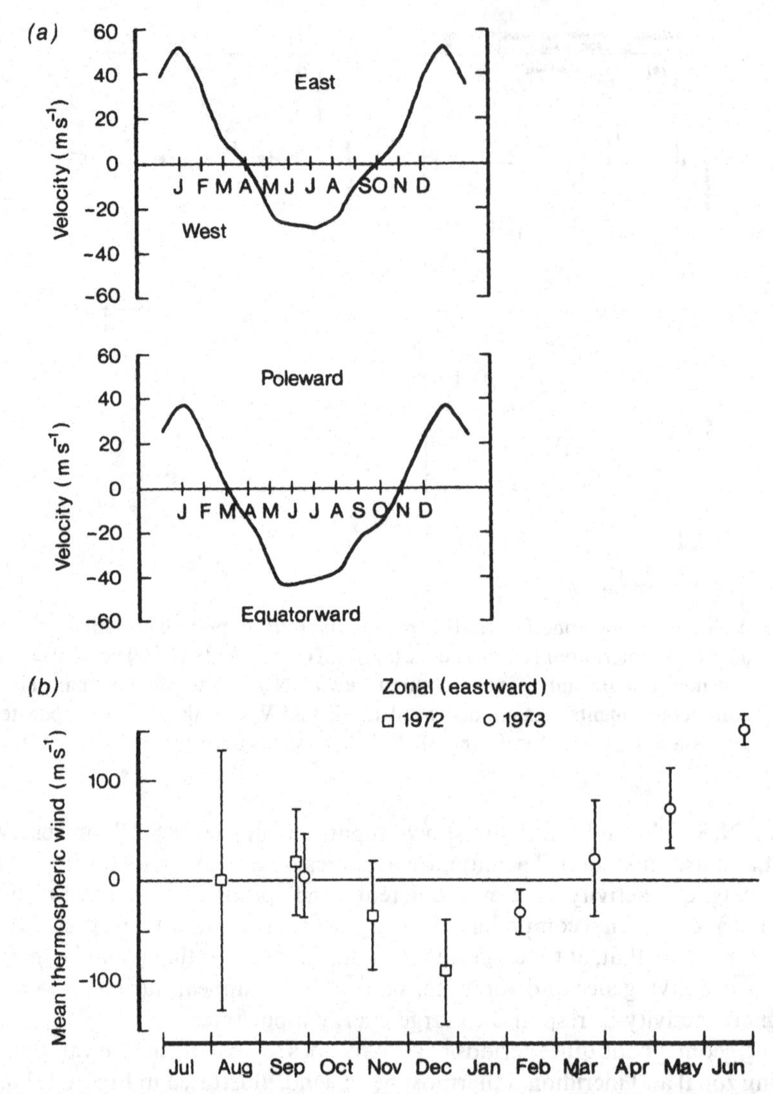

Fig. 8.2.8 (*a*) Annual variation of the prevailing zonal and meridional wind in the F-region (~ 250 km) during magnetically quiet conditions at Fritz Peak, Colorado (40° N Latitude, 105° W Longitude). (T.J. Fuller-Rowell and D. Rees, *J. Atmosph. Sci.*, **37**, 2545, 1980.) (*b*) Seasonal variation of the zonal wind in the southern hemisphere in the interval 21–03 h LMT on geomagnetically quiet nights, $AP \leqslant 10$. (F. Jacka, A.R.D. Bower and P.A. Wilksch, *J. Atmos. Terr. Phys.*, **41**, 397–407, 1979.)

where δ_θ is the declination in degrees, and d is the day number of the year with January $1 = 1$. Applying the four input parameters to the data set yields empirical values of temperature that are compared with the experimental results in Figure 8.2.10. The applicability of Equation (8.2.1) to F-region temperatures at different locations has not been tested. Perturbations attendant upon large geomagnetic storms are probably not predictable by a universal empirical formula. Indeed, effects of large energy input rates

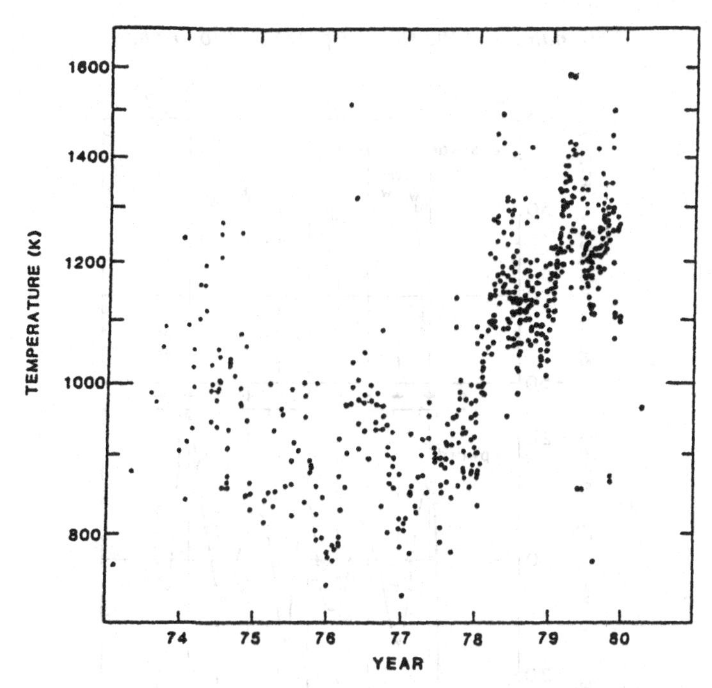

Fig. 8.2.9 Nightly mean thermospheric temperatures measured at Fritz Peak Observatory, Colorado (40° N, 105° W). (G. Hernandez, *J. Geophys. Res.*, **87**, 1623, 1982.)

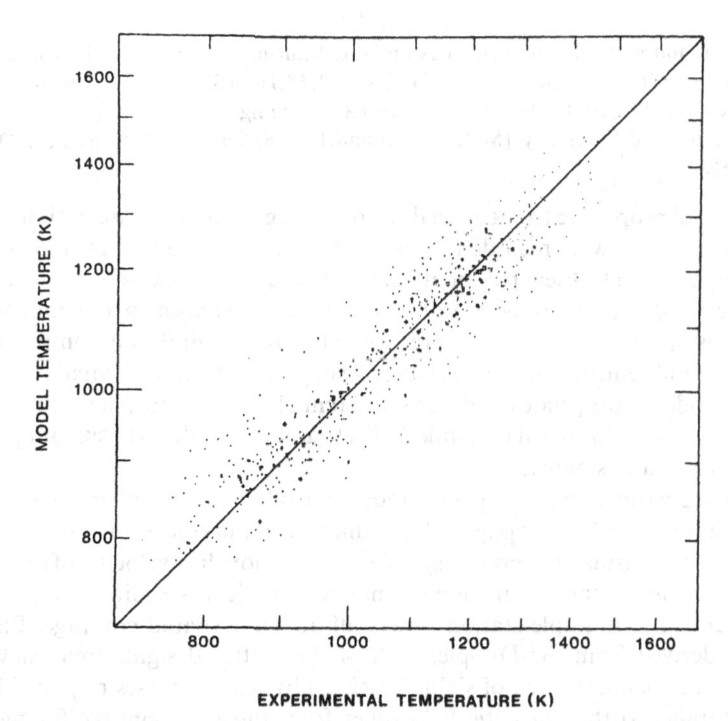

Fig. 8.2.10 Fit of the experimental data to four parameters: solar radio flux, geomagnetic activity, annual variation of the solar declination and a semiannual variation. The correlation coefficient is 0.91. (G. Hernandez, *J. Geophys. Res.*, **87**, 1623, 1982.)

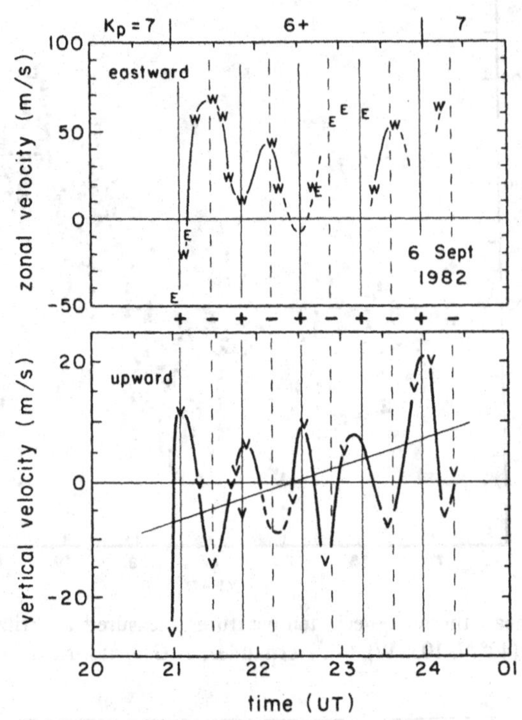

Fig. 8.2.11 Zonal and vertical wind velocities measured during the geomagnetic storm ($\sum K_p = 62+$) of 6 September, 1982 at Natal, Brazil, 6° S and 35° W. The 3 h K_p indices are given at the top of the figure. The error bars, ranging from ± 5 to $\pm 15\,\mathrm{m\,s^{-1}}$, have been omitted for clarity. (M.A. Biondi and D.P. Sipler, *Planet. Space Sci.*, **33**, 817, 1985.)

in the auroral oval propagate to equatorial regions. Large amplitude oscillations in the zonal and vertical wind were recorded during one storm at a 6° S latitude location, as shown in Figure 8.2.11. Elevated neutral temperatures of 300–400 K were also recorded. Localized perturbations are observed even in the absence of geomagnetic storms. Regions of convergence and divergence of horizontal air flow accompanied by a vertical wind and temperature perturbations suggest that vertical circulation cells may be set up under appropriate conditions. Current global thermospheric circulation models can account for localized dynamical effects as small as the grid size adopted in the model, but not any smaller.

Incoherent scattering of high frequency radio waves by the ionospheric plasma is a technique that has yielded geophysical parameters about the structure (electron density), energetics (plasma temperatures) and dynamics (ion drift velocity) of the upper atmosphere. Gating of the radar signals makes altitude discrimination possible. Steerable antennae or multiple station arrays afford some spatial coverage. Plasma velocities are derived from the Doppler shift of the scattered signal frequency, the measured quantity being the line of sight velocity. Physical processes responsible for ion motion parallel to the magnetic field differ from those responsible for motion perpendicular to the magnetic field as discussed in Sections 5.2 and 5.3. The measurements are therefore resolved into the orthogonal components $V_\parallel$ and $V_\perp$.

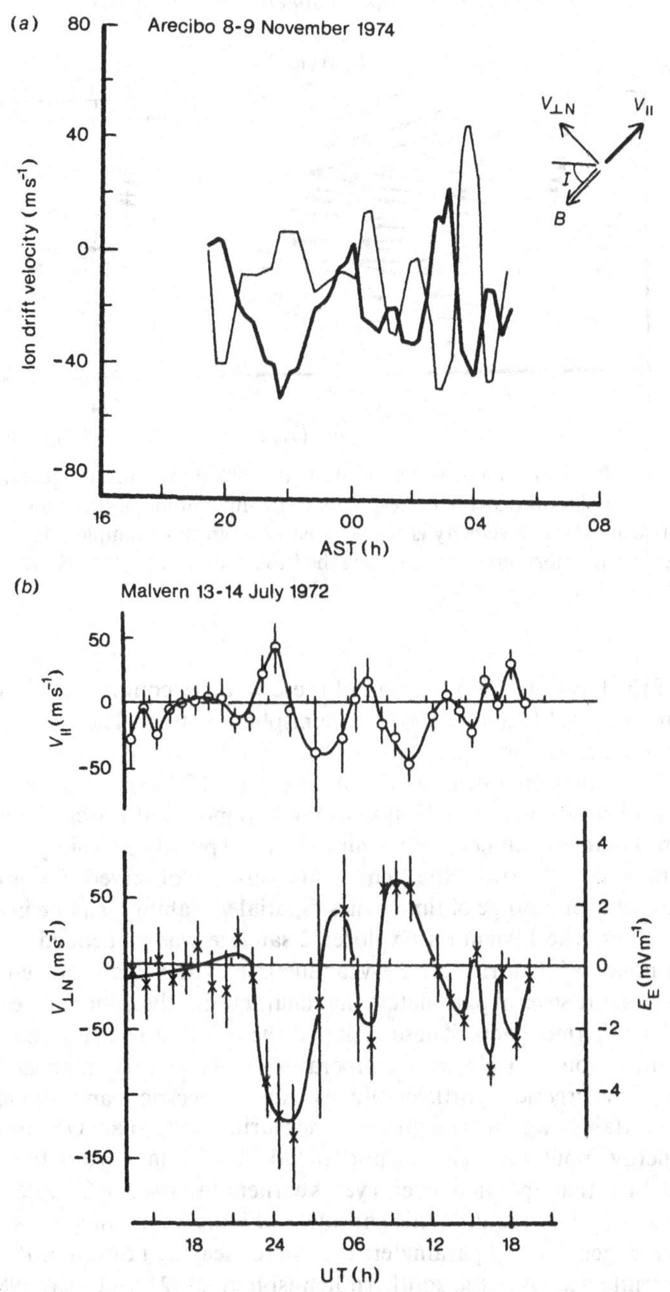

Fig. 8.2.12 (*a*) Plasma velocities in the magnetic meridian plane, measured at Arecibo; the heavy curve is $V_{\parallel}$, the light curve $V_{\perp N}$, defined as in the sketch (north to the left, dip angle $I = 50°$). A marked anticorrelation between $V_{\parallel}$ and $V_{\perp N}$ is seen. (H. Rishbeth, S. Ganguly and J.C.G. Walker, *J. Atmos. Terr. Phys.*, **40**, 767, 1978.) (*b*) Plasma velocities in the magnetic meridian plane, measured at Malvern; $V_{\parallel}$ above, $V_{\perp N}$ below. A scale of eastward electric field E_E corresponding to $V_{\perp N}$ is shown at the right. There is anticorrelation between $V_{\parallel}$ and $V_{\perp N}$. Sign convention as in (*a*). (G.N. Taylor, *J. Atmos. Terr. Phys.*, **36**, 267, 1974.)

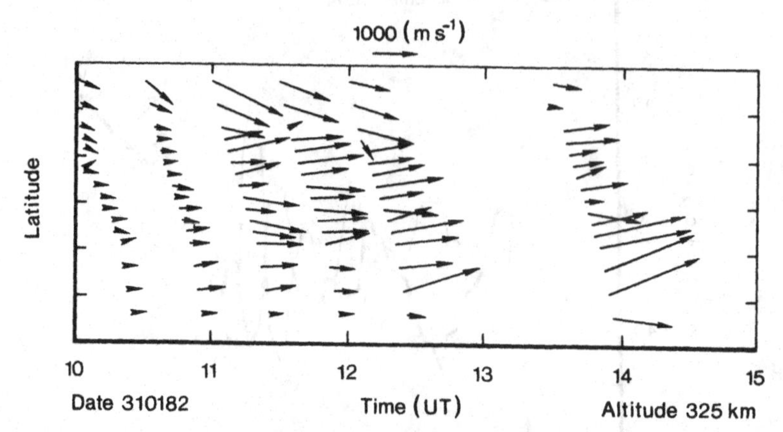

Fig. 8.2.13 Ion drift velocity vectors at an altitude of 325 km vs. time and geographic latitude. The length of the arrow corresponds to the magnitude of the velocity (see scale for 1000 m s^{-1}). The velocity is mainly eastwards in this example. Measurements were acquired in northern Norway using the EISCAT radar facility. (K. Schlegel, *J. Atmos. Terr. Phys.*, **46**, 517, 1984.)

Examples of the local diurnal variation of these velocity components in the F-region are shown in Figure 8.2.12 at two northern hemisphere stations. The diurnal variations of the F-region horizontal ion velocity over a 10° latitude range is shown in Figure 8.2.13. A contour display of the parallel ion velocity in Figure 8.2.14 shows the magnitude of $V_{\parallel}$ as a function of local time and altitude in the E-region and lower F-region.

The term dynamics implies both temporal and spatial variability. The examples presented thus far illustrate the temporal changes observed by ground based instrumentation over a range of time scales. Spatial variability may be investigated by orbiting satellites. The Dynamics Explorer 2 satellite was embedded in the thermosphere in a polar orbit during its $2\frac{1}{2}$-year lifetime. The vehicle carried instruments designed to measure several parameters that characterize the dynamics, energetics and structure of the thermosphere. These included the ion drift velocity, the neutral wind velocity, neutral, ion and electron temperatures, ion density, neutral composition and density, energetic particle fluxes and electric and magnetic field vectors. Regrettably, an instrument for measuring the solar UV irradiance, the principal energy input source, was not included. UV data from the Atmosphere Explorer satellites that operated several years earlier (discussed in Chapter 2) have to be adopted for analyses that require this parameter. Figure 8.2.15 illustrates relationships between several geophysical parameters that were measured during a 19-min interval when the satellite was over the southern hemisphere on 21 October, 1981. The high speed of the satellite, several kilometres per second, limits the spatial resolution along the track to the time interval required for obtaining a data point; this is different for the various instruments. *In situ* measurements on board satellites do not yield any information about the temporal evolution of geophysical parameters at a point in space. Satellite measurements and ground based observations are therefore complementary in providing a limited picture of thermospheric and ionospheric dynamics. Figure 8.2.15 illustrates how the ion and neutral velocity vectors influence the ion

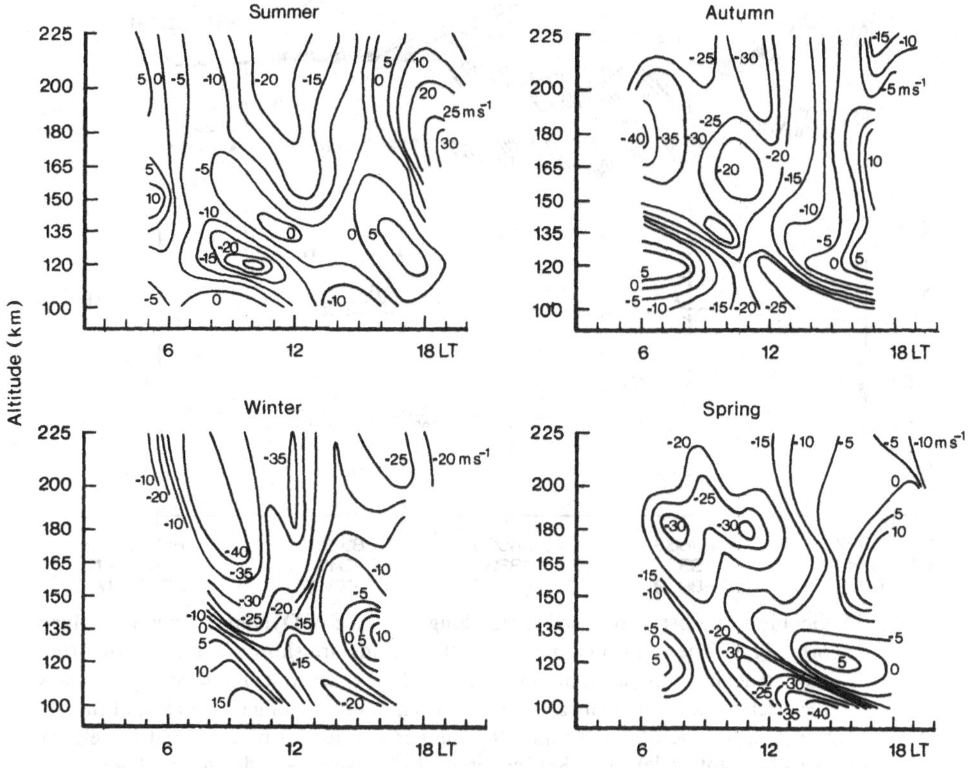

Fig. 8.2.14 Mean seasonal (1971–2 data) isocontours of parallel ion velocity deduced from St. Santin-Nançay measurements between 100 and 225 km. Numbers in the diagram indicate the value of $V_{\parallel}$ (in m s^{-1} positive upward) projected on the vertical direction. (P. Amayene, *Radio Science*, **9**, 281, 1974.)

temperature. According to Equation (6.3.1), frictional heating is a function of $(\mathbf{v}_i - \mathbf{v}_j)^2$ where $\mathbf{v}_i$ and $\mathbf{v}_j$ are the velocity vectors of two constituents. If these refer to ions and neutrals the heating rate will be large at locations where the vectorial difference in velocities is large, resulting in enhanced ion and neutral temperatures. This effect is clearly shown in Figure 8.2.15 at about UT 85160s. The relationship between the ion and neutral velocities by itself shows the coupled nature of dynamic processes in the thermosphere.

8.3 Mathematical description of thermospheric and ionospheric dynamics

Before writing down the equations that describe mathematically the dynamic behaviour of thermospheric and ionospheric geophysical parameters it is useful to discuss individually various forcing influences and the manner in which the atmosphere responds and adjusts to them. From a global perspective the unequal heating between day and night and between the winter and summer hemispheres sets up pressure gradients that force the gas to respond by moving horizontally from hot to cold regions. While the diurnal and seasonal cycles are due to the solar UV irradiance, an

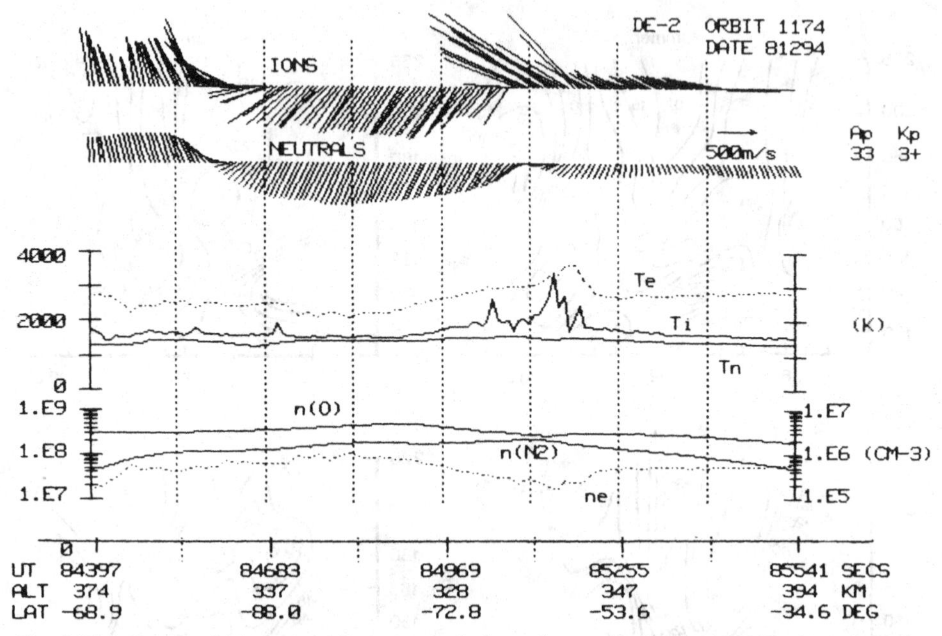

Fig. 8.2.15 Geophysical observables measured along the track of Dynamics Explorer 2 during orbit 1174. The ion drifts and the neutral winds are shown in the top two traces plotted against time, altitude, and latitude of the spacecraft. The second panel shows the electron, ion, and neutral temperatures measured along the track, and the third panel shows the atomic O and N_2 number densities (left scale) and the electron density (right scale). (T.L. Killeen, *et al.*, *J. Geophys. Res.*, **89**, 7495, 1984.)

additional substantial heat source at high geomagnetic latitudes is frictional heating of ions and neutrals due to the differential motion between them. This was discussed in Section 6.3 where it was pointed out that the driving force is a large scale electric field of magnetospheric origin that maps downward into the high latitude ionosphere. The electric field imparts directed motion to the ions which then drift in the geomagnetic field subject to the $\mathbf{v}_i \times \mathbf{B}$ Lorentz force. The partially ionized gas is anisotropic and the mobility of the ions depends on their direction of motion with respect to the direction of the **E** and **B** vectors. The mobility $\mathscr{H}_0$, defined in Section 4.5, relates the charged particle drift velocity to the electric field, $v_d = \mathscr{H}_0 E$. In an anisotropic medium the mobility becomes a tensor quantity, analogous to the electron thermal conductivity tensor, $\bar{\bar{\lambda}}_e$, given by Equation (6.5.28).

$$\bar{\bar{\mathscr{H}}} = \begin{vmatrix} \mathscr{H}_\perp & \mathscr{H}_H & 0 \\ \mathscr{H}_H & \mathscr{H}_\perp & 0 \\ 0 & 0 & \mathscr{H}_0 \end{vmatrix} \tag{8.3.1}$$

$\mathscr{H}_0$ is related to the diffusion coefficient by the Einstein relation (Equation (4.5.7)). The ion mobility coefficient is

$$\mathscr{H}_0^{in} = eD_{in}/kT_i \tag{8.3.2}$$

and may be written in terms of the collision frequency

$$\mathscr{H}_0^{in} = \frac{e}{m_i \nu_{in}} \tag{8.3.3}$$

The electron mobility is given by

$$\mathscr{K}_0^{\text{en}} = -e/m_e\nu_{\text{en}} \tag{8.3.4}$$

$\mathscr{K}_\perp$ is the mobility coefficient for the drift velocity component perpendicular to **B** and parallel to **E**

$$\mathscr{K}_\perp^{\text{in}} = \mathscr{K}_0^{\text{in}} \frac{\nu_{\text{in}}^2}{\omega_{\text{ci}}^2 + \nu_{\text{in}}^2} = \frac{e}{m_{\text{i}}} \frac{\nu_{\text{in}}}{\omega_{\text{ci}}^2 + \nu_{\text{in}}^2} \tag{8.3.5}$$

while $\mathscr{K}_\text{H}$ is the mobility coefficient for the Hall drift velocity component perpendicular to both **E** and **B**,

$$\mathscr{K}_\text{H}^{\text{in}} = \mathscr{K}_0^{\text{in}} \frac{\nu_{\text{in}}\omega_{\text{ci}}}{\omega_{\text{ci}}^2 + \nu_{\text{in}}^2} = \frac{e}{m_{\text{i}}} \frac{\omega_{\text{ci}}}{\omega_{\text{ci}}^2 + \nu_{\text{in}}^2} \tag{8.3.6}$$

Analogous expressions apply to the electron mobility. The ion and electron gyrofrequencies are given by

$$\omega_{\text{ci}} = eB/m_{\text{i}}c \text{ and } \omega_{\text{ce}} = -eB/m_{\text{e}}c \tag{8.3.7}$$

We note, in passing, that the mobility coefficient for positive ions and electrons has the same sign for the $\mathscr{K}_\text{H}$ tensor component causing the two oppositely charged species to drift in the same direction. This is not the case for $\mathscr{K}_0$ and $\mathscr{K}_\perp$.

The concept of a particle flux, $\mathbf{\Phi} = n\mathbf{v}$, was briefly defined in the continuity Equation (Equation (5.1.1)) at the beginning of Chapter 5 and n and $\mathbf{v}$ are the particle concentration and velocity, respectively. It was found unnecessary to use the parameter $\mathbf{\Phi}$ explicitly in subsequent developments in Chapter 5 but it was introduced in the definitions of ion and electron diffusion, Equations (5.3.5) and (5.3.6). Thus, while the physical processes governing ion (and electron) dynamics are readily described in terms of the mobility and drift velocity, a literature that uses the concepts of particle flux, current and conductivity is firmly established. Current is carried by a flux of charged particles,

$$\mathbf{J}_{\text{i}} = en_{\text{i}}\mathbf{v}_{\text{i}} \text{ and } \mathbf{J}_{\text{e}} = -en_{\text{e}}\mathbf{v}_{\text{e}} \tag{8.3.8}$$

positive ions and electrons carry current (the fraction of negative ions is negligible above 80 km). By convention, the ion flow direction is adopted as the current direction and the net current is the difference between the ion and electron fluxes. In addition to all the forces acting on neutral particles the ion drift velocity is governed by the Lorentz force. Excerpting this term for clarity by neglecting the partial pressure force, gravitational force and the force per unit mass on the fluid as a whole,

$$(\mathbf{v}_{\text{i}} - \mathbf{v}) = \frac{e}{m_{\text{i}}\nu_{\text{in}}}\left(\mathbf{E} + \frac{1}{c}\mathbf{v}_{\text{i}} \times \mathbf{B}\right) \tag{8.3.9}$$

The ion drift velocity component parallel to **B** is

$$(\mathbf{v}_{\text{i}} - \mathbf{v})_\parallel = \frac{e}{m_{\text{i}}\nu_{\text{in}}}\mathbf{E}_\parallel \tag{8.3.10}$$

To obtain the component perpendicular to **B**, Equation (8.3.9) is rewritten in the equivalent form

$$(\mathbf{v}_i - \mathbf{v}) = \frac{e}{m_i \nu_{in}} \left[\frac{1}{c}(\mathbf{v}_i - \mathbf{v}) \times \mathbf{B} + (\mathbf{E} + \frac{1}{c}\mathbf{v} \times \mathbf{B}) \right] \tag{8.3.11}$$

Taking the cross product of this expression with **B** and eliminating terms in $(\mathbf{v}_i - \mathbf{v}) \times \mathbf{B}$ yields

$$(\mathbf{v}_i - \mathbf{v})_\perp = \frac{e}{m_i} \left[\frac{\nu_{in}}{\omega_{ci}^2 + \nu_{in}^2} \left(\mathbf{E}_\perp + \frac{1}{c}\mathbf{v} \times \mathbf{B} \right) + \frac{\omega_{ci}}{\omega_{ci}^2 + \nu_{in}^2} \left(\mathbf{E} + \frac{1}{c}\mathbf{v} \times \mathbf{B} \right) \times \hat{\mathbf{b}} \right] \tag{8.3.12}$$

where $\hat{\mathbf{b}}$ is a unit vector in the direction of **B** and ω_{ci} is the gyrofrequency already defined by Equations (8.3.7) and (6.5.27) (Problem 42). The current density can now be written as

$$\mathbf{J} = \sigma_0 \mathbf{E}_\parallel + \sigma_\perp \left(\mathbf{E}_\perp + \frac{1}{c}\mathbf{v} \times \mathbf{B} \right) + \sigma_H \left(\mathbf{E} + \frac{1}{c}\mathbf{v} \times \mathbf{B} \right) \times \hat{\mathbf{b}} + en_i \mathbf{v} \tag{8.3.13}$$

where the σs are components of the tensor electrical conductivity which has the same form as the mobility tensor,

$$\bar{\bar{\sigma}} = \begin{vmatrix} \sigma_\perp & \sigma_H & 0 \\ \sigma_H & \sigma_\perp & 0 \\ 0 & 0 & \sigma_0 \end{vmatrix} \tag{8.3.14}$$

The conductivity coefficients are

$$\sigma_0 = \sum_i en_i \mathscr{K}_0^{in} = \sum_i n_i e^2 / m_i \nu_{in} \tag{8.3.15}$$

$$\sigma_\perp = \sum_i \sigma_0 \frac{\nu_{in}^2}{\omega_{ci}^2 + \nu_{in}^2} \tag{8.3.16}$$

$$\sigma_H = \sum_i \sigma_0 \frac{\nu_{in} \omega_{ci}}{\omega_{ci}^2 + \nu_{in}^2} \tag{8.3.17}$$

The summation applies to a multicomponent ion gas. The three components of the current vector are the parallel or Birkeland current, $J_\parallel$, the Pedersen current, $J_\perp$, and the Hall current, J_H, given by the first three terms, respectively, of Equation (8.3.13). Near-charge-neutrality prevails in the ionosphere and the bulk motion term, $en_i\mathbf{v}$, in the current equation is always negligibly small relative to the other terms. Equations (8.3.15)–(8.3.17) assume that current is carried by several species of ions; however, the electron mobility parallel to the magnetic field is larger than the ion mobility and the complete conductivity coefficient for the Birkeland current parallel to the magnetic field is

$$\sigma_0 = \frac{e^2 n_e}{m_e(\nu_{en} + \nu_{ei})} + \sum_i \frac{e^2 n_i}{m_i \nu_{in}} \tag{8.3.18}$$

which includes the electron–ion collision frequency, ν_{ei}. Electron–ion collisions become important at high ionospheric levels where the degree of ionization is large.

Assuming charge neutrality, $n_e = \Sigma_i n_i$, the net current density is given by

$$\mathbf{J} = en_e(\mathbf{v}_i - \mathbf{v}_e) \tag{8.3.19}$$

and the appropriate expressions for the Pedersen and Hall conductivity coefficients are, respectively,

$$\sigma_\perp = \sum_i \frac{en_i}{B}\left(\frac{\nu_{en}\omega_{ce}}{\omega_{ce}^2 + \nu_{en}^2} + \frac{\nu_{in}\omega_{ci}}{\omega_{ci}^2 + \nu_{in}^2}\right) \tag{8.3.20}$$

and

$$\sigma_H = \sum_i \frac{en_i}{B}\left(\frac{\omega_{ce}^2}{\omega_{ce}^2 + \nu_{en}^2} - \frac{\omega_{ci}^2}{\omega_{ci}^2 + \nu_{in}^2}\right) \tag{8.3.21}$$

(Problem 43). Electron–ion collisions do not contribute to $\sigma_\perp$ and σ_H at any altitude level. Equation (8.3.13) is a generalized Ohm's law that relates the current density **J** to an applied electric field **E** measured in a reference frame moving with the bulk velocity of the gas,

$$\mathbf{J} = \bar{\boldsymbol{\sigma}}[\mathbf{E} + (1/c)\mathbf{v} \times \mathbf{B}] \tag{8.3.22}$$

where $\bar{\boldsymbol{\sigma}}$ is the conductivity tensor.

Ionospheric currents have been studied extensively by the perturbations which they cause in the geomagnetic field. Indeed, the relative simplicity and low cost of magnetometers allowed the establishment of an extensive world wide network of magnetic observatories, providing a large data base for the study of horizontal ionospheric current systems in the equatorial, midlatitude and polar thermosphere. The existence of field-aligned, or Birkeland, currents was proposed many decades prior to their detection by satellite borne magnetometers. The early scepticism concerning the reality of field-aligned currents was based on oversimplified concepts, and today they are invoked as the backbone of electrodynamic coupling between the ionosphere and magnetosphere, in addition to their role in the generation of plasma instabilities, waves, and acceleration mechanisms.

This chapter focuses on the dynamics of the coupled neutral and ionized components of the thermospheric gas. We therefore view ion motion not in terms of currents but of ion drift velocities, which properly juxtaposes the neutral wind. The momentum equations describe the velocity field of the neutral and ionized species in the thermosphere. The momentum equation is given in Chapter 1 (Equation (1.2.2)), highlighting its importance as one of the basic conservation equations that govern the dynamics of the thermosphere. The equation is rewritten here in more detail. If $\mathbf{v}_j$ is the velocity of the jth species then

$$\frac{\mathrm{D}\mathbf{v}_j}{\mathrm{D}t} + \frac{1}{\rho_j}\nabla\cdot\bar{\mathbf{P}} = \mathbf{g} + \frac{1}{\rho_j}(\mathbf{J} \times \mathbf{B}) \tag{8.3.23}$$

Each term represents a force per unit mass or an acceleration. Identifying the terms in sequence we have the convective derivative $\mathrm{d}\mathbf{v}_j/\mathrm{D}t = \partial\mathbf{v}_j/\partial t + (\mathbf{v}\cdot\nabla)\mathbf{v}_j$ taken along the trajectory of the gas parcel and including advection. Next is the divergence of the pressure tensor which includes the scalar pressure gradient and a viscous force (Equation (6.5.7)). Gravitational acceleration is next and the last term is the acceleration due to the relative drift between ions and neutrals, which will become evident upon writing the components of the momentum equation. Solution of the momentum equation requires the specification of several parameters, including the concentration n_j, the mass density ρ_j and the temperature T that is related to the pressure by the equation of state of the gas. The concentrations of various species are obtained from the continuity equations (Equation (1.2.1)) and the temperatures from the energy equations (Equation (1.2.3)). A correct description of thermospheric dynamics therefore requires a coupled solution of the three conservation equations for all species and an equation of state. The solution to the full problem without approximations would yield a self-consistent result for the three-dimensional wind field, ion drift field, temperature field and the composition and density, all as a function of time. Current research is directed toward achieving this goal.

Equation (8.3.23) and the conservation equations in Chapter 1 are applicable in an inertial reference frame. Ground based observations of wind, temperature, and composition are made on a rotating Earth, however, so that measurements are more conveniently interpreted if the governing equations are formulated in a coordinate system that rotates with the angular velocity of the Earth. Since the atmosphere is thought to corotate with the Earth, the variation in the velocities at various locations and as a function of time emerges more readily in a rotating frame. The velocities are related by

$$\mathbf{v}_{\text{inertial}} = \mathbf{v}_{\text{rot}} + \mathbf{\Omega} \times \mathbf{r} \tag{8.3.24}$$

where $\mathbf{\Omega}$ is the angular velocity and $\mathbf{r}$ is the distance from the Earth's centre to the thermospheric level where measurements are made. The acceleration term in the momentum equation becomes

$$\left.\frac{\mathrm{D}\mathbf{v}_j}{\mathrm{D}t}\right|_{\text{inertial}} = \left.\frac{\mathrm{D}\mathbf{v}_j}{\mathrm{D}t}\right|_{\text{rot}} + 2\mathbf{\Omega} \times \mathbf{v}_{\text{rot}} - \Omega^2\mathbf{r} \tag{8.3.25}$$

where the second term on the right is the Coriolis acceleration and the third term is the centripetal acceleration (Problem 44). The direction of the centripetal acceleration is parallel to the gravitational acceleration, its magnitude is very small in the thermosphere and ionosphere and the two terms are combined into an effective gravity. Rewriting Equation (8.3.23) in the rotating coordinate system we note that the pressure tensor, gravity, and the electromagnetic force remain unchanged,

$$\frac{\mathrm{D}\mathbf{v}}{\mathrm{D}t} + 2\mathbf{\Omega} \times \mathbf{v} + \frac{1}{\rho}\nabla\bar{\bar{\mathbf{P}}} = \mathbf{g} + \frac{1}{\rho}(\mathbf{J} \times \mathbf{B}) \tag{8.3.26}$$

The subscripts referring to the rotating frame and the species have been dropped (Problem 45).

There are important differences between neutral and ion forcing influences in the atmosphere and these are reflected in the respective equations of motion. We focus first on the neutrals. In the research literature observations of neutral winds (Section 8.2) are presented in components, zonal (E–W), meridional (N–S) and vertical. The respective velocity components are u, v, and w with E, N and up taken as positive. The equations used to analyse the observations are, therefore, written in the appropriate components.

The velocity vector in the rotating coordinate system is

$$\mathbf{v} = \hat{\mathbf{i}}u + \hat{\mathbf{j}}v + \hat{\mathbf{k}}w \tag{8.3.27}$$

where $u = \mathrm{d}x/\mathrm{d}t$, $v = \mathrm{d}y/\mathrm{d}t$ and $w = \mathrm{d}z/\mathrm{d}t$, and the acceleration term in spherical coordinates is given by

$$\frac{\mathrm{D}\mathbf{v}}{\mathrm{D}t} = \hat{\mathbf{i}}\left(\frac{\mathrm{D}u}{\mathrm{D}t} - \frac{uv\tan\phi}{r} + \frac{uw}{r}\right) + \hat{\mathbf{j}}\left(\frac{\mathrm{D}v}{\mathrm{D}t} + \frac{u^2\tan\phi}{r} + \frac{vw}{r}\right)$$
$$\times\,\hat{\mathbf{k}}\left(\frac{\mathrm{D}w}{\mathrm{D}t} - \frac{u^2 + v^2}{r}\right) \tag{8.3.28}$$

where r is the radius of the sphere and ϕ is the latitude (Problem 46). The geometry is illustrated in Figure 8.3.1.

On a global scale the neutral composition is non-uniform in latitude, longtitude and altitude but the horizontal changes are small, gradual, and sluggish by comparison with the vertical variability. The horizontal components of the momentum equation may therefore be written in terms of a mass averaged velocity. The vertical acceleration in the momentum equation is small compared with the two major forcing terms in a neutral gas, the pressure gradient and gravity, and vertical momentum balance is approximately described by the hydrostatic equation (Section 5.1),

$$\partial p/\partial z = -\rho g \tag{5.1.5}$$

Compositional variation with altitude is obtained from the steady state vertical diffusion equation of a binary gas mixture. It is derived (see references to this chapter) in

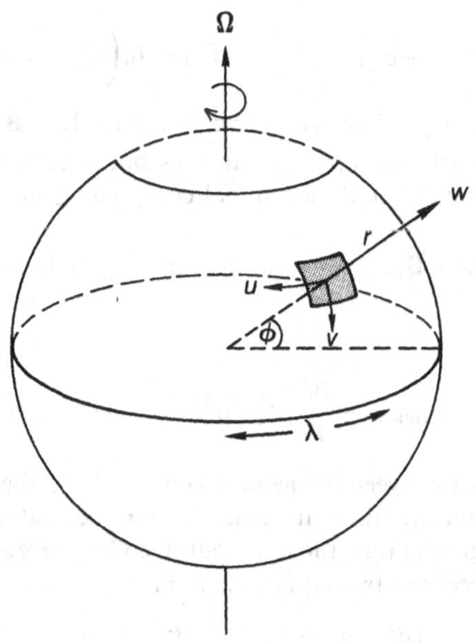

Fig. 8.3.1 Velocity components in rotating coordinates on a sphere: r = radius; ϕ = latitude; λ = longitude; Ω = rotational angular velocity; u = zonal; v = meridional; w = vertical.

terms of the change in the mean molecular mass

$$\bar{m} = \frac{m_1 n_1 + m_2 n_2}{n_1 + n_2} \tag{5.1.7}$$

In a vertical pressure coordinate system

$$p\frac{\partial \bar{m}}{\partial p} = -\frac{D_{12}}{D_{12} + D_{\mathrm{E}}} \frac{(\bar{m} - m_2)(\bar{m} - m_1)}{\bar{m}} \tag{8.3.29}$$

where D_{12} is the molecular diffusion coefficient and D_{E} is the eddy or turbulent diffusion coefficient (Section 5.2). Near the lower boundary of the thermosphere $D_{\mathrm{E}} > D_{12}$, the vertical gradient of $\bar{m}$ is small and the atmosphere tends to be mixed. With increasing altitude or decreasing pressure the mean molecular mass decreases and eddy diffusion becomes small compared with molecular diffusion (Problem 47).

Components of the pressure tensor are given by Equations (6.5.7) and (6.5.8). The scalar pressure gradient is an important forcing term in all directions, zonal, meridional and vertical, but the viscous stress is due almost solely to vertical shear of the horizontal wind, specified by the off-diagonal terms in P_{xz} and P_{yz}, $-\eta\, \partial u/\partial z$ and $\eta\, \partial v/\partial z$, where η is the coefficient of viscosity. Horizontal wind velocities dominate over vertical winds and vertical gradients are larger than horizontal gradients. The term $(1/\rho)(\nabla \cdot \bar{\mathbf{P}})$ in the momentum equation reduces to $-(1/\rho)(\partial/\partial z)(\eta\, \partial u/\partial z)$ and $-(1/\rho)(\partial/\partial z)(\eta\, \partial v/\partial z)$, in addition to the scalar pressure gradient

$$\frac{1}{\rho}\nabla p = \frac{1}{\rho}\left(\hat{\mathbf{i}}\frac{1}{r\sin\phi}\frac{\partial p}{\partial \lambda} + \hat{\mathbf{j}}\frac{1}{r}\frac{\partial p}{\partial \phi} + \hat{\mathbf{k}}\frac{\partial p}{\partial z}\right)$$

where the latitude is ϕ and the longitude is λ.

The electromagnetic forcing term is written in component form,

$$\frac{1}{\rho}(\mathbf{J} \times \mathbf{B}) = \frac{1}{\rho}\left[\sigma_0 \mathbf{E}_{\parallel} + \sigma_{\perp}\left(\mathbf{E}_{\perp} + \frac{1}{c}\mathbf{v} \times \mathbf{B}\right) + \sigma_{\mathrm{H}}\left(\mathbf{E} + \frac{1}{c}\mathbf{v} \times \mathbf{B}\right) \times \hat{\mathbf{b}}\right] \times \mathbf{B} \tag{8.3.30}$$

Carrying out the vector product with $\mathbf{B}$ we note that $\mathbf{E}_{\parallel} \times \mathbf{B} = 0$, by definition. This implies that the parallel forcing term vanishes but does not deny the existence of parallel currents. The zonal and meridional components are given by

$$\frac{1}{\rho}(\mathbf{J} \times \mathbf{B})_{\mathrm{zonal}} = \frac{B^2}{\rho}[\sigma_{\perp}(u_{\mathrm{i}} - u) + \sigma_{\mathrm{H}} \sin I (v_{\mathrm{i}} - v)] \tag{8.3.31}$$

and

$$\frac{1}{\rho}(\mathbf{J} \times \mathbf{B})_{\mathrm{meridional}} = \frac{B^2}{\rho}[\sigma_{\perp} \sin^2 I (v_{\mathrm{i}} - v) - \sigma_{\mathrm{H}} \sin I (u_{\mathrm{i}} - u)] \tag{8.3.32}$$

where I is the dip angle between the geomagnetic field and the local horizontal. These terms in the momentum equations are called the ion–neutral drag. In the F-region the ion gyrofrequency is greater than the ion–neutral collision frequency, $\omega_{\mathrm{ci}} \gg \nu_{\mathrm{in}}$, and the coefficients in the preceding two equations reduce to

$$(B^2/\rho)\sigma_{\perp} \to \nu_{\mathrm{in}}, \qquad (B^2/\rho)\sigma_{\mathrm{H}} \to \omega_{\mathrm{in}} \tag{8.3.33}$$

(Problem 48).

Depending on the relative magnitudes of the ion drift velocity and the neutral wind each can be a driver or a drag on the other. At high latitudes the externally applied convection electric field is frequently sufficiently large to impart to the ions a velocity that is larger than the neutral wind and the ion drag term is then a momentum source for the neutral gas. This is illustrated in Figures 8.2.4 and 8.2.6. The convection electric field does not extend to low latitudes, however, where the neutral wind frequently becomes a momentum source for the ions. The magnetic dip, I, causes large differences in the magnitude of the ion drag terms as a function of geomagnetic latitude. For example, at the magnetic equator where $I = 0$ the meridional component of ion drag vanishes and only the Pedersen zonal component is retained. The ion drag momentum source is most effective at high latitudes.

Collecting terms in the momentum equation the zonal component becomes

$$\underbrace{\frac{\partial u}{\partial t}}_{\text{acceleration}} + \underbrace{u\frac{\partial u}{\partial x} + v\frac{\partial u}{\partial y} + w\frac{\partial u}{\partial z}}_{\text{advection}} \underbrace{- \frac{uv\tan\phi}{r} + \frac{uv}{r}}_{\text{curvature}}$$

$$\underbrace{- 2\Omega\sin\phi v}_{\text{Coriolis}} + \underbrace{\frac{1}{\rho}\frac{1}{r\sin\phi}\frac{\partial p}{\partial\lambda}}_{\text{pressure}} \underbrace{- \frac{1}{\rho}\frac{\partial}{\partial z}\left(\eta\frac{\partial u}{\partial z}\right)}_{\text{viscosity}}$$

$$= \underbrace{\frac{B^2}{\rho}[\sigma_\perp(u_i - u) + \sigma_H \sin I(v_i - v)]}_{\text{ion–neutral drag}} \tag{8.3.34}$$

The meridional component of the momentum equation is

$$\frac{\partial v}{\partial t} + u\frac{\partial v}{\partial x} + v\frac{\partial v}{\partial y} + w\frac{\partial v}{\partial z} + \frac{u^2\tan\phi}{r} + \frac{vw}{r} + 2\Omega\sin\phi u + \frac{1}{\rho}\frac{1}{r}\frac{\partial p}{\partial\phi}$$

$$- \frac{1}{\rho}\frac{\partial}{\partial z}\left(\eta\frac{\partial v}{\partial z}\right) = \frac{B^2}{\rho}[\sigma_\perp \sin^2 I(v_i - v) - \sigma_H \sin I(u_i - u)] \tag{8.3.35}$$

It was stated earlier in this section that the vertical component of the momentum equation can be approximated by the hydrostatic equation, implying the absence of a vertical wind. The experimental data summarized in Section 8.2 show, however, that vertical neutral winds do occur, lasting a few hours and reversing direction, when measured at one observing site (Figure 8.2.7). Vertical acceleration can result from unbalanced forces or from convergence or divergence of horizontal winds. These processes are described by the vertical component of the momentum equation,

$$\frac{\partial w}{\partial t} + u\frac{\partial w}{\partial x} + v\frac{\partial w}{\partial y} + w\frac{\partial w}{\partial z} - 2\Omega\cos\phi u + \frac{1\partial p}{\rho\partial z} = g + (w_i - w)\nu_{in} \tag{8.3.36}$$

where ν_{in} is the ion–neutral collision frequency. The ion–neutral drag term could be due to a vertical component of electric field producing a force per unit mass on the ions, eE_z/m_i.

The term wind is used to describe the velocity field of the neutral air in the thermosphere but the velocity field of the ions is, by convention, called the ion drift or the plasma drift. Wind velocities are specified in an Eulerian coordinate system in which the velocity components u, v and w represent the speed of a parcel of neutral air as it passes a given point of the latitude–longitude grid as a function of time. Plasma drift, on the other hand, is described in Lagrangian coordinates in which the trajectories of ionization parcels are followed as a function of time. The reason for the two approaches is that different forcing influences govern the motion of neutrals and ions. The neutral air is assumed to corotate with the Earth. In corotating Eulerian coordinates the measured or computed neutral velocities represent the wind system. The plasma, however, is influenced not only by corotation but also by the ionospheric dynamo effect at low latitudes and by a convection electric field of magnetospheric origin at high latitudes. Plasma drift trajectories are identified with motion of flux tubes and the ionospheric properties of ion composition, densities, and temperatures may be investigated within a flux tube as a function of time, as presented in Chapters 5 and 6. The advantage of following flux tube trajectories lies in reducing both the continuity equations describing ion composition and the energy equations describing the temperatures to a one-dimensional set at mid and high latitudes where the field lines are essentially straight in the thermosphere. At low latitudes, interhemispheric transport along the curved field lines requires a two-dimensional solution.

The perpendicular (to **B**) component of the ion drift velocity, Equation (8.3.9), may be written as

$$(\mathbf{v}_i - \mathbf{v}_n)_\perp = \frac{e}{m_i \nu_{in}} \left(\mathbf{E}_\perp + \frac{1}{c} \mathbf{v}_{i\perp} \times \mathbf{B} \right) \tag{8.3.37}$$

In a reference frame moving with the neutrals and therefore defining an effective electric field,

$$\mathbf{E}_\perp = \mathbf{E}_\perp + \frac{1}{c} \mathbf{v}_n \times \mathbf{B} \tag{8.3.38}$$

the ion drift velocity becomes

$$\mathbf{v}_{i\perp} = \frac{e}{m_i \nu_{in}} \left(\mathbf{E}'_\perp + \frac{1}{c} \mathbf{v}_{i\perp} \times \mathbf{B} \right) \tag{8.3.39}$$

which may be solved for the drift velocity of the flux tubes,

$$\mathbf{v}_{i\perp} = \left(\frac{1}{1 + \nu_{in}^2/\omega_{ci}^2} \right)^{1/2} \frac{c\mathbf{E}'_\perp}{B} \tag{8.3.40}$$

Equation (8.3.40) holds under the same assumptions as Equation (8.3.9), that the Lorentz force governs $\mathbf{v}_{i\perp}$ (Problem 49).

The ionospheric plasma is assumed to corotate with the Earth, just as the neutral atmosphere. This assumption is valid in the E-region at all latitudes where the plasma and the neutral atmosphere are collisionally strongly coupled, and it is valid in the low latitude F-region where convection is small, but the assumption is poor in the high latitude F-region dominated by magnetospheric convection electric fields. The corotational electric field, $\mathbf{E}_{cor}$, is defined in terms of the corotational velocity,

$$\Omega r \cos\phi\hat{\lambda} = \frac{c\mathbf{E}_{cor} \times \mathbf{B}}{B^2} \tag{8.3.41}$$

where ϕ is the latitude.

The atmospheric dynamo also gives rise to an electric field. The neutral wind blows the plasma across magnetic field lines thereby generating a field, $(1/c)\mathbf{v} \times \mathbf{B}$, where $\mathbf{v}$ is the bulk velocity of the neutrals and ions. Ambipolar diffusion causes charge separation between ions and neutrals, as discussed in Section 5.3, setting up a polarization electric field. The polarization electric field and the $\mathbf{v} \times \mathbf{B}$ field produce a current (Equation (8.3.22)) that generates an electromagnetic force, $\mathbf{J} \times \mathbf{B}$ (Equation (8.3.3)), to drive the neutral wind. Self-consistent coupling between the neutral wind and the plasma is identified as a hydromagnetic dynamo. The dynamo is most effective in the E-region where ions and neutrals are strongly coupled. The electric field maps to the F-region along the highly conducting magnetic field and produces an $(1/c)(\mathbf{E} \times \mathbf{B})/B^2$ ion drift. Since E-region neutral winds are highly variable in time and space the resulting dynamo electric field and ion drift are likewise variable. At high latitudes the magnetospheric electric field is also variable so that only the corotational field that satisfies Equation (8.3.41) is a predictable input to dynamical studies of the thermosphere.

It has already been remarked on several occasions that plasma drift and neutral winds influence each other, suggesting that the thermospheric/ionospheric dynamic problem requires coupled solutions of the various governing equations. This has only recently been accomplished due to difficulties in overcoming the traditionally different formulations that have been adopted for neutral winds and for plasma transport. The neutral wind velocity vectors are obtained by solving the three-dimensional, time-dependent momentum equations for a single species gas together with the thermodynamic equation and the mass density continuity equation. The principal parameters of interest are the wind and the temperature, although recently a variable mean molecular mass has also become an important computed output. By contrast, ionospheric models solve the one-(or two-)dimensional continuity equations, including diffusive transport for half a dozen species of ions, together with energy equations for the plasma temperatures. The tubes of ionization are forced to move under the influence of various effective electric fields that prevail at different locations. The output parameters of interest in the models are the ion densities and composition as well as the plasma temperatures while the ion drift velocities are determined by the assumed high latitude convection field and by the assumed neutral winds. These input parameters are derived from experimental measurements and models. Since the high latitude convection electric field has its origin in the magnetosphere, a fully self-consistent solution of the thermospheric/ionospheric dynamics problem should include aspects of magnetospheric dynamics.

8.4 Geometrical aspects of thermospheric and ionospheric dynamics

A significant source of complexity in the global morphology of ionospheric and thermospheric parameters is the asymmetric offset between the geomagnetic and

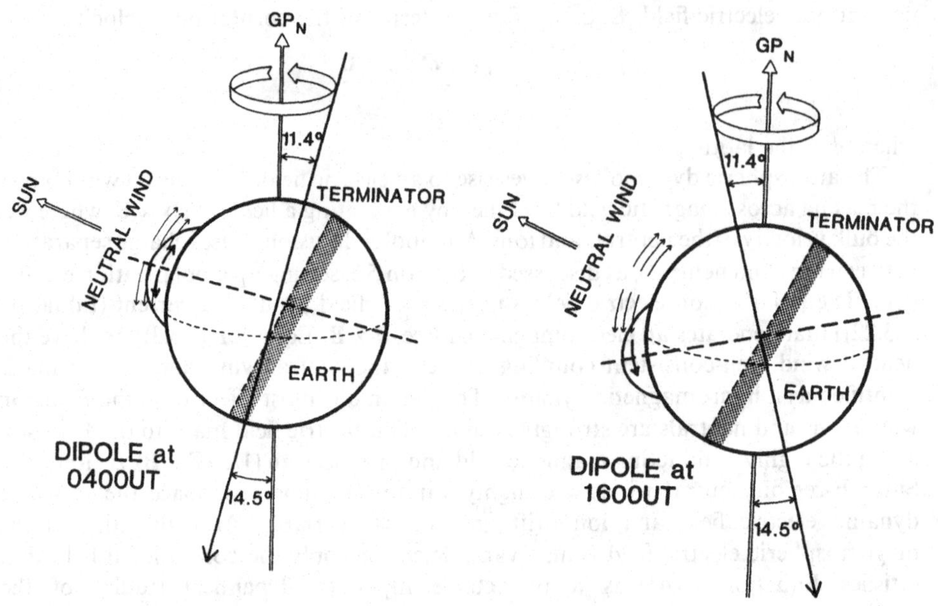

Fig. 8.4.1 Magnetic dipole geometry at two universal times. The offset of the magnetic axis from the geographic axis is 11.4° in the northern hemisphere and 14.5° in the southern hemisphere. The neutral wind at noon blows away from the subsolar point and induces a field-aligned ion drift that depends on the magnetic field configuration. (J.J. Sojka and R.W. Schunk, *J. Geophys. Res.*, **90**, 5285, 1985.)

geographic axes of the Earth. The geometry is illustrated in Figure 8.4.1. Corotation is about the geographic axis while the solar zenith angle is symmetric about the subsolar point. At low latitudes the neutral wind also has approximate subsolar symmetry but the electric field is governed by the topology of the dipole magnetic field and therefore the geomagnetic reference frame. The high latitude convection electric field is driven by the solar wind and maps into the thermosphere along the magnetic field lines. Figure 8.4.1 shows that the combined effect of the tilt of the Earth's axis of rotation with respect to the ecliptic plane and the rotation of the geomagnetic axis about the geographic axis (or vice versa) leads to a local and a universal time dependence of all the parameters that characterize the structure, dynamics, and energetics of the thermosphere and ionosphere. In addition, the different offsets between the geomagnetic and geographic axes in the northern and southern hemispheres leads to asymmetric responses to momentum sources and to energy or heating sources. The two hemispheres are not decoupled, however, and complex feedback is expected to occur. The dependence of various momentum and energy sources on different reference frames vitiates the possibility of finding a 'natural' coordinate system for describing the global morphology of the coupled thermosphere and ionosphere. The relationship between the geographic coordinates and a magnetic latitude–magnetic local time coordinate system is illustrated in Figure 8.4.2 at two universal times, 03 UT and 15 UT.

One possible approach to combining the corotation ion drift with magnetospheric convection ion drift is to transform the corotation velocity from a geographic inertial

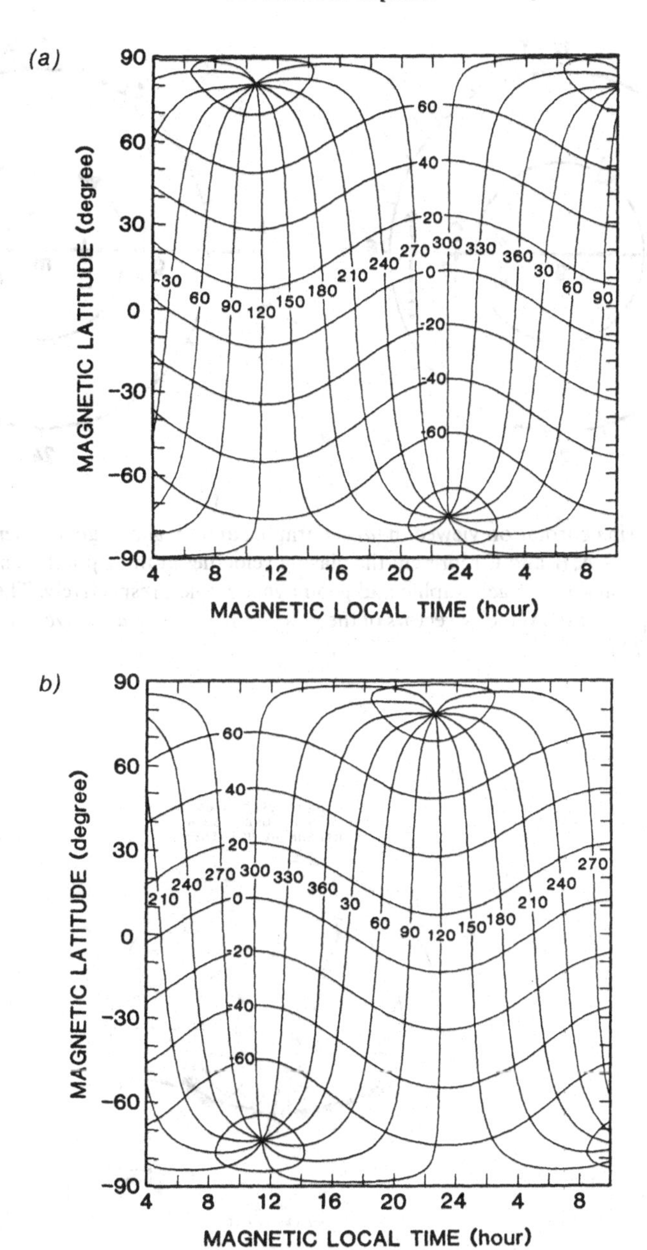

Fig. 8.4.2 Geographic coordinate system displayed in a dipole magnetic latitude, MLT coordinate system: (*a*) 03 UT, and (*b*) 15 UT. The geographic latitude is labelled at 20° intervals, and the geographic longitude at 30° intervals. (J.J. Sojka and R.W. Schunk, *J. Geophys. Res.*, **90**, 5285, 1985.)

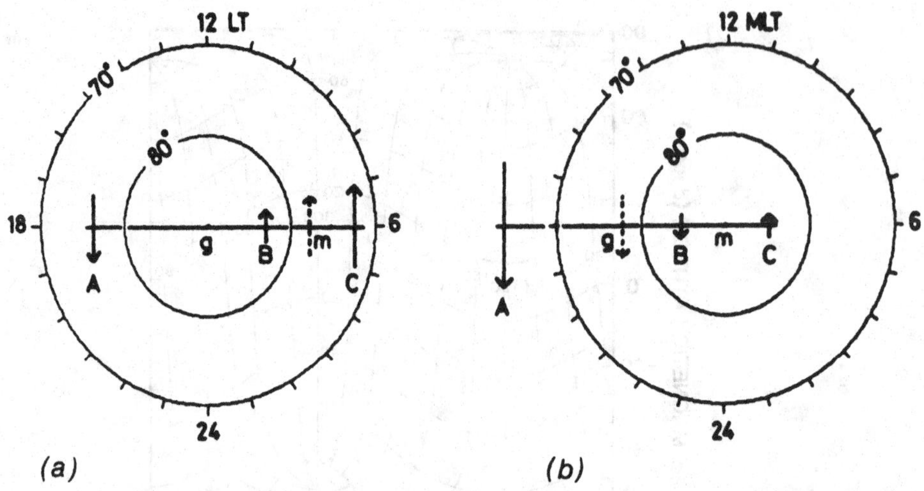

Fig. 8.4.3 Plasma corotation viewed in (*a*) geographic and (*b*) geomagnetic inertial frames. The arrows A, B, and C represent the plasma velocities at three points. The symbols g and m represent the geographic and geomagnetic poles, respectively. The dashed arrows represent the velocity vectors of the poles. (J.J. Sojka, *et al.*, *J. Geophys. Res.*, **84**, 5493, 1979.)

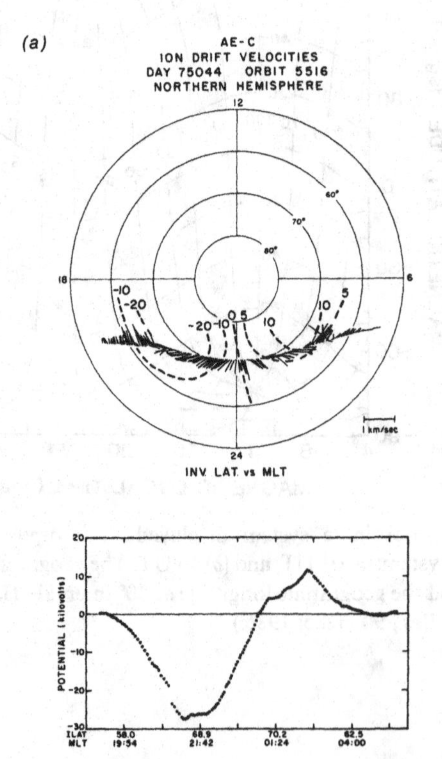

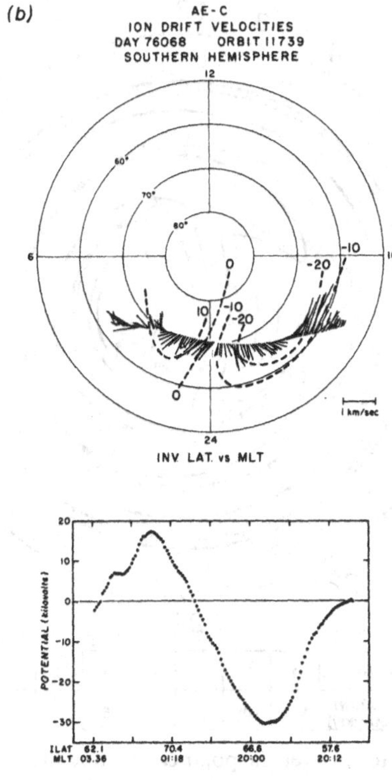

Fig. 8.4.4 High latitude passes of the Atmosphere Explorer C satellite over (*a*) the northern and (*b*) the southern hemisphere. The electrostatic potential distribution is derived from the ion drift velocities. (R.A. Heelis and W.B. Hanson, *J. Geophys. Res.*, **85**, 1995, 1980.)

frame to a geomagnetic inertial frame. This is illustrated in Figure 8.4.3 which shows the changes in speed and direction of three representative velocity vectors. While all velocities change in magnitude, and some even reverse direction, the basic corotation pattern still occurs as the plasma motion is viewed in the geomagnetic inertial frame. This implies that corotation is independent of universal time in a geomagnetic inertial reference frame.

The total potential difference across the polar caps that is established by the interaction of the solar wind and the magnetosphere is variable in magnitude ($\sim$15–150 kV) and the topology depends on the vectorial interplanetary magnetic field configuration. Average patterns have been constructed from ion drift measurements obtained through numerous rocket release experiments and from thousands of satellite passes over the northern and southern polar caps, as explained in Section 8.2. The electric field is derived from the measured drift velocity, $\mathbf{v}_i = (1/c)\mathbf{E} \times \mathbf{B}$, and the electrostatic potential, V, for the divergence free field, is $\mathbf{E} = \nabla V$. These relationships are illustrated in Figure 8.4.4 for a northern and southern hemisphere pass of the Atmosphere Explorer C satellite. The inferred potential contours are sketched in to fit the measurements obtained along the trajectory (Problem 50). The potentials across the polar cap, in this example, are at least 40 kV and 48 kV in the northern and southern hemispheres, respectively. Most model computations assume simplified two-cell

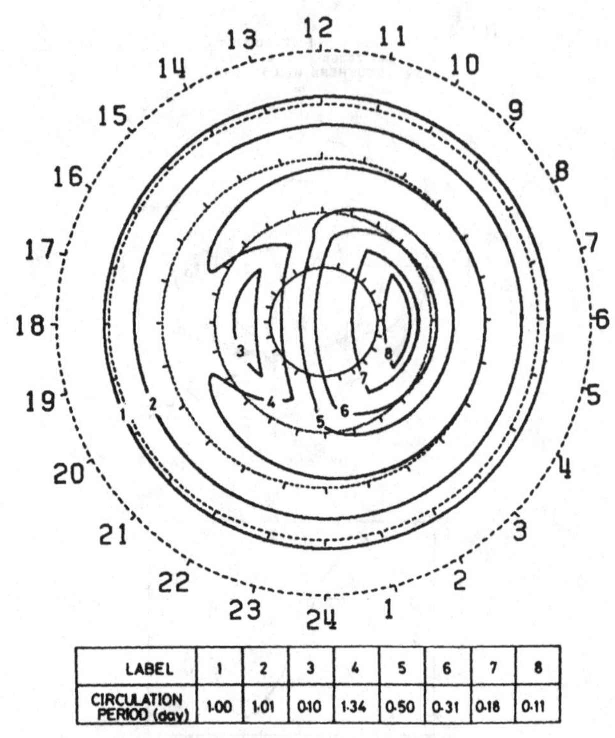

LABEL	1	2	3	4	5	6	7	8
CIRCULATION PERIOD (day)	1.00	1.01	0.10	1.34	0.50	0.31	0.18	0.11

Fig. 8.4.5 Plasma drift trajectories over the polar cap viewed in a magnetic inertial frame. Each of the eight trajectories shown has a different circulation period, tabulated in the figure. For these calculations the total potential across the polar cap was maintained at 64 kV. (J.J. Sojka, *et al.*, *J. Geophys. Res.*, **84**, 5943, 1979.)

convection patterns. The combined corotational and convective electric field produces a complex plasma flow pattern that may be examined by tracing representative trajectories. Eight drift trajectories are shown in Figure 8.4.5 in a magnetic inertial frame; the concentric circles are the magnetic latitudes while magnetic local time marks are indicated on the 40° latitude circle. The electric field is assumed to remain constant for the time interval required for flux tubes to complete a full trajectory by returning to the starting point. This assumption is rarely fully satisfied, given the variability of the solar wind. Flux tubes that follow the trajectory labelled 1 essentially corotate counterclockwise on the diagram in one day but, with increasing latitude, the convection field assumes greater dominance. Flux tubes that follow trajectory 4 approach a stagnation point at about 17 MLT and reverse direction several times before the trajectory is closed. The horizontal ion drift is governed by the vectorial sum of the convection electric field and the corotational electric field (Equation (8.3.39)). The flow reversals noted on the dusk side of trajectory 4 in Figure 8.4.5 are due to opposing directions and changes in magnitude of the two electric field components. A flux tube on trajectory 4 requires 1.34 days to complete the path. Flux tubes on trajectory 3 remain on the dusk side of the polar cap while those on trajectory 8 remain in the dawn sector. Only about one tenth of a day is required to complete the drift trajectory 3 or 8.

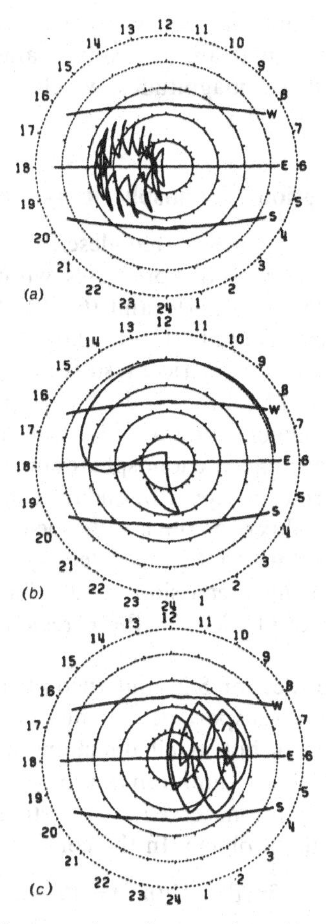

Fig. 8.4.6 Plasma drift trajectories over the polar cap viewed in a geographic inertial frame for nonaligned geomagnetic and geographic axes. The trajectories in (*a*) (*b*) and (*c*) correspond to those shown in Figure 8.4.5 and labelled 3, 4, and 7, respectively. These trajectories have been followed for a 24 h period. The positions of the terminator for winter solstice (W), equinox (E), and summer solstice (S) are shown. (J.J. Sojka *et al.*, *J. Geophys. Res.*, **84**, 5943, 1979.)

Rotation of the geomagnetic pole about the geographic pole introduces a universal time dependence when plasma drift is transformed from the geomagnetic to the geographic reference frame. The consequences of this are shown for the three trajectories labelled 3, 4 and 7, in Figure 8.4.6 and displayed in a geographic inertial frame over a 24 h period. The rotation of the poles introduces a wobble on trajectories 3 and 7. The position of the terminator is shown for equinox and solstice conditions, winter and summer. Plasma on trajectory 3 will be in complete daylight at summer solstice and in complete darkness during winter solstice. At the equinoxes, the flux tubes drift from sunlit to non-sunlit regions. A similar pattern is followed along trajectory 7 but trajectory 4 maps into a more complex pattern and the flux tube has not returned to the starting point in a 24 h period. Trajectories 3 and 7 are confined to the dusk and dawn periods, respectively. Flux tubes in the two trajectories are therefore immersed in segments of the auroral oval that are generally characterized by different charged particle energy spectra and flux magnitudes, which cause different ion density and plasma temperature height profiles.

8.5 Mathematical description of ionospheric dynamics

The continuity and energy equations that describe the physical processes in the ionosphere take account of transport processes within a flux tube. The relevant equations are Equations (5.1.1), (6.5.19) and (6.5.40). Solution of these differential equations requires the adoption of suitable boundary conditions. Indeed, the boundary conditions largely determine the numerical results in regions of the ionosphere where transport dominates over local sources and sinks of plasma or energy. At latitudes where the dipole field configuration provides continuity of flux tubes between hemispheres the structure of conjugate ionospheres must be analysed self-consistently, avoiding the need for an arbitrary upper boundary condition. There is observational evidence that an appreciable flux of conjugate photoelectrons flows between sunlit and dark ionospheres, contributing to the ion density in the dark hemisphere and producing enhancement of twilight emissions in OI 6300 Å and 5577 Å radiations from dissociative recombination of O_2^+ ions in the F-region, $O_2^+ + e \rightarrow O(^1D, ^1S) + O$ (see Section 4.4).

Observational evidence (Section 8.2) and theoretical analysis (Section 8.3) have shown that ion and energy transport occur not only along the magnetic field, as assumed in Chapters 5 and 6, but in all directions in response to various forcing influences. Thus, the simplifying assumptions adopted in Chapter 5, that ion transport occurs only by ambipolar diffusion, served as a starting point to a more realistic but more complex analysis of the problem. In the continuity equation

$$\partial n_i/\partial t + \nabla \cdot (n_i \mathbf{v}_i) = P_i - L_i \tag{8.5.1}$$

the plasma velocity in the low latitude F-region includes several terms,

$$\mathbf{v}_i = \mathbf{v}_{em} + \mathbf{v}_{diff} + (v \cos I + w \sin I)\hat{\mathbf{b}} \tag{8.5.2}$$

where $\mathbf{v}_{em}$ is the electromagnetic drift resulting from the dynamo electric field and $\mathbf{v}_{diff}$ is the plasma diffusion velocity along the magnetic field. The third term is the neutral velocity component tangential to the magnetic field, it being assumed that the ions are dragged along with the neutral wind. A meridional cross section of the ionosphere spanning the equatorial region is shown in Figure 8.5.1. The vertical component of the neutral wind is w and v is the meridional component. A simple transequatorial wind blowing from the summer to the winter hemisphere forces ions up and down the field as shown in the figure. More complex neutral wind patterns such as convergence or divergence and zonal components would produce correspondingly different ion velocity vectors.

Figure 8.5.1 also illustrates the consequences of the dynamo electric field which maps from the E- to the F-region. The magnetic field is northward, by convention, so that an eastward directed electric field causes an upward ion flow. This mechanism provides a substantial enhancement of the ion density in the upper F-region. Ambipolar diffusion has already been discussed.

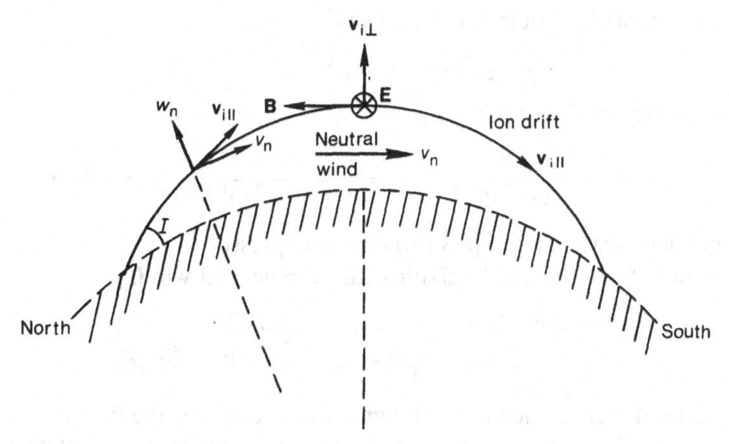

Fig. 8.5.1 Cross section of the equatorial ionosphere showing the relationship between neutral wind and ion drift.

At low and middle latitudes the Earth's magnetic field closely approximates to that of a dipole. The analysis of electric fields, potentials and currents, i.e., ionospheric electrodynamics, is most conveniently carried out in dipole coordinates. An orthogonal dipole coordinate system, a, λ, b, may be defined in terms of the spherical coordinates r, θ, λ

$$a = r/\sin^2\theta; \quad b = \frac{\cos\theta}{r}; \quad \lambda = \lambda \tag{8.5.3}$$

where θ is measured with respect to the north geomagnetic pole, λ is the geomagnetic east longitude, and r is the radius vector, as shown in Figure 8.5.2. The distance

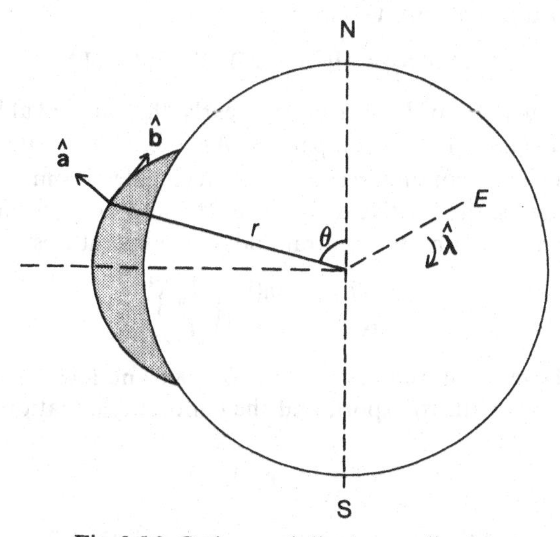

Fig. 8.5.2 Orthogonal dipole coordinates.

between two points in dipole coordinates is

$$ds = [(h_a \, da)^2 + (h_\lambda \, d\lambda)^2 + (h_b \, db)^2]^{\frac{1}{2}} \tag{8.5.4}$$

where the metric coefficients, h_j, are

$$h_a = \frac{\sin^3\theta}{(1+3\cos^2\theta)^{1/2}}; \quad h_b = \frac{r^3}{(1+3\cos^2\theta)^{1/2}}; \quad h_\lambda = r\sin\theta \tag{8.5.5}$$

The current density (Equation (8.3.13)) can be expressed in dipole coordinates, given the electrostatic field, the conductivities, and the neutral wind.

$$\begin{aligned} \mathbf{J} = \hat{\mathbf{a}}\,[\sigma_\perp(E_a + v_\lambda B) - \sigma_{\mathrm{H}}(E_\lambda - v_a B)] \\ + \hat{\lambda}[\sigma_\perp(E_\lambda - v_a B) + \sigma_{\mathrm{H}}(E_a + v_\lambda B)] + \hat{\mathbf{b}}\sigma_0 E_b \end{aligned} \tag{8.5.6}$$

In the equatorial region, the dipolar magnetic field confines the flow of current to the ionosphere. With increasing geomagnetic latitude, however, the field lines extend out into the magnetosphere, introducing additional physical processes that must be included in the electrodynamic description of the high latitude current system: (1) the magnetospheric plasma is no longer collision dominated and field-aligned currents are not governed by ohmic conductance; (2) the energetic magnetospheric plasma becomes an additional source of current; (3) the magnetic field configuration in the magnetosphere becomes distorted and is no longer represented by a dipolar field. Since the magnetosphere is the upper boundary of the ionosphere, its characteristics must be taken into account in the mathematical description of high latitude electrodynamics.

At high latitudes in the polar cap, where flux tubes extend into the deep magnetosphere, thermal ions that are not gravitationally bound to the planet diffuse upward along the magnetic field and may escape from the top of the ionosphere. In the terrestrial ionosphere H^+ and He^+ are in this category (Problem 51). H^+ and He^+ are minor ions and their motion is influenced by the major ion through Coulomb collisions and by the polarization electrostatic field set up by ambipolar diffusion of the major ion and electrons. H^+ ions are produced in the ionosphere almost solely by accidental resonant charge exchange with O^+ ions,

$$O^+({}^4S_{\frac{3}{2}}) + H({}^2S_1) \leftrightarrows O({}^3P_{2,1,0}) + H^+ \tag{8.5.7}$$

The ionization potentials of H and O are nearly the same (Table 2.2.1) and direct photoionization of H is small, by comparison. Assuming that in the production region the transport term in the continuity equation is negligible in comparison with the local source and sink terms, the relative densities of H^+ and O^+ ions are governed by the densities of H and O and by the neutral and ion temperatures,

$$\frac{n(H^+)}{n(O^+)} = \frac{9n(H)}{8n(O)}\left(\frac{T_n}{T_i}\right)^{\frac{1}{2}} \tag{8.5.8}$$

(Problem 52). Above an altitude of about 800–1000 km, local sources and sinks are small by comparison with transport, and the continuity equation reduces to

$$\frac{\partial}{\partial z}(n_{mi} w_{mi}) = 0 \tag{8.5.9}$$

under steady state conditions and for a vertical magnetic field. The minor ion density is labelled n_{mi} and the velocity is w_{mi}. A constant minor ion flux implies an increase in the speed as the density decreases with increasing altitude. As long as speed remains subsonic the momentum equation for minor ion flow is

$$w_{mi} = w_i - D_{mi}\left[\frac{\partial(\ln n_{mi})}{\partial z} + \frac{m_{mi}g}{kT_{mi}} + \frac{1}{T_{mi}}\frac{\partial(T_{mi}+T_e)}{\partial z} + \frac{T_e}{n_e T_{mi}}\frac{\partial n_e}{\partial z}\right] \tag{8.5.10}$$

where the diffusion coefficient is defined in terms of the collision frequency

$$D_{mi} = kT_{mi}/m_{mi}\nu_{mi,i} \tag{8.5.11}$$

and all quantities with the subscript mi refer to the minor ion. The major ion carries the minor ions along but the minor ions also have their own diffusion terms (Problem 53). Boundary conditions to the transport equations determine the detailed altitude profiles of ion densities and velocities. At some altitude level, H^+ or He^+ ions are no longer minor ions but become the major ionic species. In addition, the flow may become supersonic and Equation (8.5.9) will no longer be valid. A detailed analysis of the field-aligned outflow of ions, called the polar wind effect is available in references listed with this chapter. It appears highly probable that O^+ ions detected in the magnetosphere originate in the polar cap ionosphere; a detailed formulation of the mechanisms is a current research topic.

The physical coupling between thermospheric dynamics and ionospheric electrodynamics has been expounded in Sections 8.3, 8.4 and this section from several points of view such as common forcing terms, collisional interactions, currents, and topological considerations. Attention was paid to feedback mechanisms between the neutral wind and ion drifts and quantitative relationships between various parameters were derived. Considerable insight into the nature of the coupled processes has thereby been acquired. The equations of motion for neutral and ion species are beginning to be solved self-consistently. However, equations that relate the electric and magnetic fields and the currents are not included as yet; these are Maxwell's equations.

To achieve, in principle, a fully self-consistent solution, coupled solutions to the hydromagnetic equations are required for each species. These are: the neutral and ion momentum equations

$$\frac{D\mathbf{v}}{Dt} + 2\mathbf{\Omega}\times\mathbf{v} + \frac{1}{\rho}\nabla\cdot\bar{\mathbf{P}} = \mathbf{g} + \frac{1}{\rho}(\mathbf{J}\times\mathbf{B}) \tag{8.3.26}$$

the energy equations

$$\frac{\mathcal{N}}{2}k\frac{D(nT)}{Dt} + \frac{5}{2}nkT\nabla\cdot\mathbf{v} + \nabla\cdot\mathbf{q} + \bar{\mathbf{P}}:\nabla\mathbf{v} = Q - L \tag{6.5.1}$$

the continuity equations

$$\partial n/\partial t + \nabla\cdot(n\mathbf{v}) = P - L \tag{1.2.1}$$

Ohm's law

$$\mathbf{J} = \bar{\boldsymbol{\sigma}}[\mathbf{E} + (1/c)\mathbf{v}\times\mathbf{B}] \tag{8.3.22}$$

and Maxwell's equations

$$\nabla \times \mathbf{E} = -\frac{1}{c}\frac{\partial \mathbf{B}}{\partial t}; \quad \nabla \times \mathbf{B} = \frac{4\pi}{c}\mathbf{J} \tag{8.5.12}$$

As the data base of simultaneously measured thermospheric and ionospheric parameters becomes more comprehensive, the need for better modelling of the observed quantities becomes more important. This can best be met by solving the governing equations self-consistently and applying the appropriate boundary conditions.

8.6 Global dynamics of the thermosphere

Ground based observatories and space borne instruments yield measurements that have limited spatial and temporal coverage. A global picture may be built up with satellite measurements accumulated over long periods of time, perhaps years, and with more than one spacecraft. Although this chapter focuses on atmospheric and ionospheric dynamics, characterized by winds and ion drifts, one cannot ignore the global energy budget, characterized by temperatures, and the global neutral and ion composition and densities. Indeed the physical processes that govern the dynamics, energetics, structure and composition of the thermosphere and ionosphere demonstrate the interdependence of the various geophysical parameters. It is therefore highly desirable for an observational programme to include simultaneous measurements of as many parameters as possible. This has been the aim of several satellite missions dedicated to exploring the near-Earth space environment by including in the payload instruments for measuring the solar UV flux, energetic charged particle fluxes, neutral and ion densities and composition, neutral and plasma temperatures, electric fields and magnetic field perturbations, ion drifts, and neutral winds. Not all spacecraft have carried a complete complement of instruments, however. It has also been a recent goal to cluster ground based instruments to yield complementary geophysical data. Various instruments and the measured parameters are briefly described in Section 8.2. There have been several campaigns that combined ground based, rocket, and satellite observations, in order to study simultaneously spatial and temporal variations of geophysical parameters. While the observational approach is essential to new discoveries and for testing one's understanding of the physics and chemistry of the upper atmosphere, the ability to formulate equations that describe and predict the global behaviour of the system has provided considerable insight into the relative importance of various processes. Further, it has provided guidance for planning critical observations.

Numerical models that generate global neutral wind, temperature and composition fields are called thermospheric general circulation models (TGCM). (Current work on this topic is described in references listed with this chapter.) The momentum, energy, mass conservation and the equation of state for the neutral and ionized gas are solved self-consistently as a function of time but, as already noted, the high latitude convection electric field and auroral energetic particle precipitation are arbitrary and adjustable inputs to the model. Several parameters are generated by the models and the variety of cases that have been examined covers all seasons, levels of solar activity, and auroral contributions. Examples of computed neutral wind fields and perturbation temperatures (from a global mean) at constant pressure surfaces corresponding to the E- and F-regions are shown in Figure 8.6.1 for equinox conditions. The results are displayed in geographic coordinates at 12 UT (noon on the Greenwich meridian) on a mercator projection. The assumed magnitude of the cross polar cap potential is 60 kV; the resulting ion drag and joule heating substantially enhance the neutral wind and temperature at high latitudes. Comparison between model computations and observations requires displaying the data in compatible formats. For example, observations of

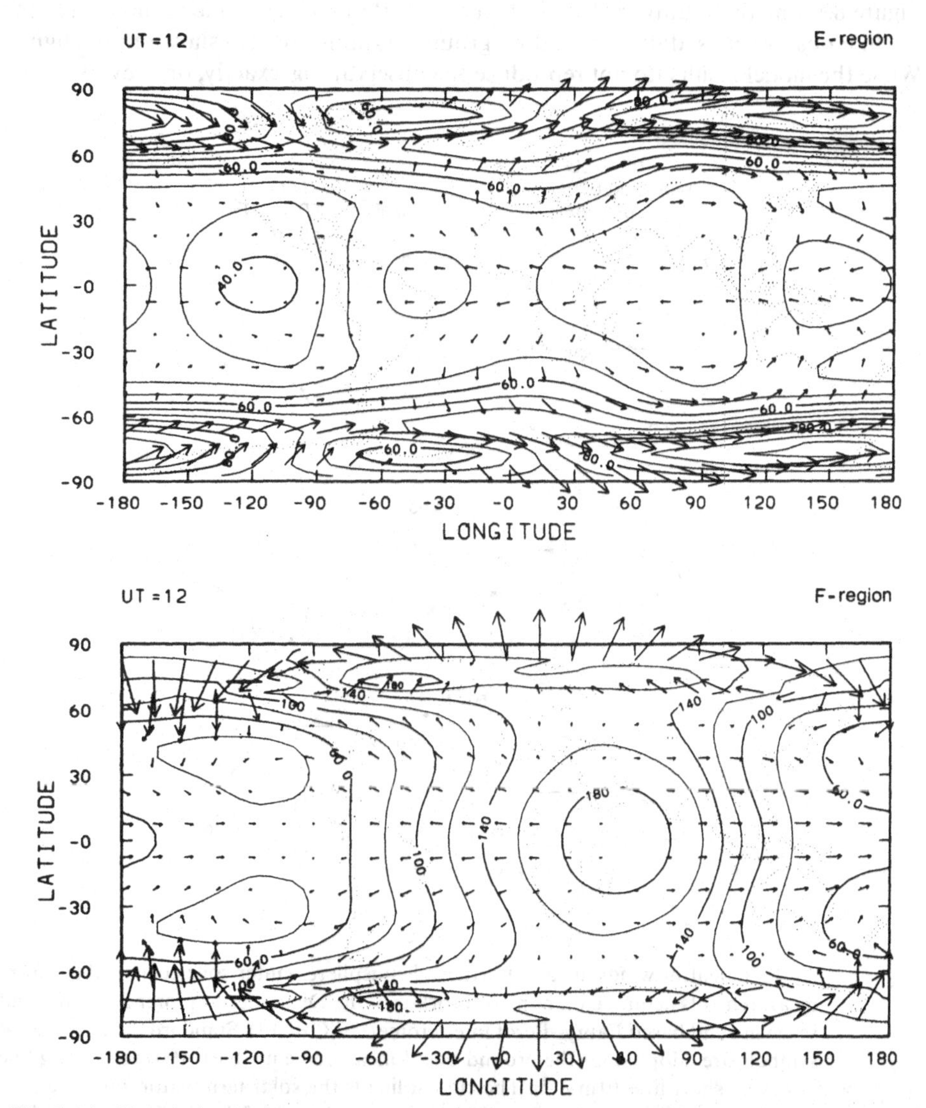

Fig. 8.6.1 Global circulation and temperature structure computed with a TGCM. Arrows represent the wind speed with the maximum length in the F-region equal to 336 m s^{-1} and in the E-region equal to 79 m s^{-1}. Temperature contours are deviations from the global mean. (R.E. Dickinson *et al.*, *J. Atmosph. Sci.*, **41**, 205, 1984.)

neutral winds were acquired along the satellite track by instruments on the Dynamics Explorer 2 (Figure 8.2.15) which required about 20 min to traverse the polar region. The data and model can be compared in geographic polar coordinates, latitude and solar local time, which requires specifying the universal time. During a 40 day period at the end of 1981 observations acquired on 122 dawn/dusk north polar passes of the Dynamics Explorer 2 satellite and by seven ground stations were used to generate the average neutral wind pattern for northern hemisphere winter solstice at several universal times. The observations are compared with TGCM predicted wind fields in Figure 8.6.2 at three universal time intervals. Although the display emphasizes the satellite observations, data acquired by ground stations are also shown in the figure. While the model results do not reproduce the observations exactly, or in every detail,

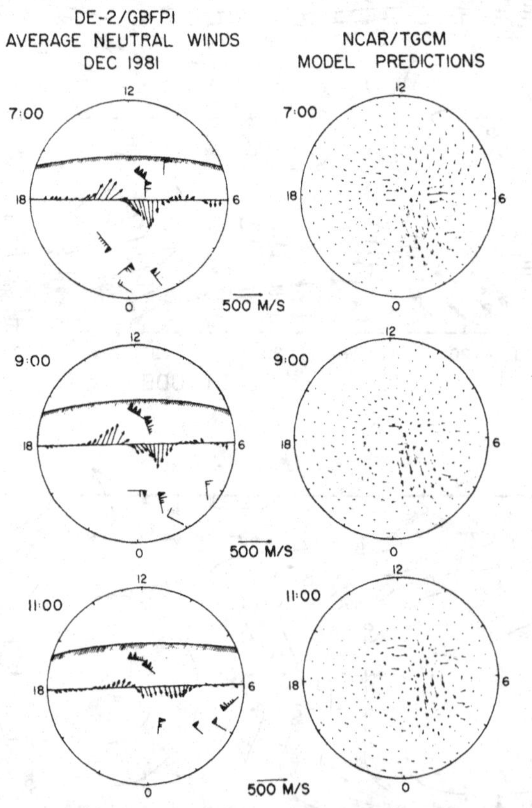

Fig. 8.6.2 Average neutral winds around northern hemisphere winter solstice assembled from measurements acquired on numerous orbits of the Dynamics Explorer 2 satellite and seven ground based Fabry–Perot interferometers (GBFPI). Standard meteorological symbols are adopted for the ground based measurements (barb $100\,\mathrm{m\,s^{-1}}$, long line $50\,\mathrm{m\,s^{-1}}$, short line $10\,\mathrm{m\,s^{-1}}$). The curved line is the solar terminator. Observations and model predictions extend to 40° latitude. The length of the wind vectors used for the TGCM results refer to the same wind speed scale as the satellite observations. Results for three universal time periods are shown: 07, 09 and 11. (T.L. Killeen, *et al.*, *J. Geophys. Res.*, **91**, 1633, 1986.)

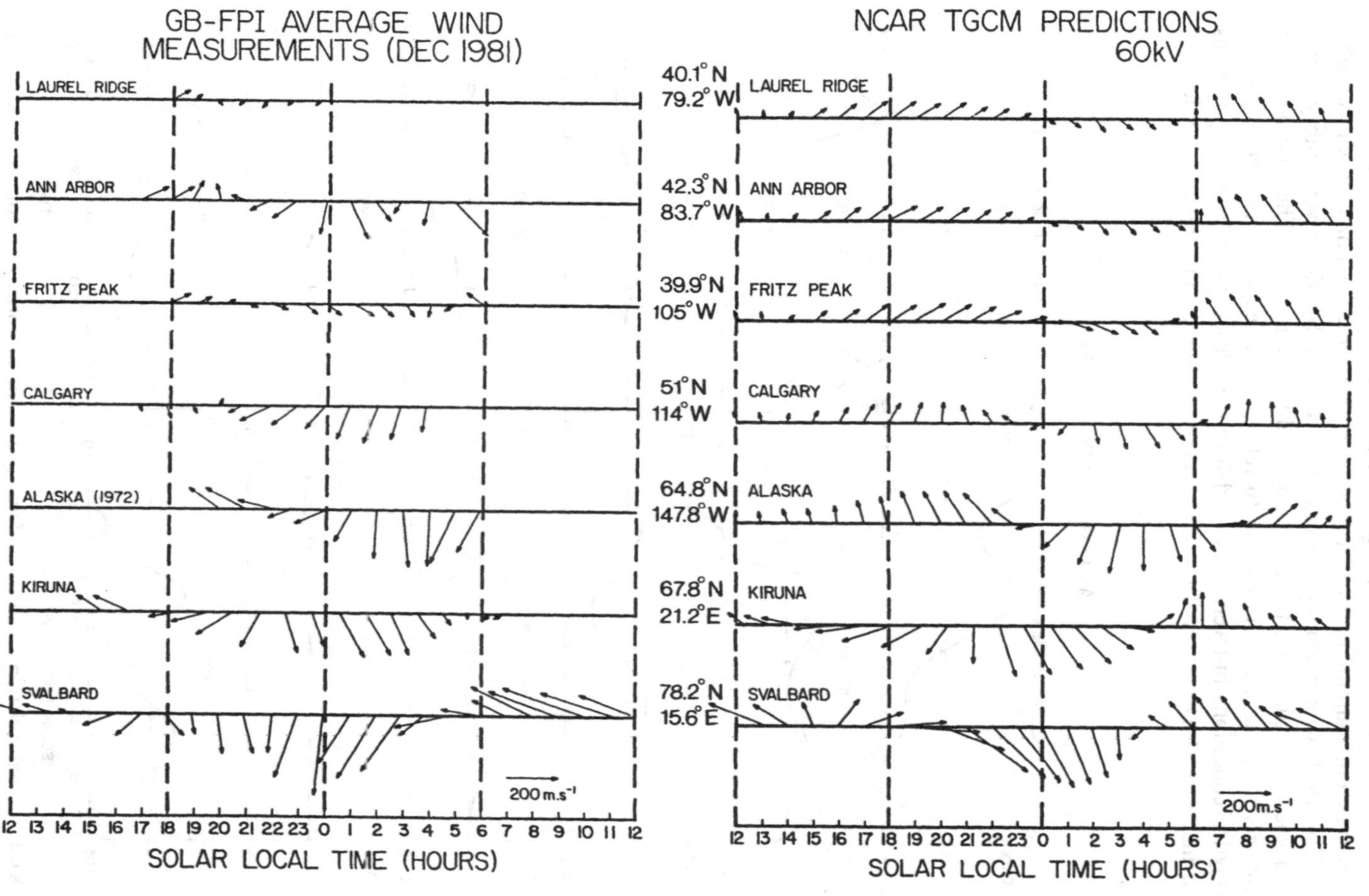

Fig. 8.6.3 Averaged neutral wind measurements from the referenced GBFPI stations plotted as a function of solar local time according to the wind scale in the lower right hand corner. The TGCM predictions for the individual GBFPI stations are plotted in the same format. (T.L. Killeen, *et al.*, *J. Geophys. Res.*, **91**, 1633, 1986.)

they do provide a global view of the wind field that cannot be obtained observationally. Ground based measurements are acquired at one location as a function of local solar time. The 1981 winter averaged data set acquired by the seven Fabry–Perot interferometers are compared to the TGCM results in Figure 8.6.3. The arbitrary input parameters to the model were chosen to reflect values expected for the prevailing geophysical conditions, but were not varied during the 40 day period, the computations being made for 4 December 1981. The extent of agreement between model and

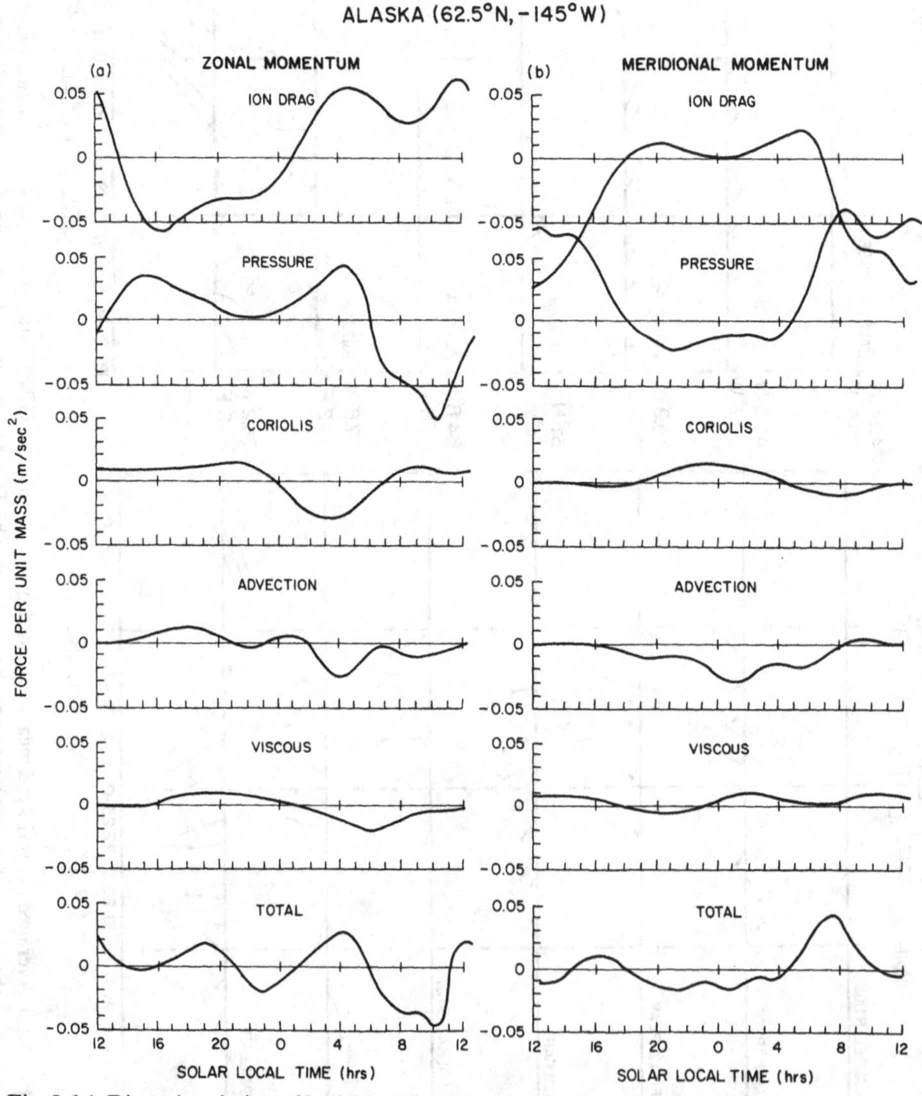

Fig. 8.6.4 Diurnal variation of individual forcing terms in the horizontal momentum equations at one location. The results apply to a pressure surface corresponding to about 300 km. (T.L. Killeen and R.G. Roble, *J. Geophys. Res.*, **91**, 11 291, 1986.)

observations suggests that the major physical processes are included in the model and that it is possible to choose arbitrary input parameters that are appropriate to the event being investigated.

To evaluate the relative importance of various physical processes that establish the neutral wind field, the individual forcing terms in the horizontal momentum equations (Equations (8.3.34) and (8.3.35)) can be displayed. In Figure 8.6.4 these are shown as a function of local solar time at a selected geographic position. Ion–neutral drag and the pressure gradient forces appear to dominate at high latitudes and all the forcing terms show large variations with time. These results apply to the F-region (about 300 km

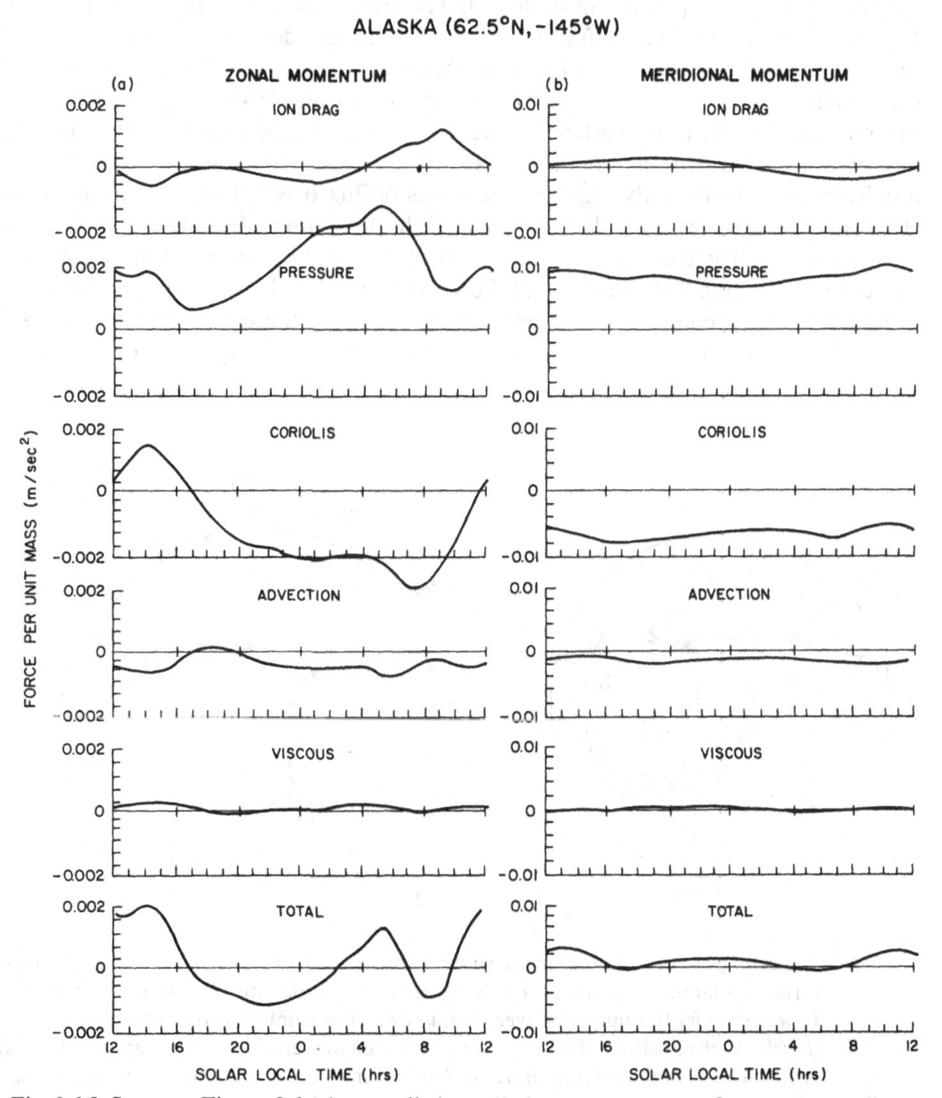

Fig. 8.6.5 Same as Figure 8.6.4 but applied to a constant pressure surface corresponding to about 130 km. (T.L. Killeen and R.G. Roble, *J. Geophys. Res.*, **91**, 11 291, 1986.)

altitude). A similar decomposition into various forcing terms in the E-region (about 130 km) is shown in Figure 8.6.5 to emphasize the large differences that prevail in the two regions. Insufficient observations of E-region winds are as yet available to allow meaningful comparisons with model predictions.

The observations and model results shown in Figures 8.6.2–8.6.5 are all displayed as a function of local solar time, and therefore refer to different parcels of gas moving into and out of the field of view of the detector. The measured wind, however, is not determined by local forcing but by the forces experienced by the gas parcel as it travels along its trajectory. By tracing the trajectory backward in time, the regions that it traversed can be identified: the time spent in the sunlit or dark region, the auroral oval, the dayside polar cusp, etc. can be found. The trajectory of one parcel is shown in Figure 8.6.6 in two reference frames, geographic longitude/latitude and local solar time/latitude, corresponding to the wind fields as displayed in Figure 8.6.1 and in subsequent figures. Knowledge of the trajectories of air parcels exposes details in the models that help in understanding the model predictions and thereby the observations, provided that there is reasonable agreement between them. Similar insight emerged from analysing the trajectories of flux tubes of ionization driven by electromagnetic forces. The drift velocity of the flux tubes (Equation (8.3.40)) was assumed to be divergence free so that their motion could be described in Lagrangian coordinates (Section 8.3). Neutral air flow, on the other hand, contains divergent components due primarily to Joule heating and the computation of parcel trajectories cannot be carried out without solving the system of governing equations given in

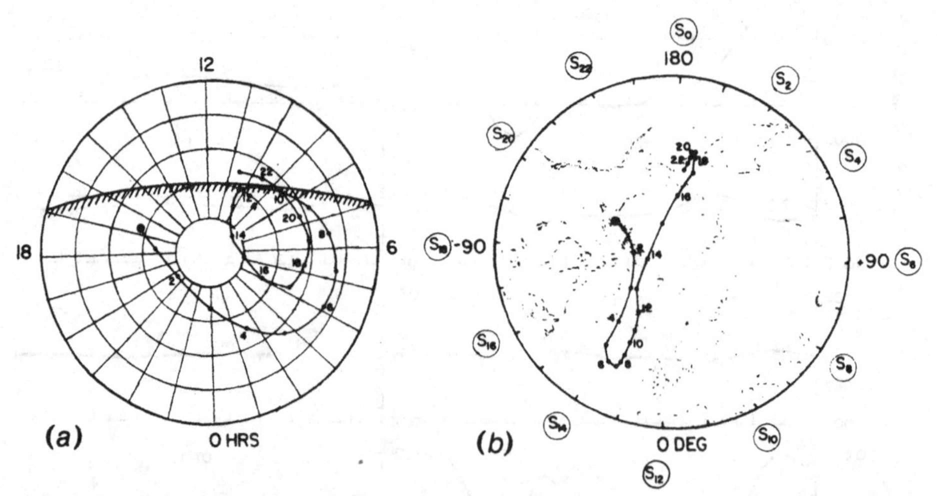

Fig. 8.6.6 Trajectory of a parcel of gas in the winter hemisphere F-region in two reference frames, (*a*) local solar time/latitude and (*b*) geographic longitude/latitude. The numbers on the trajectory refer to hours of universal time while the numbers around the periphery of (*b*) refer to the position of the sun at the indicated universal time. The curved line in (*a*) is the location of the solar terminator. A 60 kV cross polar potential has been assumed in the model computation and all other inputs are appropriate to 4 December, 1981. (T.L. Killeen and R.G. Roble, *J. Geophys. Res.*, **91**, 11 291,1986.)

Section 8.3. The trajectories shown in Figure 8.6.6 were obtained for horizontal displacements only, while vertical transport was neglected. The effects of a high latitude convection electric field on ion drift and neutral winds in the F-region are somewhat different from the consequences in the E-region due to varying contributions of coupling by ion drag momentum exchange and by Joule heating. Contours of an assumed electrostatic potential associated with magnetospheric convection are shown in Figure 8.6.7(*a*) in geographic latitude/longitude polar coordinates at 06 UT, with displaced geomagnetic and geographic poles. Ion drift vectors computed for a 60 kV cross polar potential, including the effects of ion–neutral collisions, are shown in Figure 8.6.7(*b*) for the F-region (~ 300 km) and for the E-region (~ 130 km). Evidently, in the F-region the ion drift motion is little affected by collisions with neutrals and the

Fig. 8.6.7 (*a*) and (*b*) for caption see p. 228.

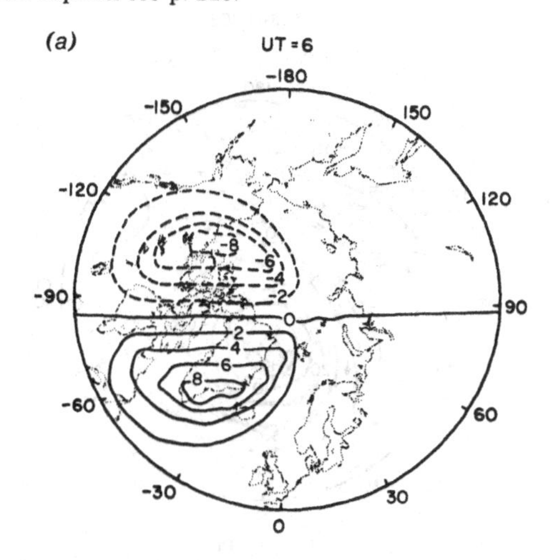

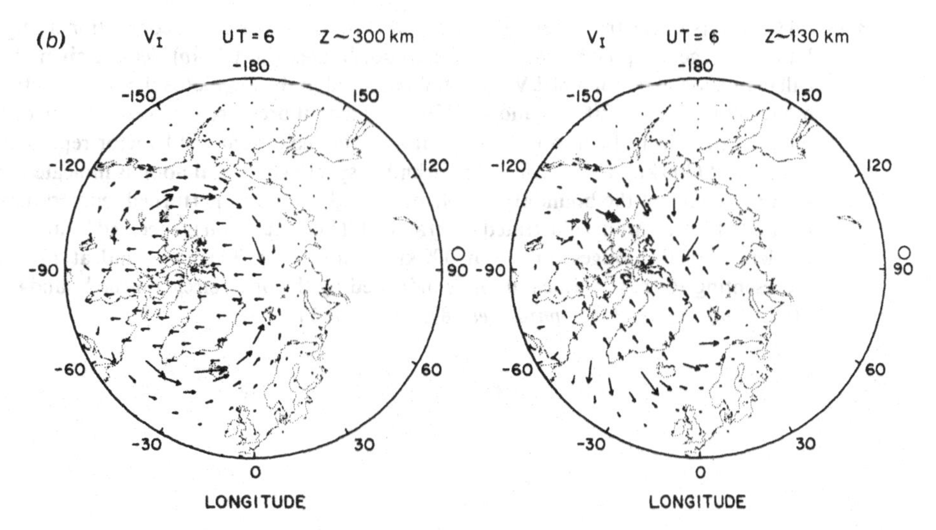

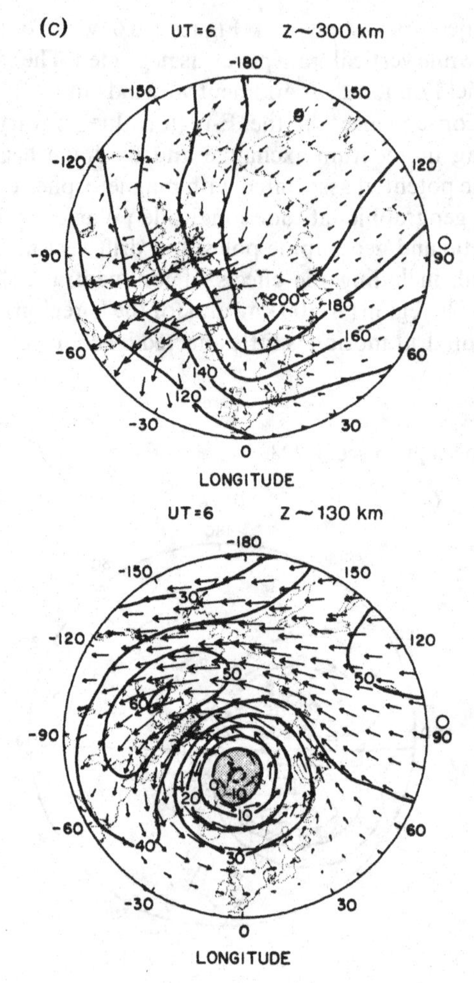

Fig. 8.6.7 (*a*) Contours of electric potential associated with magnetospheric convection at high latitudes in geographic latitude/longitude coordinates at 6 UT. (*b*) Vectors giving the calculated ion drift, for a 60 kV cross-tail potential in the magnetospheric convection model, along the $z \sim 300$ km and $z \sim 130$ km constant pressure surfaces. The ion drift includes the effect of collisions with neutrals. The maximum wind vector represents $1010\,\mathrm{m\,s^{-1}}$ (300 km) and $573\,\mathrm{m\,s^{-1}}$ (130 km), respectively. Local noon is indicated by the open circle on the boundary. (*c*) Neutral wind fields and perturbed temperatures computed for the model illustrated in (*a*) and (*b*). The circulation is specified by arrows, the length of the longest arrow at 300 km representing $395\,\mathrm{m\,s^{-1}}$ and at 130 km representing $100\,\mathrm{m\,s^{-1}}$. Local noon is indicated by the open circle on the boundary. (R.G. Roble, *et al., J. Geophys. Res.*, **87**, 1599, 1982.)

trajectories follow the $(1/c)\mathbf{E} \times \mathbf{B}$ drift pattern. Since $(1/c)\mathbf{E} \times \mathbf{B}$ drift is non-divergent, the ion drift pattern in the F-region is specified largely by a rotational non-divergent velocity. The ion drift pattern in the E-region is characterized by vectors that are rotated clockwise about 45° from the $(1/c)\mathbf{E} \times \mathbf{B}$ vectors. Thus, both rotational non-divergent and divergent irrotational components contribute (about equally) to ion drift in the E-region. With reference to Equation (8.3.13), the difference between the E- and F-region ion drift patterns may be interpreted in terms of the variation with altitude of the ratio of the Hall to Pedersen conductivities (Problem 54). The neutral wind patterns in the E- and F-regions corresponding to the 60 kV cross polar potential are shown in Figure 8.6.7(*c*). The model computation takes account of all the momentum and energy sources, including ion drag and Joule heating. In the F-region, the wind is generally enhanced by ion drag and the circulation pattern tends to follow the ion drift vortices. In the E-region, however, circulation is governed principally by the topology of cold low pressure and warm high pressure regions. This is illustrated in Figure 8.6.7(*c*) by including contours of the perturbation temperature (from the global mean), negative values indicating regions of below average temperature at the particular pressure level. The divergent wind field and large pressure variations are caused principally by Joule heating.

It was noted in the introduction to this chapter that the development of models that solve the coupled ionospheric and thermospheric governing equations self-consistently is a current research activity in upper atmospheric physics. Structure, composition, dynamics, and energetics of the neutral and ionized gases must be computed interactively as a function of time. Solar and geomagnetically controlled processes in a rotating fluid must be properly coupled, with allowance for a wide range of time constants. References are given in the bibliography to the first published results (1987) of a coupled global model. From the large number of geophysical parameters that are computed by such a model as a function of time and altitude (or pressure level) a small sample is shown in Figure 8.6.8(*a*)–(*c*). The offset between the geographic and geomagnetic poles requires the specification of a universal time. Photoionization is the principal source of the electron density on the dayside while auroral ionization contributes to the nightside enhancement. The pattern associated with an auroral oval is grossly distorted, however, by the drift of flux tubes of ionization driven by the convection electric field and modified by the neutral wind and the ion–neutral chemistry. The electron density is equal to the sum of the various positive ions in the ionosphere. The ion composition varies over the globe, as does the neutral composition, due primarily to pressure gradients and heating. A measure of the variability in the composition is the mean molecular weight. The neutral temperature is governed by local heating and cooling, radiative cooling as well as dynamical effects. The sparse observations available to date support the model predictions, suggesting that the major physical and chemical processes that operate in the thermosphere and ionosphere are correctly taken into account.

In spite of the complexity of modelling global thermospheric dynamics, a method has been developed that leads to an analytic solution to the governing equations. The method involves application of linear tidal theory and the equations are cast in terms of small amplitude disturbances of the wind and pressure fields. The advantage of such a perturbation method lies in allowing variations of geophysical parameters to be expressed by periodic components, semidiurnal, diurnal, semiannual, etc. Simplifying

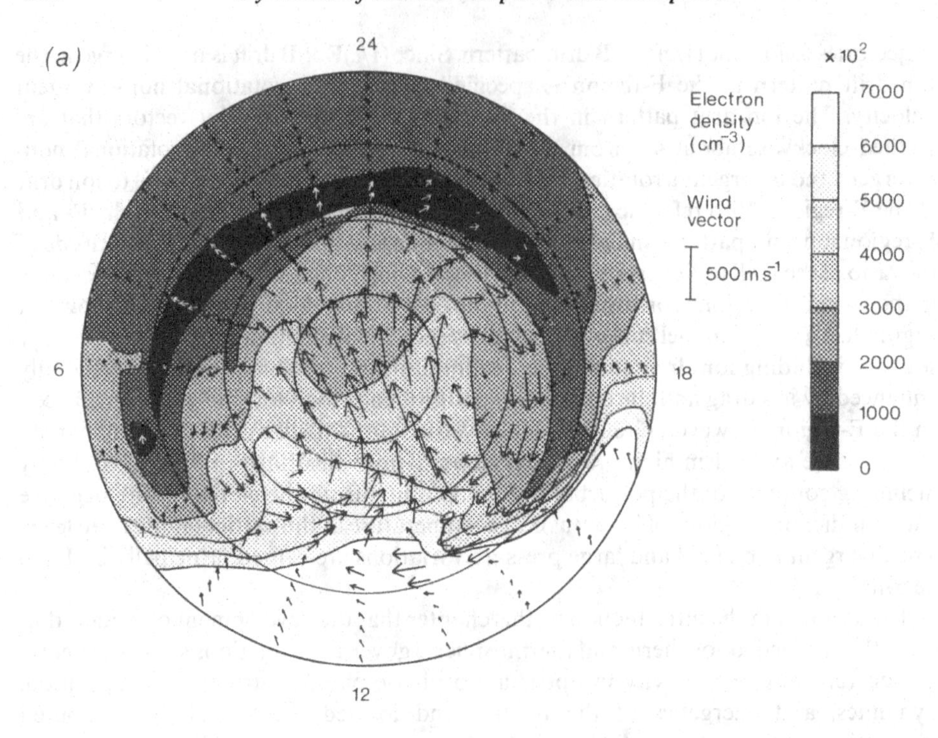
(a)
24
6
18
12
Electron density (cm⁻³)
Wind vector
500 m s⁻¹
× 10²
7000
6000
5000
4000
3000
2000
1000
0

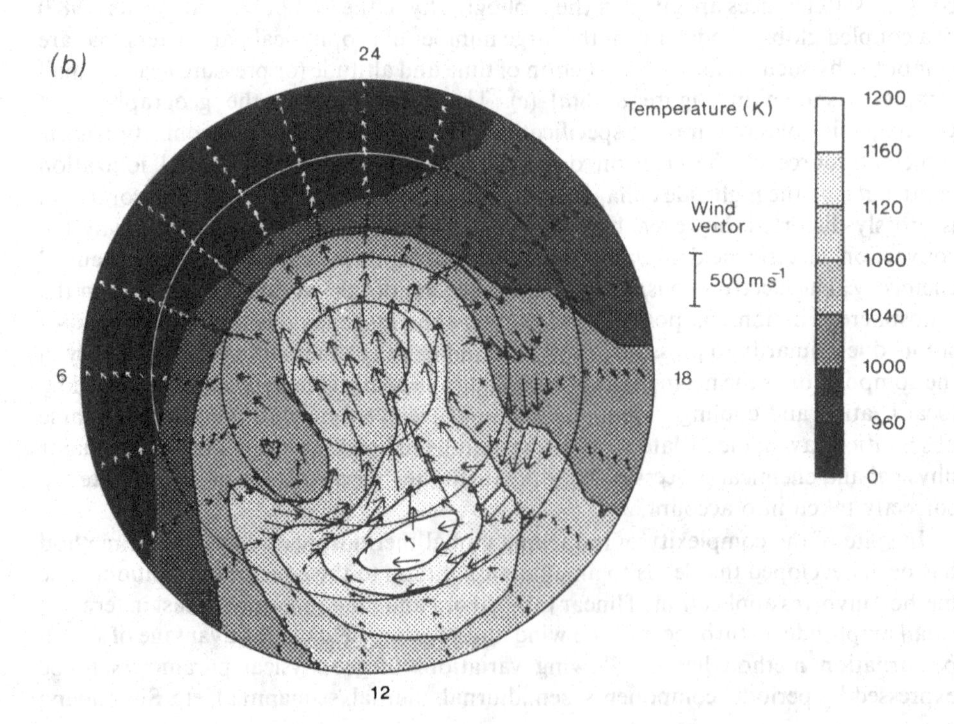
(b)
24
6
18
12
Temperature (K)
Wind vector
500 m s⁻¹
1200
1160
1120
1080
1040
1000
960
0

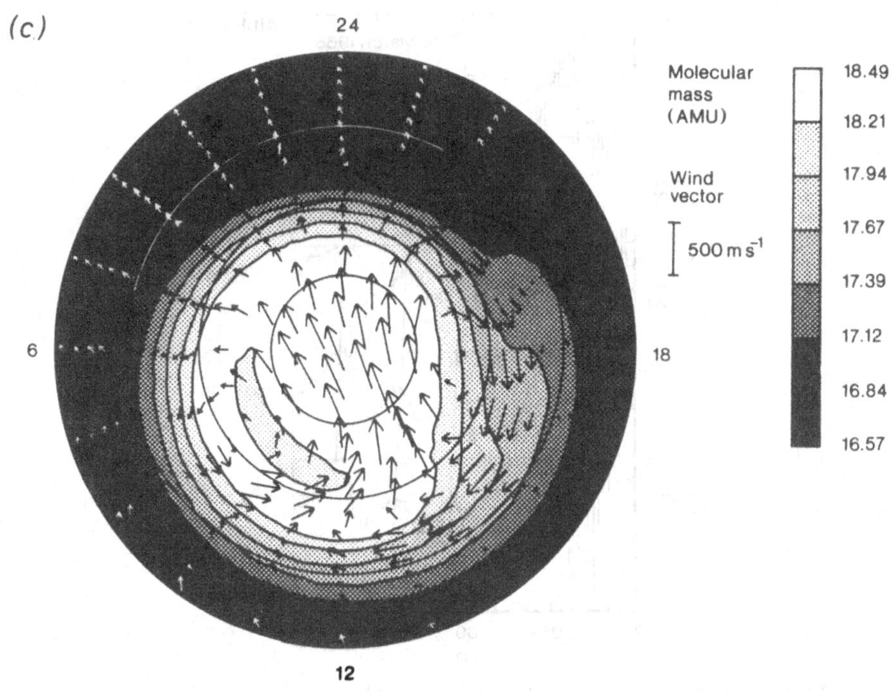

Fig. 8.6.8 (*a*) Northern hemisphere region in geographic latitude (50°–90°) local time coordinates for 18 UT. The two geophysical parameters shown for an altitude of about 320 km are contours of the electron density, an ionospheric property, and vectors of the neutral wind, a thermospheric characteristic. (*b*) Contours of the neutral temperature and wind vectors for the model computations shown in (*a*). (*c*) Contours of the mean molecular mass and wind vectors for the model computations shown in (*a*). (Courtesy of T.J. Fuller-Rowell, 1987.)

assumptions must be adopted with regard to viscosity, heat conduction, and the Coriolis effect. Coupling between ions and neutrals by the Lorentz force and by the dynamo effect is included. A solution is obtained in terms of a series of products of certain harmonic functions. Each element is identified with a wave mode that describes the space/time variation of a geophysical parameter, pressure, current, etc. For mathematical reasons, coupling between modes is neglected, although physical arguments suggest that this simplification may not be correct in the thermosphere. Publications that describe this approach to modelling thermospheric dynamics are listed in the bibliography of this chapter. While harmonic analysis cannot be expected to describe global thermospheric structure and dynamics in the detail possible with numerical models, the periodic character of several forcing functions naturally gives rise to wave motions in the atmosphere. These are described briefly in the following section.

8.7 Waves in the atmosphere and ionosphere

The quasi-periodic variations in the zonal and vertical neutral wind velocities illustrated in Figure 8.2.11 suggest that wave phenomena contribute to thermospheric

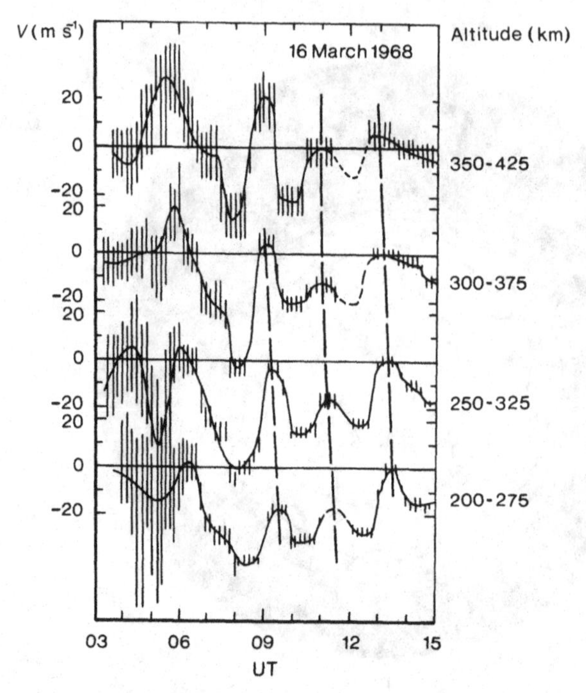

Fig. 8.7.1 Ion velocity as a function of time at various heights during the passage of a large scale TID at St-Santin, (44.7° N, 2.2° E). The velocity $V_\parallel$ is 0.87 V_0 where $V_\parallel$ is the ion velocity parallel to the magnetic field. (J. Testud, Doctoral thesis, University of Paris, 1973.)

dynamics. Figure 8.7.1 shows that variations also occur in the vertical ion velocity. The most commonly observed manifestation of waves in the thermosphere–ionosphere system occurs in the electron density profiles. Figure 8.7.2 shows contours of constant electron density as a function of altitude and time measured with four spaced ionosondes. The observed pattern has been aptly called a travelling ionospheric disturbance (TID). TIDs propagate from high to low latitudes and are associated with auroral activity.

Optical observations of wave patterns have been reported from different locations spanning a wide range of latitude. For example, using all-sky images acquired with an intensifying camera and a filter that selected the nightglow emission at 5577 Å (Equations (7.2.36) and (7.2.37)), a well-defined wave structure was observed to move across the field of view almost due south (poleward). A 5577 Å photometer recorded the changes in emission rate as the wave passed a fixed location. Substantial amplitude modulation was observed, as shown in Figure 8.7.3. Assuming an airglow emission altitude of 95–100 km, the all-sky images yielded a wavelength of about 250 km and a wave speed of about 70 m s^{-1} giving a period of about 1 h, in agreement with the photometer data. The observed intensity fluctuation is a consequence of density and temperature variations produced by poleward propagating gravity waves that originated in the lower atmosphere equatorward of the observing station.

The questions of concern deal with the sources of wave-like perturbations, whether these are internal or external to the thermosphere, and the mechanisms that produce

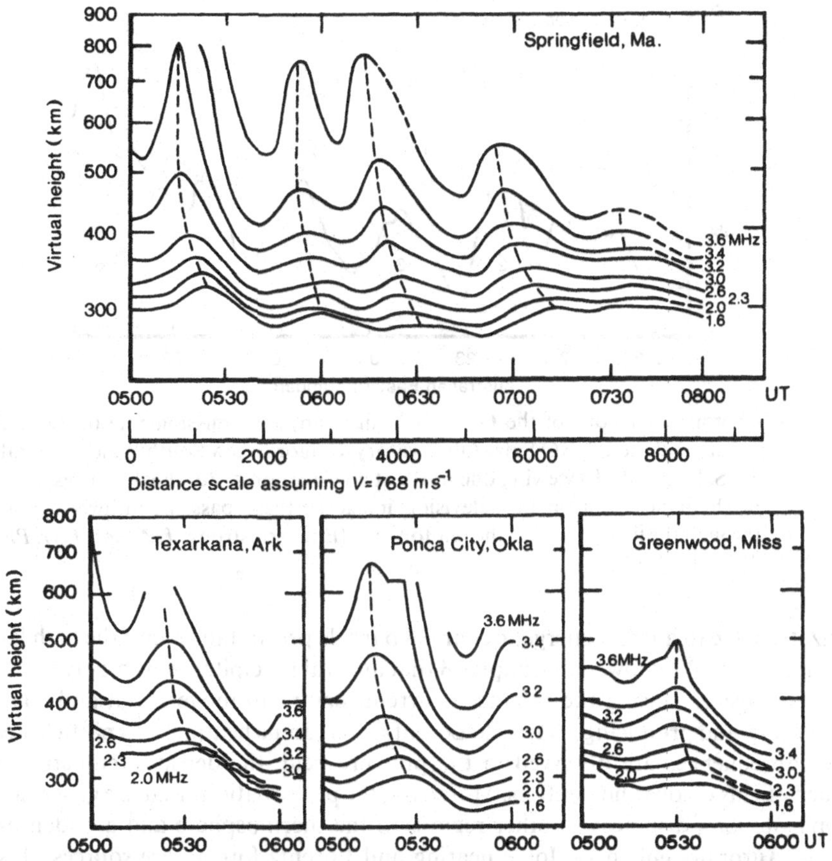

Fig. 8.7.2 Contours of constant electron density as a function of virtual height and time acquired with four spaced ionosondes. The virtual height assumes a free space propagation velocity for the electromagnetic wave reflected by the ionized layer. The contours are labelled with the local plasma frequency, f_p, where the electron density is given by $n(e) = 1.24 \times 10^{-8} f_p^2 \text{ cm}^{-3}$, with f_p in MHz. The wave structure shows the passage of a large scale TID. (T.M. Georges, *J. Atmos. Terr. Phys.*, **30**, 735–46, 1968.

the observed effects. The propagation and dissipation of waves in the atmosphere has spawned an extensive literature dealing with the mathematical analysis of the phenomenon (selected works are given in the bibliography). Since the disturbances are observed in the neutral as well as the ionized components of the atmosphere, the coupling of energy and momentum between species needs to be considered.

Wave phenomena are embedded in the overall dynamic morphology of the thermosphere and ionosphere, and their analysis involves the same physical and chemical processes discussed in the preceding sections of this chapter. The same conservation equations are adopted, momentum, energy and continuity, together with an equation of state for the gas. It should only be necessary to specify the forcing and heating sources corresponding to the physical processes that give rise to wave phenomena. For example, ionospheric currents can be substantially enhanced by

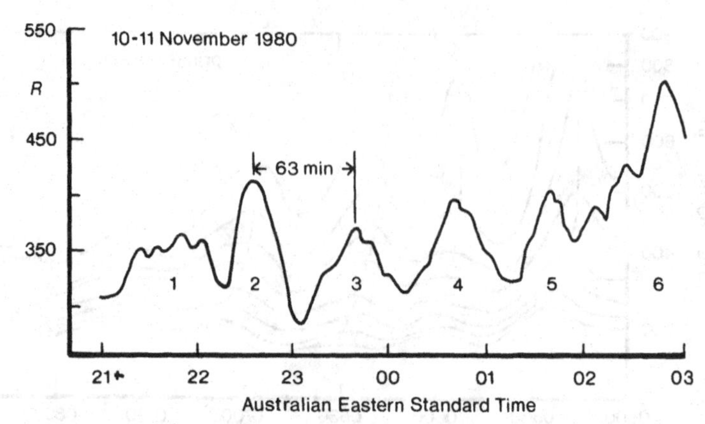

Fig. 8.7.3 Photometer record of the OI 5577 Å night airglow emission rate (in rayleighs, R) acquired at the CSIRO Solar Observatory, Culgora, New South Wales, Australia (lat. 30° S, long. 149° E) viewing due north at an elevation of 30°. Peaks correspond to the bright bands seen on the television image as these passed southward (poleward) through the field of view of the photometer. (E.B. Armstrong, *J. Atmos. Terr. Phys.*, **44**, 325–336, 1982.)

ionization due to auroral energetic charged particle precipitation into the high latitude thermosphere. We showed in Chapter 3 that auroral precipitation is highly variable in time and space, causing the enhanced current regions to move about and change in magnitude, and producing a variable Lorentz force, Joule heating, and particle heating. The consequences are observed in the pressure field, the density, temperature and velocity fields of ions and neutrals. Moreover, the perturbations that are generated may propagate long distances to other regions of the thermosphere and are identified as waves. Aurorally enhanced Joule heating and Lorentz forcing are sources of waves, internal to the thermosphere, which are confined principally to high latitudes. The diurnal variation of the solar irradiance is a predictable periodic variation of thermospheric heating. As discussed in Chapter 2, dissociation of O_2 in the Schumann Runge continuum is a major source of heating in the lower thermosphere while photoionization results in heating throughout the upper E- and lower F-regions. By far the largest fraction of the sun's energy, however, is deposited at levels below the thermosphere, in the stratosphere and troposphere. Several processes have been identified that generate waves in the troposphere and stratosphere which can propagate upward into the thermosphere. Transfer of energy and momentum at the lower boundary of our 'arbitrarily defined region' therefore needs to be included in our study of thermospheric dynamics.

The diurnal and semidiurnal variations in the global wind and temperature fields (e.g., Figure 8.6.1) are identified with atmospheric tides which are due to the gravitational attraction by the sun and moon and to solar heating. The diurnal tide is predominantly excited locally in the thermosphere by solar heating whereas the semidiurnal tidal oscillations are excited both within the thermosphere and at lower altitudes, whence they propagate upward. The long period and the planetary scale size of tidal waves requires analysis in a global spherical geometry and inclusion of the Coriolis effect. The equations set forth in Section 8.3 therefore specify the response of various geophysical parameters. Although the 24 h and 12 h periods are the dominant tidal components, atmospheric properties (temperature, wind) can also oscillate with shorter periods, higher frequencies that are integral multiples of the forcing frequency and are due to temperature gradients and horizontal winds.

In addition to the large scale, long period tidal oscillations there are atmospheric waves with periods ranging from tens of minutes to a few hours and wavelengths of a few hundred to perhaps one thousand kilometres. While the physical processes that govern the behaviour of all waves are the same, the smaller scale and shorter period perturbations allow the adoption of Cartesian coordinates and the neglect of the Coriolis effect; these waves are known as acoustic-gravity waves. Possible sources of acoustic-gravity waves external to (and below) the thermosphere include winds blowing across orographic features, instabilities and perturbations in the general tropospheric circulation system, and major impulsive perturbations such as volcanic eruptions and nuclear explosions. Joule heating and the Lorentz force associated with auroral currents have been identified as sources of gravity waves within the thermosphere.

The simplest possible situation is invoked in our presentation to illustrate the generation and propagation of waves. It is assumed that there are no dissipative forces and that the atmosphere is isothermal. (These assumptions are not made in the current research literature referenced with this chapter.) A parcel of air is displaced vertically by a source mechanism such as a thunderstorm in an unperturbed atmosphere at rest with respect to the adopted Cartesian coordinates. Gravitational acceleration acts in the vertical z direction and hydrostatic balance (discussed in Chapter 5) holds in the unperturbed atmosphere,

$$\partial p/\partial z = -\rho g \tag{5.1.5}$$

Displacements with velocity $\mathbf{v}$ result in perturbations of the pressure and density in the displaced air parcel which experiences additional forces to balance the inertial force. For emphasis, perturbed quantities are denoted by primes and average quantities by overbars. The perturbations are assumed to be small by comparison with average field quantities allowing the neglect of non-linear acceleration terms. In a two-dimensional atmosphere the force equations under the above assumptions are

$$\bar{\rho}\partial w'/\partial t + \partial p'/\partial z + \rho' g = 0 \tag{8.7.1}$$

$$\bar{\rho}\partial u'/\partial t + \partial p'/\partial x = 0 \tag{8.7.2}$$

Conservation of mass yields

$$\partial \rho'/\partial t + \partial(\bar{\rho}u')/\partial x + \partial(\bar{\rho}w')/\partial z = 0 \tag{8.7.3}$$

and conservation of energy is applied to an adiabatic change of state

$$\frac{R}{\gamma - 1}\left(\frac{\partial T'}{\partial t} + w'\frac{\partial T'}{\partial z}\right) = 0 \tag{8.7.4}$$

where γ is the ratio of specific heats, c_p/c_v, and R is the universal gas constant. (An invariant composition is assumed.) The equation of state is

$$p'/\bar{p} = T'/\bar{T} + \rho'/\bar{\rho} \tag{8.7.5}$$

Solution of the coupled set of equations yields expressions for the pressure and density perturbations, p'/p and ρ'/ρ, and the velocities u and w. Mathematical manipulation leads to wave equations for which oscillatory solutions exist. The desired parameters may all be expressed in the form

$$p'/p \propto \exp(-\beta z)\exp[\mathrm{i}(kx + lz + \omega t)] \tag{8.7.6}$$

in the two-dimensional Cartesian space with

$$\beta = -\frac{1}{2}(\omega_{\mathrm{B}}^2/g + g/c_{\mathrm{s}}^2) \tag{8.7.7}$$

where c_s is the speed of sound

$$c_s^2 = \gamma RT/M$$

for a mean molecular mass, M. The parameter ω_{B} is the Brunt–Väisälä frequency which, for an adiabatic change of state, is

$$\omega_{\mathrm{B}}^2 = (\gamma - 1)\frac{g^2}{c_{\mathrm{s}}^2} - \frac{g}{\rho}\frac{\partial \rho}{\partial z} \tag{8.7.8}$$

The proportionality factors in Equation (8.7.6) contain the phase and magnitude of $p'/p, \rho'/\rho$, u and w in terms of the wave frequency ω, the wave numbers k and l and several (assumed) constants. A dispersion equation relates the frequency to the wave numbers

$$\omega^2\left[l^2 + \frac{1}{4}\left(\frac{g}{c_{\mathrm{s}}^2} - \frac{\omega_{\mathrm{B}}^2}{g}\right)^2\right] = \left(k^2 - \frac{\omega}{c_{\mathrm{s}}^2}\right)(\omega_{\mathrm{B}}^2 - \omega^2) \tag{8.7.9}$$

which has two solutions in ω^2

$$\omega_{\mathrm{a}}^2 = \frac{c_{\mathrm{s}}^2\zeta^2}{2}\left[1 + \left(1 - \frac{4k^2\omega_{\mathrm{B}}^2}{c_{\mathrm{s}}^2\zeta^4}\right)^{1/2}\right] \tag{8.7.10}$$

$$\omega_{\mathrm{g}}^2 = \frac{c_{\mathrm{s}}^2\zeta^2}{2}\left[1 - \left(1 - \frac{4k^2\omega_{\mathrm{B}}^2}{c_{\mathrm{s}}^2\zeta^4}\right)^{1/2}\right] \tag{8.7.11}$$

where

$$\zeta^2 = k^2 + l^2 + \beta^2 \tag{8.7.12}$$

The higher frequency waves, Equation (8.7.10), are called acoustic waves while the lower frequency waves are gravity waves.

The vertical wave number l determines the type of solution obtained for the altitude dependence. For $l^2 < 0$, the solution is periodic and the waves are called internal waves. Acoustic waves generated in the lower atmosphere are efficiently dissipated by viscosity and thermal conduction before reaching thermospheric levels. Dissipation also produces a filtering effect on the gravity waves, with the high frequencies being attenuated before reaching the ionosphere. Waves that reach the thermosphere have periods ranging from some tens of minutes to several hours. The phase surfaces of propagating gravity waves in a stratified, compressible atmosphere have a forward tilt that is illustrated by their effect on the ionospheric electron density contours shown in Figure 8.7.2. Energy propagation of internal gravity waves occurs obliquely upwards while phase surfaces propagate downwards. These results are schematically illustrated

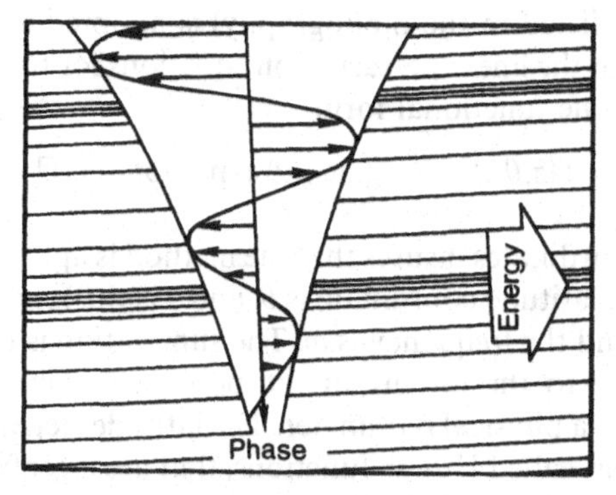

Fig. 8.7.4 Pictorial representation of internal atmospheric gravity waves. Instantaneous velocity vectors are shown, together with their instantaneous and overall envelopes. Density variations are depicted by a background of parallel lines lying in surfaces of constant phase. Phase progression is essentially downwards in this case, and energy propagation obliquely upwards; gravity is directed vertically downwards. (C.O. Hines, *Canadian J. Phys.*, **38**, 1441–81, 1960.)

in Figure 8.7.4. Were it not for dissipative forces, the wave amplitude would increase with altitude. However, molecular viscosity and turbulence dissipate wave energy and heat conduction tends to destroy the vertical wave structure. Ion–neutral drag also extracts energy from the waves, setting the ions and electrons in motion. The available energy is the sum of the perturbation kinetic energy, $\rho v'^2/2$, and the available potential energy of the perturbed flow.

The effect of waves on the ion gas is analyzed with the continuity equation that governs the ionization density (Equation (8.5.1)),

$$\partial n_i/\partial t + \nabla\cdot(n_i \mathbf{v}_i) = P_i - L_i \tag{8.5.1}$$

Perturbations of the production term, P_i, are negligible and only in the lower thermosphere is there a small perturbation in the loss rate, L_i, due to changes in chemical reactions. The principal effect of waves is upon the transport term. The ion gyrofrequency is larger than the collision frequency in the upper F-region and the perturbation ion velocity is given by

$$\mathbf{v}_i' = (u' \cos I + w' \sin I)\hat{\mathbf{b}} \tag{8.7.13}$$

where u' and w' are the perturbed velocity components of the neutral gas. If the perturbed parameters vary as $\exp[i(\mathbf{k}\cdot\mathbf{r} - \omega t)]$ then the ion density perturbation, n_i', at the peak of the F-region is given by

$$n_i'/\bar{n} = \mathbf{v}_i' \cdot \hat{\mathbf{k}}/v_p \tag{8.7.14}$$

where $\hat{\mathbf{k}} = \mathbf{k}/k$ and $v_p = \omega/k$ is the phase velocity of the wave with ω the frequency and k the wave number. Propagation of the ion density perturbations are identified with TIDs.

Internal gravity waves and semidiurnal tidal oscillations are generated in the lower atmosphere, and propagate upward to the lower boundary of the thermosphere where they become forcing parameters for the overlying gas. Well-developed classical tidal theory (references are listed in the bibliography) is adopted to specify the boundary conditions. Solutions to the linear perturbation equations for the physical variables ρ, p, T and w are given in the functional form

$$f(z,\theta,t) = \sum_n f_n(z)G_n(\theta) \exp[i(\omega t + s\lambda)] \tag{8.7.15}$$

Adopting spherical coordinates insures that the method is applicable to both small and large scale waves. The latitude coordinate is θ, and s is the longitudinal wave number. The wavelength is λ and the frequency is ω. The summation is taken over all the wave modes that contribute to the forcing at the boundary. Thus, $f_n(z)$ is the altitude-dependent nth mode of a physical parameter. The latitude-dependent quantities, $G_n(\theta)$, are harmonic functions, called Hough functions, that are eigenfunctions of the vertical structure differential equation. The Hough functions are expansions of the associated Legendre polynomials and approach spherical harmonics at thermospheric altitudes. The important point is that the physical parameters oscillate with time at the lower boundary and the effect on the thermosphere is the same as that produced by wave propagation from below.

Numerical computations have shown that gravity waves can propagate over long horizontal distances in the thermosphere. Taking quantitative account of the structure and composition of the real atmosphere and ionosphere leads to a complex mathematical formulation of the propagation of waves and the coupling of energy and momentum with the neutral and ionized gas. The problem is a current research topic stimulated by a diversity of recent observational evidence.

The dynamical response of the neutral and ionized components of the thermosphere to various energy sources and forcing mechanisms has, so far, been assumed to be stable, while dissipating the applied energy and momentum. Once the sources are turned off, the neutral gas and plasma relax to the unperturbed state. The response of this system does not allow for positive feedback leading to growth of the perturbation. While the stable operational mode accounts for most of the large scale dynamical behaviour of the thermosphere, there are observations, especially of the ionized components, that clearly demonstrate the occurrence of an unstable response to perturbations. Observations of wave phenomena have been useful for identifying the nature and source of the instabilities. The variety of plasma instabilities is large, and an exposition of the mechanisms that produce them is the subject of several specialized monographs. We only note here that the hydromagnetic equations given at the end of Section 8.5 are the governing equations for low frequency phenomena, but it is the boundary conditions in time and space that determine solutions which may be associated with one or more observed instabilities.

Bibliography

Section 8.2

Sample measurements of the dynamic behaviour of the neutral and ionized gases given in this section were acquired by several observational techniques. Release of chemical tracers from rockets has been described by

J.P. Heppner and M.L. Miller, Thermospheric winds at high latitudes from chemical release observations, *J. Geophys. Res.*, **87**, 1633–47, 1982.

Several groups scattered worldwide are using interferometric spectroscopy to measure neutral and ion winds and temperatures. The more recent articles give references to previous publications about different instruments and analysis techniques.

D. Rees, P. Charleton, M. Carlson and P. Rounce, High latitude thermospheric circulation during the Energy Budget Campaign, *J. Atmos. Terr. Phys.*, **47**, 195–232, 1985.

M.A. Biondi, D.P. Sipler and M. Weinschenker, Multiple aperture exit plate for field widening a Fabry–Perot interferometer, *Appl. Optics*, **24**, 232–6, 1985.

T.L. Killeen, B.C. Kennedy, P.B. Hays, D.A. Symanow and D.H. Ceckowski, Image plane detector for the Dynamics Explorer Fabry–Perot Interferometer, *Appl. Optics*, **22**, 3503–13, 1983.

G. Hernandez, Measurement of thermospheric temperature and winds by remote Fabry–Perot spectrometry, *Opt. Engin.*, **19**, 518–32, 1980.

F. Jacka, Application of Fabry–Perot spectrometers for measurement of upper atmosphere winds and temperature, *Handbook for the Middle Atmosphere Program*, Vol. 13, Ed. R.A. Vincent, 1985.

F. Jacka, A.R.D. Bower, D.F. Creighton and P.A. Wilksch, A large aperture high resolution Fabry– Perot spectrometer for airglow studies, *J. Phys. E: Scient. Instrum.*, **13**, 562–8, 1980.

T.D. Cocks, D.F. Creighton and F. Jacka, Application of a dual Fabry–Perot spectrometer for daytime airglow studies, *J. Atmos. Terr. Phys.*, **42**, 499–511, 1980.

G.G. Shepherd, *et al.*, Optical Doppler imaging of the Aurora Borealis, *Geophys. Res. Lett.*, **11**, 1003–6, 1984.

The principles of operation of incoherent scatter radar and the extraction of geophysical parameters are explained by

J.V. Evans, Ionospheric movements measured by incoherent scatter: A review: *J. Atmos. Terr. Phys.*, **34**, 175, 1972; *Proc. IEEE*, **57**, 496, 1969; Incoherent scatter contributions to studies of the dynamics of the lower thermosphere, *Rev. Geophys.*, **16**, 195–216, 1978.

The most recently completed incoherent scatter radar facility is in northern Scandinavia. A complete special issue of a journal is devoted entirely to a description of the facility and first results,

EISCAT Science: Results from the first year's operation of the European Incoherent Scatter Radar, *J. Atmos. Terr. Phys.*, **46**, No. 6/7, 1984.

There have been several satellite missions flown that provided data relevant to thermospheric and ionospheric dynamics. The Dynamics Explorer 2 satellite has yielded a wealth of measurements, winds, ion drifts, temperatures and densities. The instruments are described in a special issue of a journal,

The Dynamics Explorer Mission, *Space Sci. Instrum.*, 5, No. 4, 1981.

Scientific results have been reported in many articles and only a few recent contributions are listed which provide the reader with references to previous publications,

T.L. Killeen, *et al.*, Mean neutral circulation in the winter polar F-region, *J. Geophys.* Res., **91**, 1633–49, 1986.

D. Rees, *et al.*, The composition, structure, temperature and dynamics of the upper thermosphere in the polar regions during October to December 1981, *Planet. Space Sci.*, **33**, 617–66, 1985.

D. Rees, *et al.*, A comparison of wind observations of the upper thermosphere from the Dynamic Explorer satellite with the predictions of a global time-dependent model, *Planet. Space Sci.*, **31**, 1299–314, 1983.

R.G. Roble *et al.*, Thermospheric circulation, temperature and compositional structure of the southern hemisphere polar cap during October–November 1981, *J. Geophys. Res.*, **89**, 9057–68, 1984.

Sections 8.3–8.6

Theoretical analysis of thermospheric dynamics has evolved into three-dimensional time-dependent numerical modelling. Two independent groups of investigators have contributed to this advance and we list the two papers that set forth the theoretical models adopted by each group,

T.J. Fuller-Rowell and D. Rees, A three-dimensional and time-dependent global model of the thermosphere, *J. Atmos. Sci.*, **37**, 2545–67, 1980.

R.E. Dickinson, E.C. Ridley and R.G. Roble, A three-dimensional general circulation model of the thermosphere, *J. Geophys. Res.*, **86**, 1499–512, 1981.

Continuing development has focused on removing arbitrary constraints and assumptions, adopting more realistic input parameters, and in achieving self-consistent solutions. The following papers indicate a few of the steps.

T.J. Fuller-Rowell and D. Rees, Derivation of a conservation equation for mean molecular weight for a two-constituent gas within a three-dimensional, time-dependent model of the thermosphere, *Planet. Space Sci.*, **31**, 1209, 1222, 1983.

R.E. Dickinson, E.C. Ridley and R.G. Roble, Thermospheric general circulation with coupled dynamics and composition, *J. Atmos. Sci.*, **41**, 205–19, 1984.

R.G. Roble, R.E. Dickinson and E.C. Ridley, Global circulation and temperature structure of the thermosphere with high-latitude plasma convection, *J. Geophys. Res.*, **87**, 1599–614, 1982.

M.F. Smith, D. Rees and T.J. Fuller-Rowell, The consequences of high latitude particle precipitation on global thermospheric dynamics, *Planet. Space Sci.*, **30**, 1259–67, 1982.

Plasma dynamics has been investigated theoretically by several groups with emphasis partitioned between high and low latitude phenomena. A few of the articles that develop the basic concepts are listed.

D.N. Anderson, Modelling the ambient, low latitude F-region ionosphere – a review,, *J. Atmos. Terr. Phys*, **43**, 753–61, 1981.

J.J. Sojka, W.J. Raitt and R.W. Schunk, A theoretical study of the high latitude winter F-region at solar minimum for low magnetic activity, *J. Geophys. Res.*, **86**, 609–21, 1981.

S. Quegan *et al.*, A theoretical study of the distribution of ionization in the high latitude ionosphere and the plasmasphere: first results on the mid-latitude trough and the light ion trough, *J. Atmos. Terr. Phys.*, **44**, 619–42, 1982.

A global perspective of the dynamics of the plasma is developed by

J.J. Sojka and R.W. Schunk, A theoretical study of the global F-region for June solstice, solar maximum and low magnetic activity, *J. Geophys. Res.*, **90**, 5285–98, 1985.

The first fully self-consistent analysis of neutral and plasma dynamics and composition at high latitudes has been developed by

T.J. Fuller-Rowell, D. Rees, S. Quegan, R.J. Moffett and G.J. Bailey, Interactions between neutral thermospheric composition and the polar ionosphere using a coupled ionosphere–thermosphere model, *J. Geophys. Res.*, **92**, 7744–48, 1987.

and the first self-consistent global mean model has been presented by

R.G. Roble, E.C. Ridley and R.E. Dickinson, On the global mean structure of the thermosphere, *J. Geophys. Res.*, **92**, 8745–58, 1987.

The following articles expound on the conceptual aspects of thermospheric and ionospheric dynamics, providing further insight to the physical processes at work.

H. Rishbeth, Thermospheric winds and the F-region: a review, *J. Atmos. Terr. Phys.*, **34**, 1–47, 1972; Electromagnetic transport processes in the ionosphere, in *Physics and Chemistry of the Upper Atmosphere*, Ed. B.M. McCormac, D. Reidel Publ. Co., Dordrecht, 1973, pp. 43–53; Drifts and winds in the polar F-region, *J. Atmos. Terr. Phys.*, **39**, 111–6, 1977.

H.G. Mayr, I. Harris and N.W. Spencer, Some properties of upper atmosphere dynamics, *Rev. Geophys. Space Phys.*, **16**, 539–65, 1978.

H. Volland and L. Grellman, A hydromagnetic dynamo of the atmosphere, *J. Geophys. Res.*, **83**,

3699–708, 1978; Magnetospheric electric fields and currents and their influence on large scale thermospheric circulation and composition, *J. Atmos. Terr. Phys.*, **41**, 853–66, 1979.

There are two sections in a recent book that concisely summarize much of our theoretical understanding of the thermosphere and ionosphere.

A.D. Richmond, Thermospheric dynamics and electrodynamics, pp. 523–607; R.W. Schunk, The terrestrial ionosphere, pp. 609–676, in *Solar Terrestrial Physics – Principles and Theoretical Foundations*, Ed. R.L. Carovillano and J.M. Forbes, D. Reidel Publ. Co., Dordrecht, 1983.

Section 8.7

One scientist, more than any other, has brought to the attention of atmospheric dynamicists the important role of wave phenomena. Colin Hines has collected his and some of his colleagues' research papers in a monograph, profusely annotated and updated to the publication deadline:

C.O. Hines, The Upper Atmosphere in Motion, *Geophysical Monograph 18*, American Geophysical Union, Washington, D.C. 1974.

There are several texts devoted to the general concepts and the basic mathematical formulation of waves in the atmosphere and energy transport.

C. Eckart, *Hydrodynamics of Oceans and Atmospheres*, Pergamon Press, Oxford, 1960.

J. Van Mieghem, *Atmospheric Energetics*, Clarendon Press, Oxford, 1973.

J.R. Holton, *The Dynamic Meteorology of the Stratosphere and Mesosphere*, Meteorological Monograph Vol 15, No. 37, American Meteorological Society, Boston, 1975.

R.G. Fleagle and J.A. Businger, *An Introduction to Atmospheric Physics*, Academic Press, New York, 1980.

An authoritative specialized monograph which provides a detailed mathematical development of the subject is:

S. Chapman and R.S. Lindzen, *Atmospheric Tides, Thermal and Gravitational*, Gordon and Breach, 1970.

A recent concise mathematical development of the subject of atmospheric waves is given by

A.D. Richmond, Thermospheric Dynamics and Electrodynamics, chapter in *Solar Terrestrial Physics*, Ed. R.L. Carovillano and J.M. Forbes, pp. 555–72, D. Reidel Publ. Co., Dordrecht, 1983.

Fundamental aspects of wave dissipation are expounded from basic principles by

C.O. Hines, Relaxational dissipation in atmospheric waves – I Basic formulation; II Application to Earth's upper atmosphere, *Planet. Space Sci.*, **25**, pp. 1045–74, 1977.

Complementary articles show that a subject may be discussed in different ways:

K.C. Yeh and C.H. Liu, Acoustic gravity waves in the upper atmosphere, *Rev. of Geophys. and Space Phys.*, **12**, 193–216, 1974.

S.H. Francis, Global propagation of Atmospheric Gravity Waves: A review, *J. Atmos. Terr. Phys.*, **37**, 1011–54, 1975.

H.G. Mayr, *et al.*, Global excitation of wave phenomena in a dissipative multi-constituent medium 1. Transfer function of the Earth's thermosphere. 2. Impulsive perturbations in the Earth's thermosphere, *J. Geophys. Res.*, **89**, 10 929–86, 1984; 3. Response characteristics for different sources in the Earth's thermosphere, *J. Geophys. Res.*, **92**, 7657–72, 1987.

R.D. Hunsucker, Atmospheric Gravity Waves generated in the high-latitude ionosphere: a review, *Rev. of Geophys. and Space Phys.*, **20**, 293–315, 1982.

Recent articles discuss coupling from the lower to the upper atmosphere and give references to many previous publications.

C.G. Fesen, R.E. Dickinson and R.G. Roble, Simulation of the thermospheric tides at equinox with the NCAR thermospheric general circulation model, *J. Geophys. Res.*, **91**, 4471–89, 1986.

J.M. Forbes, Atmospheric tides 1. Model description and results for the solar diurnal component. 2. The solar and lunar semidiurnal components, *J. Geophys. Res.*, **87**, 5222–52, 1982.

A summary of recent observations, theory and numerical modelling is given by

D. Rees, R.W. Smith, P.J. Charleton, F.G. McCormac, N. Lloyd and Å. Steen, The generation of vertical thermospheric winds and gravity waves at auroral latitudes; I observations of vertical winds; and II theory and numerical modelling of vertical winds, *Planet. Space Sci.*, **32**, 667–705, 1984.

Problems 42–54

Problem 42 Starting with the ion diffusion equation derive Equation (8.3.12) for the orthogonal ion drift velocity.

Problem 43 Derive Equations (8.3.20) and (8.3.21) for $\sigma_\perp$ and σ_H. Compute and plot altitude profiles of ν_{en}, ν_{in}, ν_{ei}, ω_{ce}/ν_{en}, ω_{ci}/ν_{in} and of the three conductivity coefficients σ_0, $\sigma_\perp$ and σ_H for a typical daytime ionosphere and neutral atmosphere. To alleviate computational complexity assume an average molecular ion mass in the E-region and O^+ ions in the F-region.

Problem 44 Derive the Relationship (8.3.25) between acceleration in a rotating reference frame and in an inertial reference frame.

Problem 45 Show that the pressure tensor and the electromagnetic force are unchanged by transformation from an inertial to a rotating reference frame.

Problem 46 Derive Equation (8.3.28) for the acceleration terms appropriate to the momentum equation in a rotating coordinate system.

Problem 47 Letting m_2 be the mass of molecular nitrogen and m_1 that of atomic O, find how the mean molecular mass, $\bar{m}$, decreases with increasing altitude, neglecting eddy diffusion.

Problem 48 Derive the zonal and meridional components of the ion–neutral drag term. Show that, for $\omega_{ci} \gg \nu_{in}$, the coefficients reduce to the collision frequency and gyrofrequency. Describe qualitatively how zonal and meridional ion drag are coupled in the high latitude F-region. What would the coefficients reduce to in the lower E-region where $\omega_{ci} \ll \nu_{in}$?

Problem 49 Derive Equation (8.3.40) for the drift velocity of a flux tube.

Problem 50 Show that the equipotentials are identical to the plasma drift trajectories for a divergence-free electric field.

Problem 51 Show that some H^+ and He^+ ions can escape from the Earth's ionosphere. Assume a plasma temperature of 2000 K.

Problem 52 Assume that the reaction rate coefficients for the forward and backward channels in the accidental resonant charge exchange process (Reaction (8.5.7)) may be expressed by Equation (4.4.17) for interacting Maxwellian gases. Noting that the fine structure levels are populated according to their statistical weights, derive the ion density ratio, Equation (8.5.8).

Problem 53 Derive Equation (8.5.10) for the (subsonic) flow of a minor ion in the upper ionosphere assuming that the forces acting on the ions are the pressure gradient, gravity, polarization field due to major ion and electron flow, and Coulomb collisions with the major ion.

Problem 54 Compute the ratio of Hall to Pedersen conductivity as a function of altitude for typical daytime densities and composition conditions. Discuss how this result accounts for the high latitude ion drift in the E- and F-regions of the ionosphere.

Appendix 1

The neutral atmosphere

Empirical models are constructed from a large data base of observations made *in situ* by satellite borne mass spectrometers and remotely from the ground. Minimum perigee for long-lived satellites is at about 150 km, depending on the available on-board propulsion system. Data points below this altitude are therefore inferred from incoherent scatter radar measurements, which also serve to follow the temporal variation of several parameters at one location. Satellites provide the global coverage.

One such empirical model is based on decades of accumulated measurements acquired by several low-altitude orbiting satellites and radar measurements. It has been updated periodically and the most recent version is the MSIS-86. The large number of individual measurements allowed binning by geographic position and universal time, solar zenith angle, solar activity and geomagnetic activity.

The parameters computed by the empirical model as a function of altitude are the concentrations of N_2, O, O_2, H, He, Ar, N, the mass density and the neutral temperature. The data sets provided here correspond to the solar minimum (July 1976) and solar maximum (February 1979) periods used as examples in Chapter 2, as well as day 23 April, 1974 on which the solar irradiance reference spectrum for the Atmosphere Explorer satellite was acquired by a rocket payload.

Reference:

A.E. Hedin, MSIS-86 Thermospheric Model, *J. Geophys. Res.*, **92**, 4649, 1987.

Concentrations of several minor species in the lower thermosphere are collected from various sources and plotted as a function of altitude. There is a large variability associated with several species due to geographic, diurnal and seasonal effects.

Table A1.1 *MSIS-86 model atmosphere for 13 July 1976*
LAT = 50.0, LON = − 150.0, ZEN = 28.
F107A = 71.8, F107 = 67.6, AP = 7.0

Alt. (KM)	N(N2) (CM** − 3)	N(O) (CM** − 3)	N(02) (CM** − 3)	N(H) (CM** − 3)	N(HE) (CM** − 3)	N(AR) (CM** − 3)	N(N) (CM** − 3)	Density (G/CM**3)	Neut. Temp. (K)
90	5.577E + 13	1.670E + 11	1.445E + 13	7.761E + 07	4.227E + 08	6.573E + 11	1.770E + 05	3.408E − 09	163.
100	7.097E + 12	2.344E + 11	1.665E + 12	1.772E + 07	8.673E + 07	7.391E + 10	2.499E + 05	4.294E − 10	202.
110	1.344E + 12	1.273E + 11	2.501E + 11	7.465E + 06	3.818E + 07	1.059E + 10	5.230E + 05	7.984E − 11	257.
120	3.202E + 11	5.391E + 10	4.394E + 10	3.481E + 06	1.838E + 07	1.797E + 09	1.147E + 06	1.877E − 11	360.
130	1.104E + 11	2.572E + 10	1.217E + 10	1.724E + 06	5.987E + 06	4.746E + 08	2.501E + 06	6.491E − 12	495.
140	5.133E + 10	1.490E + 10	4.991E + 09	9.716E + 05	2.893E + 06	1.774E + 08	5.194E + 06	3.058E − 12	595.
150	2.769E + 10	9.578E + 09	2.478E + 09	6.106E + 05	2.264E + 06	7.879E + 07	9.279E + 06	1.679E − 12	669.
160	1.628E + 10	6.575E + 09	1.362E + 09	4.250E + 05	1.994E + 06	3.866E + 07	1.406E + 07	1.006E − 12	723.
170	1.011E + 10	4.716E + 09	7.965E + 08	3.236E + 05	1.811E + 06	2.020E + 07	1.836E + 07	6.393E − 13	764.
180	6.512E + 09	3.488E + 09	4.845E + 08	2.652E + 05	1.669E + 06	1.101E + 07	2.118E + 07	4.223E − 13	795.
190.	4.301E + 09	2.639E + 09	3.029E + 08	2.299E + 05	1.552E + 06	6.177E + 06	2.218E + 07	2.870E − 13	817.
200	2.893E + 09	2.030E + 09	1.930E + 08	2.075E + 05	1.452E + 06	3.541E + 06	2.159E + 07	1.994E − 13	834.
210	1.971E + 09	1.583E + 09	1.248E + 08	1.928E + 05	1.364E + 06	2.062E + 06	1.991E + 07	1.409E − 13	846.
220	1.356E + 09	1.246E + 09	8.154E + 07	1.827E + 05	1.287E + 06	1.215E + 06	1.768E + 07	1.010E − 13	856.
230	9.397E + 08	9.896E + 08	5.369E + 07	1.755E + 05	1.217E + 06	7.225E + 05	1.528E + 07	7.323E − 14	863.
240	6.550E + 08	7.909E + 08	3.557E + 07	1.701E + 05	1.152E + 06	4.326E + 05	1.298E + 07	5.368E − 14	868.
250	4.586E + 08	6.356E + 08	2.368E + 07	1.659E + 05	1.093E + 06	2.606E + 05	1.090E + 07	3.973E − 14	872.
260	3.222E + 08	5.131E + 08	1.583E + 07	1.625E + 05	1.038E + 06	1.576E + 05	9.093E + 06	2.967E − 14	875.
270	2.271E + 08	4.157E + 08	1.061E + 07	1.596E + 05	9.859E + 05	9.571E + 04	7.557E + 06	2.235E − 14	877.
280	1.604E + 08	3.378E + 08	7.137E + 06	1.570E + 05	9.373E + 05	5.830E + 04	6.272E + 06	1.696E − 14	878.
290	1.135E + 08	2.753E + 08	4.809E + 06	1.547E + 05	8.916E + 05	3.561E + 04	5.205E + 06	1.297E − 14	880.
300	8.051E + 07	2.249E + 08	3.248E + 06	1.526E + 05	8.485E + 05	2.180E + 04	4.323E + 06	9.995E − 15	881.
310	5.718E + 07	1.798E + 08	2.197E + 06	1.505E + 05	8.077E + 05	1.338E + 04	3.595E + 06	7.641E − 15	881.
320	4.067E + 07	1.480E + 08	1.488E + 06	1.486E + 05	7.792E + 05	8.224E + 03	2.995E + 06	5.976E − 15	882.

330	2.897E + 07	1.219E + 08	1.010E + 06	1.466E + 05	7.326E + 05	5.066E + 03	2.500E + 06	4.701E − 15	882.
340	2.066E + 07	1.005E + 08	6.863E + 05	1.448E + 05	6.980E + 05	3.126E + 03	2.090E + 06	3.718E − 15	883.
350	1.475E + 07	8.285E + 07	4.670E + 05	1.431E + 05	6.651E + 05	1.932E + 03	1.750E + 06	2.956E − 15	883.
360	1.054E + 07	6.838E + 07	3.182E + 05	1.414E + 05	6.339E + 05	1.196E + 03	1.469E + 06	2.362E − 15	883.
370	7.544E + 06	5.648E + 07	2.171E + 05	1.397E + 05	6.042E + 05	7.415E + 02	1.234E + 06	1.895E − 15	883.
380	5.404E + 06	4.667E + 07	1.483E + 05	1.380E + 05	5.761E + 05	4.604E + 02	1.038E + 06	1.527E − 15	883.
390	3.875E + 06	3.859E + 07	1.014E + 05	1.364E + 05	5.493E + 05	2.863E + 02	8.750E + 05	1.235E − 15	883.
400	2.782E + 06	3.193E + 07	6.941E + 04	1.348E + 05	5.239E + 05	1.783E + 02	7.382E + 05	1.002E − 15	883.
410	1.999E + 06	2.644E + 07	4.758E + 04	1.332E + 05	4.997E + 05	1.112E + 02	6.235E + 05	8.156E − 16	883.
420	1.438E + 06	2.190E + 07	3.265E + 04	1.316E + 05	4.768E + 05	6.945E + 01	5.272E + 05	6.658E − 16	883.
430	1.035E + 06	1.815E + 07	2.243E + 04	1.301E + 05	4.549E + 05	4.344E + 01	4.462E + 05	5.450E − 16	883.
440	7.461E + 05	1.505E + 07	1.543E + 04	1.286E + 05	4.341E + 05	2.721E + 01	3.779E + 05	4.472E − 16	883.
450	5.382E + 05	1.249E + 07	1.062E + 04	1.271E + 05	4.143E + 05	1.706E + 01	3.204E + 05	3.677E − 16	883.
460	3.886E + 05	1.037E + 07	7.321E + 03	1.256E + 05	3.955E + 05	1.072E + 01	2.697E + 05	3.030E − 16	883.
470	2.809E + 05	8.614E + 06	5.051E + 03	1.242E + 05	3.776E + 05	6.740E + 00	2.293E + 05	2.501E − 16	883.
480	2.032E + 05	7.159E + 06	3.489E + 03	1.227E + 05	3.605E + 05	4.245E + 00	1.950E + 05	2.069E − 16	883.
490	1.472E + 05	5.953E + 06	2.413E + 03	1.213E + 05	3.442E + 05	2.677E + 00	1.660E + 05	1.714E − 16	883.
500	1.067E + 05	4.953E + 06	1.671E + 03	1.199E + 05	3.288E + 05	1.690E + 00	1.413E + 05	1.423E − 16	883.
510	7.740E + 04	4.124E + 06	1.158E + 03	1.186E + 05	3.141E + 05	1.069E + 00	1.203E + 05	1.183E − 16	883.
520	5.621E + 04	3.435E + 06	8.032E + 02	1.172E + 05	3.000E + 05	6.767E − 01	1.026E + 05	9.845E − 17	883.
530	4.086E + 04	2.862E + 06	5.578E + 02	1.159E + 05	2.867E + 05	4.290E − 01	8.744E + 04	8.208E − 17	883.
540	2.973E + 04	2.387E + 06	3.878E + 02	1.146E + 05	2.739E + 05	2.724E − 01	7.458E + 04	6.853E − 17	883.
550	2.165E + 04	1.991E + 06	2.699E + 02	1.133E + 05	2.618E + 05	1.731E − 01	6.364E + 04	5.731E − 17	883.
560	1.578E + 04	1.662E + 06	1.881E + 02	1.120E + 05	2.502E + 05	1.102E − 01	5.434E + 04	4.799E − 17	883.
570	1.151E + 04	1.388E + 06	1.312E + 02	1.108E + 05	2.392E + 05	7.024E − 02	4.641E + 04	4.026E − 17	883.
580	8.406E + 03	1.160E + 06	9.157E + 01	1.095E + 05	2.287E + 05	4.483E − 02	3.966E + 04	3.382E − 17	883.
590	6.144E + 03	9.695E + 05	6.399E + 01	1.083E + 05	2.187E + 05	2.865E − 02	3.391E + 04	2.846E − 17	883.
600	4.495E + 03	8.109E + 05	4.477E + 01	1.071E + 05	2.091E + 05	1.833E − 02	2.900E + 04	2.399E − 17	883.

MSIS-86 Model atmosphere, 13 July 1976 at 22 UT
Lat. 50°, Long. −150°, Zenith angle of Sun 28°
$\overline{F10.7}$ = 71.8, F10.7 = 67.6, Ap = 7.0

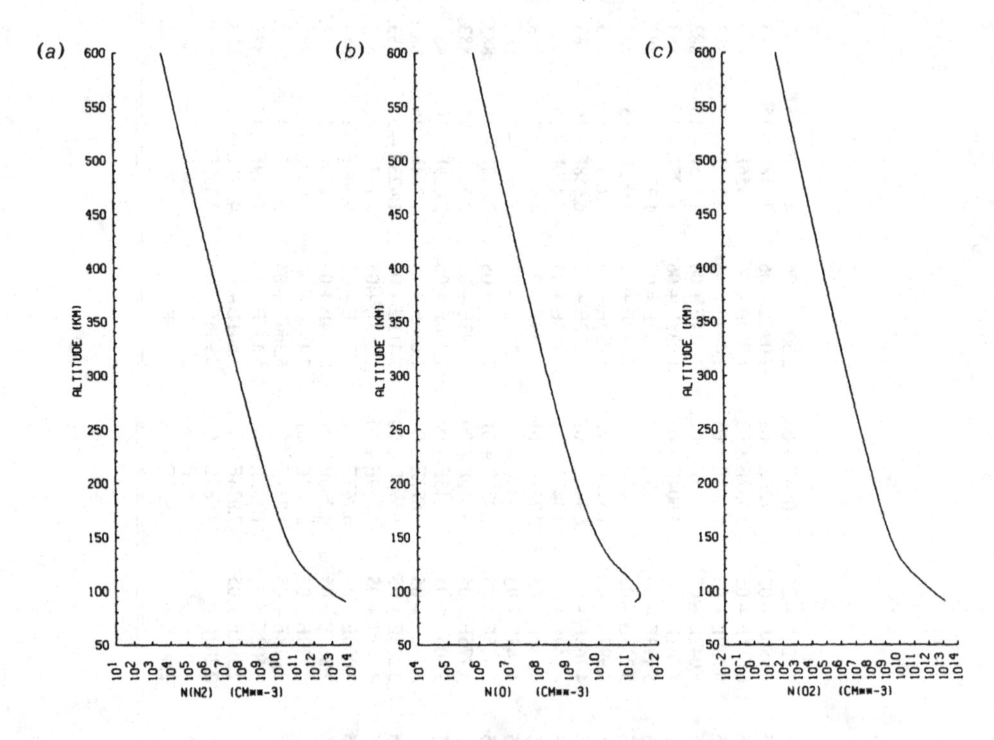

MSIS-86 Model atmosphere, 13 July 1976 at 22 UT
Lat. 50°, Long. −150°, Zenith angle of Sun 28°
$\overline{F10.7}$ = 71.8, F10.7 = 67.6, Ap = 7.0

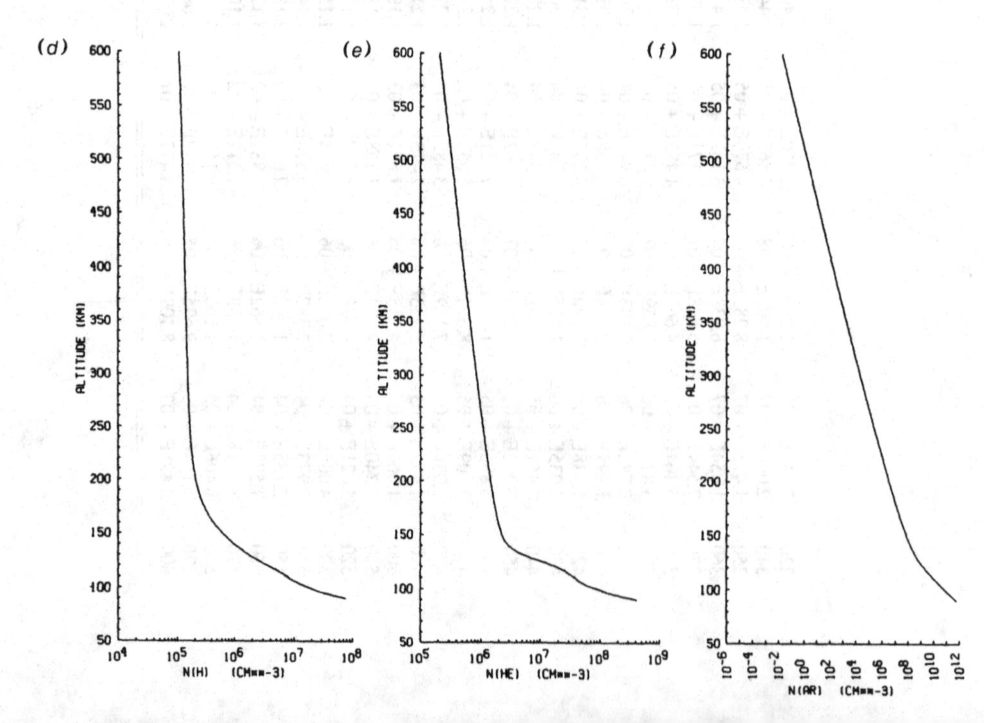

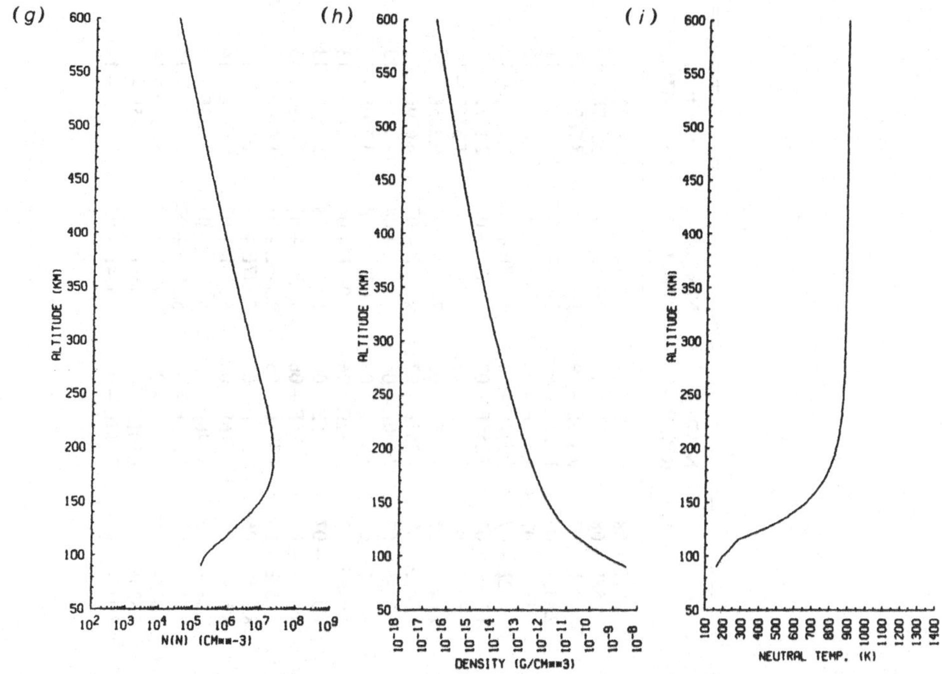

Fig. A1.1 MSIS-86 model atmosphere for 13 July, 1976 at 22 UT (lat. = 50.0°, long. = − 150.0°, zen. = 28, F107A = 71.8, F107 = 67.6, $AP = 7.0$): (*a*) N_2, (*b*) O, (*c*) O_2, (*d*) H, (*e*) He, (*f*) Ar, (*g*) N, (*h*) density, (i) neutral temperature.

Table A1.2. *MSIS-86 model atmosphere for 19 February 1979*

LAT = 50.0, LON = − 150.0, ZEN = 62.
F107A = 187.7, F107 = 243.3, AP = 15.0

Alt. (KM)	N(N2) (CM** − 3)	N(O) (CM** − 3)	N(02) (CM** − 3)	N(H) (CM** − 3)	N(HE) (CM** − 3)	N(AR) (CM** − 3)	N(N) (CM** − 3)	Density (G/CM**3)	Neut. Temp. (K)
90	4.903E + 13	2.163E + 11	1.276E + 13	5.304E + 07	3.811E + 08	5.497E + 11	3.391E + 05	2.999E − 09	194.
100	8.605E + 12	4.355E + 11	2.054E + 12	1.397E + 07	1.058E + 08	8.356E + 10	7.076E + 05	5.262E − 10	191.
110	1.429E + 12	2.227E + 11	2.752E + 11	4.967E + 06	4.636E + 07	9.879E + 09	1.442E + 06	8.760E − 11	249.
120	3.186E + 11	9.141E + 10	4.710E + 10	1.863E + 06	2.832E + 07	1.521E + 09	3.144E + 06	1.984E − 11	393.
130	1.160E + 11	4.725E + 10	1.423E + 10	7.889E + 05	2.336E + 07	4.151E + 08	7.591E + 06	7.432E − 12	547.
140	5.594E + 10	2.907E + 10	6.154E + 09	3.804E + 05	2.062E + 07	1.582E + 08	1.707E + 07	3.710E − 12	674.
150	3.131E + 10	1.979E + 10	3.206E + 09	2.084E + 05	1.796E + 07	7.255E + 07	3.282E + 07	2.157E − 12	779.
160	1.918E + 10	1.439E + 10	1.858E + 09	1.297E + 05	1.595E + 07	3.740E + 07	5.339E + 07	1.376E − 12	867.
170	1.250E + 10	1.095E + 10	1.153E + 09	9.058E + 04	1.444E + 07	2.088E + 07	7.476E + 07	9.365E − 13	940.
180	8.511E + 09	8.607E + 09	7.493E + 08	6.962E + 04	1.327E + 07	1.233E + 07	9.242E + 07	6.670E − 13	1000.
190	5.979E + 09	6.926E + 09	5.042E + 08	5.759E + 04	1.233E + 07	7.590E + 06	1.036E + 08	4.916E − 13	1051.
200	4.302E + 09	5.673E + 09	3.481E + 08	5.029E + 04	1.155E + 07	4.819E + 06	1.077E + 08	3.720E − 13	1093.
210	3.153E + 09	4.709E + 09	2.454E + 08	4.563E + 04	1.088E + 07	3.132E + 06	1.061E + 08	2.874E − 13	1128.
220	2.346E + 09	3.950E + 09	1.756E + 08	4.253E + 04	1.031E + 07	2.075E + 06	1.004E + 08	2.258E − 13	1157.
230	1.765E + 09	3.340E + 09	1.273E + 08	4.038E + 04	9.803E + 06	1.395E + 06	9.239E + 07	1.798E − 13	1181.
240	1.341E + 09	2.843E + 09	9.319E + 07	3.883E + 04	9.353E + 06	9.487E + 05	8.346E + 07	1.449E − 13	1201.
250	1.027E + 09	2.433E + 09	6.882E + 07	3.767E + 04	8.945E + 06	6.521E + 05	7.447E + 07	1.178E − 13	1218.
260	7.911E + 08	2.091E + 09	5.117E + 07	3.676E + 04	8.573E + 06	4.516E + 05	6.593E + 07	9.665E − 14	1232.
270	6.128E + 08	1.804E + 09	3.827E + 07	3.603E + 04	8.230E + 06	3.148E + 05	5.811E + 07	7.985E − 14	1244.
280	4.769E + 08	1.560E + 09	2.877E + 07	3.543E + 04	7.912E + 06	2.208E + 05	5.111E + 07	6.639E − 14	1254.
290	3.725E + 08	1.353E + 09	2.171E + 07	3.490E + 04	7.615E + 06	1.556E + 05	4.492E + 07	5.551E − 14	1262.
300	2.919E + 08	1.176E + 09	1.644E + 07	3.444E + 04	7.336E + 06	1.101E + 05	3.949E + 07	4.665E − 14	1269.
310	2.294E + 08	1.029E + 09	1.249E + 07	3.402E + 04	7.073E + 06	7.817E + 04	3.474E + 07	3.952E − 14	1275.

320	1.807E + 08	8.966E + 08	9.518E + 06	3.364E + 04	6.824E + 06	5.569E + 04	3.060E + 07	3.348E − 14	1280.
330	1.427E + 08	7.823E + 08	7.268E + 06	3.326E + 04	6.588E + 06	3.978E + 04	2.700E + 07	2.847E − 14	1284.
340	1.129E + 08	6.834E + 08	5.562E + 06	3.293E + 04	6.363E + 06	2.849E + 04	2.385E + 07	2.429E − 14	1287.
350	8.942E + 07	5.977E + 08	4.264E + 06	3.262E + 04	6.149E + 06	2.045E + 04	2.110E + 07	2.079E − 14	1290.
360	7.096E + 07	5.233E + 08	3.274E + 06	3.232E + 04	5.944E + 06	1.471E + 04	1.870E + 07	1.785E − 14	1292.
370	5.638E + 07	4.586E + 08	2.518E + 06	3.202E + 04	5.748E + 06	1.060E + 04	1.659E + 07	1.536E − 14	1294.
380	4.486E + 07	4.022E + 08	1.939E + 06	3.174E + 04	5.560E + 06	7.648E + 03	1.473E + 07	1.325E − 14	1296.
390	3.573E + 07	3.530E + 08	1.496E + 06	3.146E + 04	5.379E + 06	5.528E + 03	1.310E + 07	1.146E − 14	1298.
400	2.849E + 07	3.100E + 08	1.155E + 06	3.119E + 04	5.206E + 06	4.002E + 03	1.166E + 07	9.925E − 15	1299.
410	2.274E + 07	2.724E + 08	8.926E + 05	3.093E + 04	5.039E + 06	2.901E + 03	1.039E + 07	8.616E − 15	1300.
420	1.817E + 07	2.396E + 08	6.907E + 05	3.067E + 04	4.878E + 06	2.106E + 03	9.265E + 06	7.492E − 15	1301.
430	1.453E + 07	2.108E + 08	5.350E + 05	3.042E + 04	4.724E + 06	1.530E + 03	8.269E + 06	6.526E − 15	1301.
440	1.163E + 07	1.856E + 08	4.148E + 05	3.017E + 04	4.575E + 06	1.114E + 03	7.385E + 06	5.693E − 15	1302.
450	9.312E + 06	1.634E + 08	3.219E + 05	2.992E + 04	4.431E + 06	8.111E + 02	6.600E + 06	4.974E − 15	1302.
460	7.465E + 06	1.440E + 08	2.500E + 05	2.968E + 04	4.293E + 06	5.915E + 02	5.870E + 06	4.350E − 15	1303.
470	5.989E + 06	1.270E + 08	1.944E + 05	2.945E + 04	4.159E + 06	4.318E + 02	5.257E + 06	3.811E − 15	1303.
480	4.808E + 06	1.120E + 08	1.512E + 05	2.921E + 04	4.030E + 06	3.156E + 02	4.710E + 06	3.342E − 15	1304.
490	3.863E + 06	9.880E + 07	1.178E + 05	2.898E + 04	3.906E + 06	2.308E + 02	4.221E + 06	2.934E − 15	1304.
500	3.106E + 06	8.722E + 07	9.177E + 04	2.875E + 04	3.786E + 06	1.690E + 02	3.785E + 06	2.579E − 15	1304.
510	2.499E + 06	7.702E + 07	7.158E + 04	2.853E + 04	3.670E + 06	1.239E + 02	3.394E + 06	2.269E − 15	1304.
520	2.012E + 06	6.804E + 07	5.587E + 04	2.831E + 04	3.558E + 06	9.091E + 01	3.046E + 06	1.998E − 15	1304.
530	1.621E + 06	6.013E + 07	4.364E + 04	2.809E + 04	3.449E + 06	6.676E + 01	2.733E + 06	1.761E − 15	1305.
540	1.306E + 06	5.316E + 07	3.412E + 04	2.787E + 04	3.345E + 06	4.907E + 01	2.454E + 06	1.554E − 15	1305.
550	1.054E + 06	4.702E + 07	2.669E + 04	2.766E + 04	3.244E + 06	3.611E + 01	2.204E + 06	1.372E − 15	1305.
560	8.507E + 05	4.160E + 07	2.090E + 04	2.744E + 04	3.146E + 06	2.659E + 01	1.980E + 06	1.213E − 15	1305.
570	6.871E + 05	3.682E + 07	1.637E + 04	2.724E + 04	3.051E + 06	1.960E + 01	1.780E + 06	1.073E − 15	1305.
580	5.554E + 05	3.261E + 07	1.284E + 04	2.703E + 04	2.960E + 06	1.446E + 01	1.600E + 06	9.494E − 16	1305.
590	4.492E + 05	2.888E + 07	1.007E + 04	2.682E + 04	2.871E + 06	1.068E + 01	1.439E + 06	8.410E − 16	1305.
600	3.635E + 05	2.559E + 07	7.907E + 03	2.662E + 04	2.786E + 06	7.892E + 00	1.294E + 06	7.456E − 16	1305.

MSIS-86 Model atmosphere, 19 Feb 1979 at 22 UT
Lat. 50°, Long. −150°, Zenith angle of Sun 62°
$\overline{F10.7}$ = 187.7, F10.7 = 243.3, Ap = 15.0

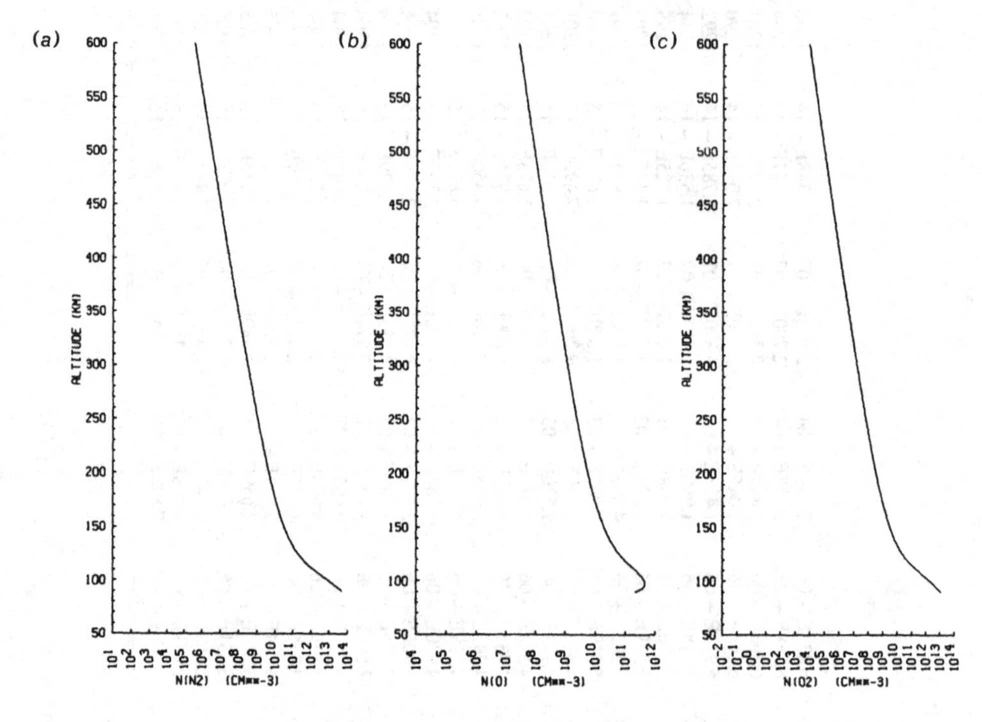

MSIS-86 Model atmosphere, 19 Feb 1979 at 22 UT
Lat. 50°, Long. −150°, Zenith angle of Sun 62°
$\overline{F10.7}$ = 187.7, F10.7 = 243.3, Ap = 15.0

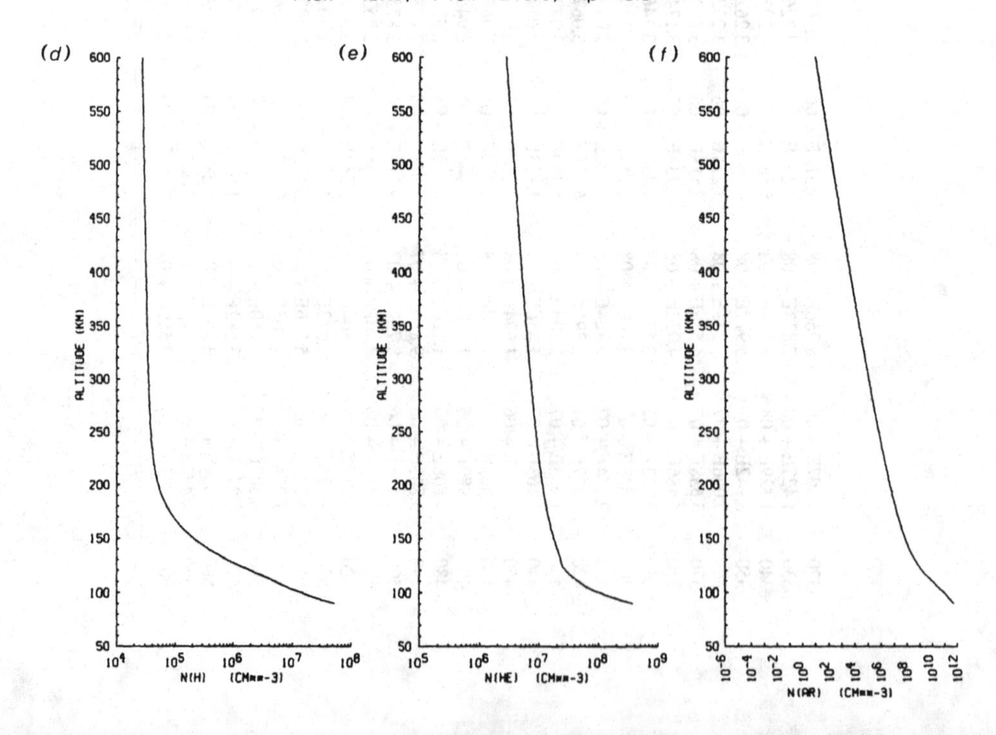

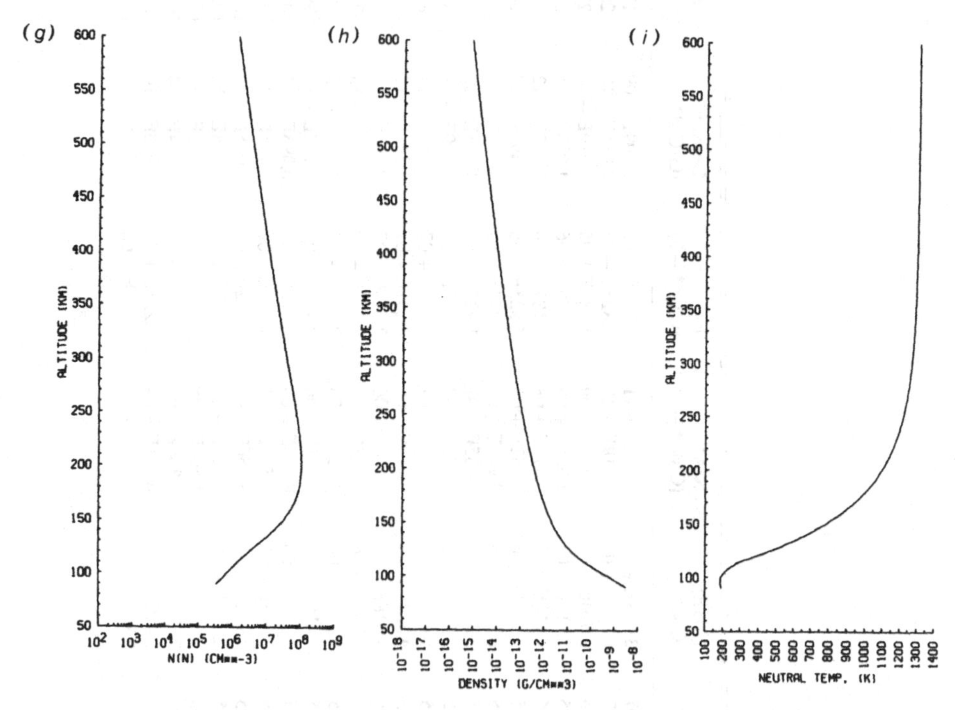

Fig. A1.2 MSIS-86 model atmosphere for 19 February, 1979 at 22 UT (lat. = 50.0°, long. = − 150.0°, zen. = 62°, F107A = 187.7, F107 = 243.3, *AP* = 15.0): (*a*) N_2, (*b*) O, (*c*) O_2, (*d*) H, (*e*) He, (*f*) Ar, (*g*) N, (*h*) density, (*i*) neutral temperature.

Table A1.3. *MSIS-86 model atmosphere for 23 April 1974*

LAT = 50.0, LON = − 150.0, ZEN = 38.
F107A = 86.1, F107 = 74.5, AP = 29.0

Alt. (KM)	N(N2) (CM** − 3)	N(O) (CM** − 3)	N(02) (CM** − 3)	N(H) (CM** − 3)	N(HE) (CM** − 3)	N(AR) (CM** − 3)	N(N) (CM** − 3)	Density (G/CM**3)	Neut. Temp. (K)
90	4.752E + 13	1.624E + 11	1.240E + 13	6.632E + 07	3.628E + 08	5.602E + 11	1.818E + 05	2.909E − 09	171.
100	6.538E + 12	2.505E + 11	1.570E + 12	1.595E + 07	8.014E + 07	6.819E + 10	2.828E + 05	3.985E − 10	204.
110	1.224E + 12	1.353E + 11	2.437E + 11	6.466E + 06	3.520E + 07	9.790E + 09	5.888E + 05	7.410E − 11	276.
120	3.187E + 11	6.136E + 10	5.009E + 10	2.999E + 06	2.016E + 07	1.888E + 09	1.376E + 06	1.923E − 11	390.
130	1.177E + 11	3.119E + 10	1.556E + 10	1.498E + 06	1.113E + 07	5.478E + 08	3.187E + 06	7.162E − 12	525.
140	5.652E + 10	1.866E + 10	6.718E + 09	8.406E + 05	7.549E + 06	2.152E + 08	6.825E + 06	3.494E − 12	632.
150	3.120E + 10	1.230E + 10	3.444E + 09	5.241E + 05	6.285E + 06	9.941E + 07	1.247E + 07	1.967E − 12	707.
160	1.874E + 10	8.634E + 09	1.948E + 09	3.618E + 05	5.567E + 06	5.065E + 07	1.927E + 07	1.208E − 12	784.
170	1.192E + 10	6.336E + 09	1.172E + 09	2.735E + 05	5.052E + 06	2.753E + 07	2.566E + 07	7.870E − 13	837.
180	7.884E + 09	4.798E + 09	7.362E + 08	2.227E + 05	4.650E + 06	1.566E + 07	3.021E + 07	5.347E − 13	879.
190	5.363E + 09	3.720E + 09	4.767E + 08	1.920E + 05	4.323E + 06	9.212E + 06	3.231E + 07	3.748E − 13	913.
200	3.726E + 09	2.936E + 09	3.158E + 08	1.725E + 05	4.048E + 06	5.554E + 06	3.214E + 07	2.691E − 13	940.
210	2.630E + 09	2.350E + 09	2.129E + 08	1.597E + 05	3.808E + 06	3.413E + 06	3.032E + 07	1.969E − 13	961.
220	1.879E + 09	1.902E + 09	1.453E + 08	1.510E + 05	3.600E + 06	2.129E + 06	2.775E + 07	1.464E − 13	978.
230	1.356E + 09	1.552E + 09	1.003E + 08	1.448E + 05	3.415E + 06	1.344E + 06	2.439E + 07	1.102E − 13	991.
240	9.855E + 08	1.276E + 09	6.975E + 07	1.401E + 05	3.247E + 06	8.566E + 05	2.122E + 07	8.398E − 14	1002.
250	7.208E + 08	1.055E + 09	4.885E + 07	1.366E + 05	3.093E + 06	5.504E + 05	1.827E + 07	6.461E − 14	1011.
260	5.299E + 08	8.767E + 08	3.439E + 07	1.337E + 05	2.951E + 06	3.556E + 05	1.562E + 07	5.015E − 14	1017.
270	3.911E + 08	7.313E + 08	2.433E + 07	1.313E + 05	2.819E + 06	2.310E + 05	1.332E + 07	3.924E − 14	1023.
280	2.897E + 08	6.120E + 08	1.727E + 07	1.292E + 05	2.695E + 06	1.507E + 05	1.134E + 07	3.093E − 14	1027.
290	2.151E + 08	5.136E + 08	1.230E + 07	1.274E + 05	2.579E + 06	9.866E + 04	9.653E + 06	2.454E − 14	1031.
300	1.602E + 08	4.320E + 08	8.784E + 06	1.257E + 05	2.470E + 06	6.481E + 04	8.226E + 06	1.960E − 14	1033.
310	1.195E + 08	3.593E + 08	6.288E + 06	1.242E + 05	2.366E + 06	4.270E + 04	7.020E + 06	1.561E − 14	1036.

320	8.935E + 07	3.040E + 08	4.510E + 06	1.227E + 05	2.268E + 06	2.820E + 04	6.001E + 06	1.262E − 14	1037.
330	6.690E + 07	2.575E + 08	3.241E + 06	1.212E + 05	2.175E + 06	1.866E + 04	5.139E + 06	1.026E − 14	1039.
340	5.017E + 07	2.184E + 08	2.333E + 06	1.199E + 05	2.086E + 06	1.238E + 04	4.409E + 06	8.373E − 15	1040.
350	3.767E + 07	1.853E + 08	1.682E + 06	1.186E + 05	2.002E + 06	8.223E + 03	3.789E + 06	6.865E − 15	1041.
360	2.832E + 07	1.574E + 08	1.214E + 06	1.174E + 05	1.921E + 06	5.473E + 03	3.262E + 06	5.651E − 15	1042.
370	2.132E + 07	1.338E + 08	8.776E + 05	1.162E + 05	1.844E + 06	3.648E + 03	2.812E + 06	4.669E − 15	1042.
380	1.607E + 07	1.138E + 08	6.351E + 05	1.150E + 05	1.771E + 06	2.436E + 03	2.428E + 06	3.872E − 15	1043.
390	1.212E + 07	9.685E + 07	4.602E + 05	1.138E + 05	1.701E + 06	1.629E + 03	2.099E + 06	3.221E − 15	1043.
400	9.152E + 06	8.248E + 07	3.339E + 05	1.126E + 05	1.633E + 06	1.090E + 03	1.817E + 06	2.687E − 15	1043.
410	6.917E + 06	7.028E + 07	2.425E + 05	1.115E + 05	1.569E + 06	7.310E + 02	1.574E + 06	2.248E − 15	1044.
420	5.233E + 06	5.992E + 07	1.763E + 05	1.104E + 05	1.508E + 06	4.908E + 02	1.365E + 06	1.886E − 15	1044.
430	3.963E + 06	5.111E + 07	1.283E + 05	1.093E + 05	1.449E + 06	3.299E + 02	1.185E + 06	1.586E − 15	1044.
440	3.003E + 06	4.362E + 07	9.345E + 04	1.082E + 05	1.393E + 06	2.220E + 02	1.030E + 06	1.336E − 15	1044.
450	2.278E + 06	3.725E + 07	6.814E + 04	1.071E + 05	1.339E + 06	1.496E + 02	8.951E + 05	1.129E − 15	1044.
460	1.730E + 06	3.182E + 07	4.974E + 04	1.061E + 05	1.287E + 06	1.009E + 02	7.730E + 05	9.549E − 16	1044.
470	1.314E + 06	2.720E + 07	3.634E + 04	1.050E + 05	1.237E + 06	6.818E + 01	6.738E + 05	8.094E − 16	1044.
480	9.994E + 05	2.326E + 07	2.658E + 04	1.040E + 05	1.190E + 06	4.611E + 01	5.876E + 05	6.873E − 16	1045.
490	7.606E + 05	1.990E + 07	1.945E + 04	1.030E + 05	1.144E + 06	3.122E + 01	5.126E + 05	5.846E − 16	1045.
500	5.794E + 05	1.703E + 07	1.425E + 04	1.020E + 05	1.101E + 06	2.116E + 01	4.474E + 05	4.979E − 16	1045.
510	4.417E + 05	1.459E + 07	1.045E + 04	1.010E + 05	1.059E + 06	1.436E + 01	3.906E + 05	4.247E − 16	1045.
520	3.370E + 05	1.250E + 07	7.672E + 03	1.000E + 05	1.019E + 06	9.759E + 00	3.412E + 05	3.628E − 16	1045
530	2.573E + 05	1.071E + 07	5.637E + 03	9.909E + 04	9.802E + 05	6.638E + 00	2.981E + 05	3.103E − 16	1045.
540	1.966E + 05	9.185E + 06	4.145E + 03	9.814E + 04	9.432E + 05	4.520E + 00	2.606E + 05	2.658E − 16	1045.
550	1.504E + 05	7.880E + 06	3.051E + 03	9.720E + 04	9.077E + 05	3.081E + 00	2.279E + 05	2.279E − 16	1045.
560	1.151E + 05	6.763E + 06	2.247E + 03	9.628E + 04	8.737E + 05	2.103E + 00	1.994E + 05	1.957E − 16	1045.
570	8.815E + 04	5.807E + 06	1.657E + 03	9.537E + 04	8.411E + 05	1.437E + 00	1.745E + 05	1.682E − 16	1045.
580	6.757E + 04	4.989E + 06	1.223E + 03	9.446E + 04	8.097E + 05	9.828E − 01	1.528E + 05	1.448E − 16	1045.
590	5.184E + 04	4.287E + 06	9.033E + 02	9.357E + 04	7.796E + 05	6.730E − 01	1.338E + 05	1.248E − 16	1045.
600	3.980E + 04	3.686E + 06	6.678E + 02	9.269E + 04	7.507E + 05	4.613E − 01	1.172E + 05	1.077E − 16	1045.

MSIS-86 Model atmosphere, 23 April 1974 at 22 UT
Lat. 50°, Long. −150°, Zenith angle of Sun 38°
$\overline{F10.7}$ = 86.7, F10.7 = 74.5 Ap = 29.0

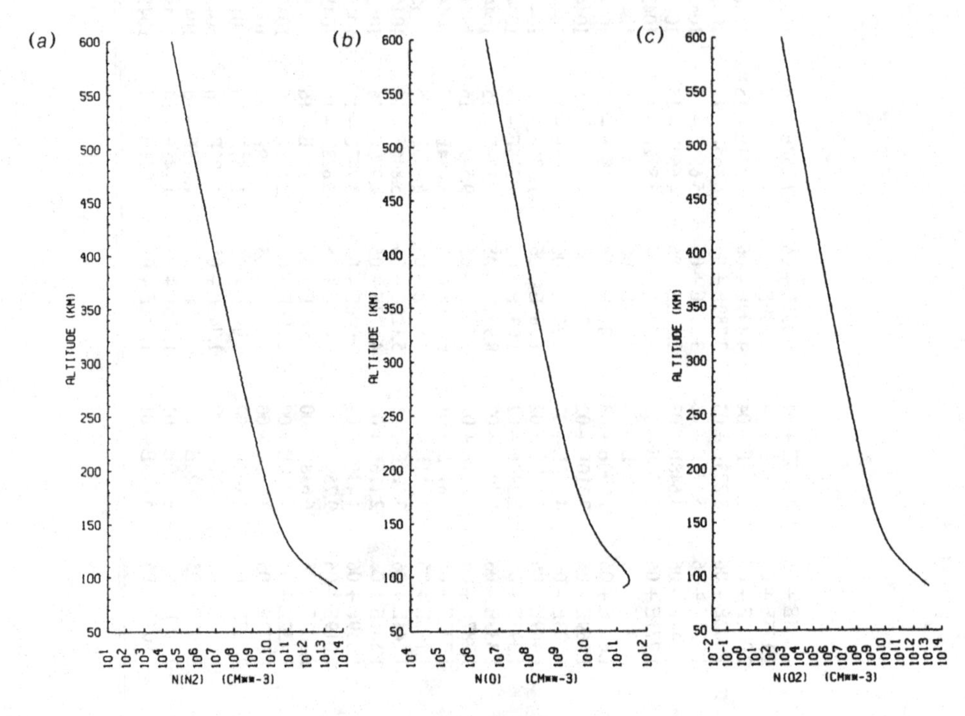

MSIS-86 Model atmosphere, 23 April 1974 at 22 UT
Lat. 50°, Long. −150°, Zenith angle of Sun 38°
$\overline{F10.7}$ = 86.7, F10.7 = 74.5 Ap = 29.0

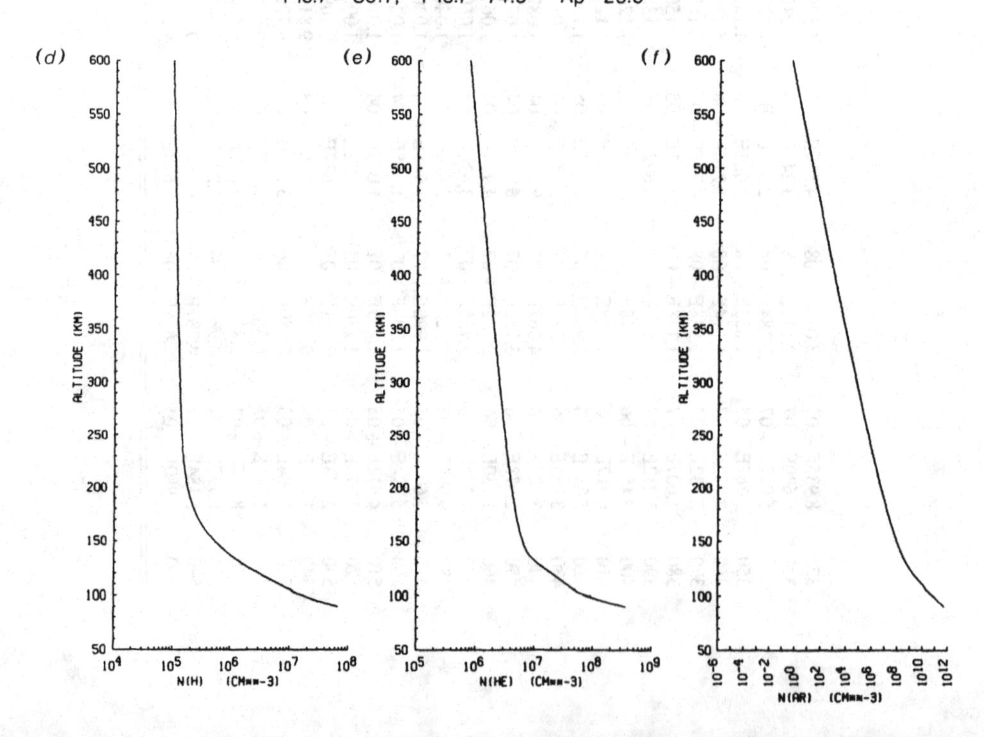

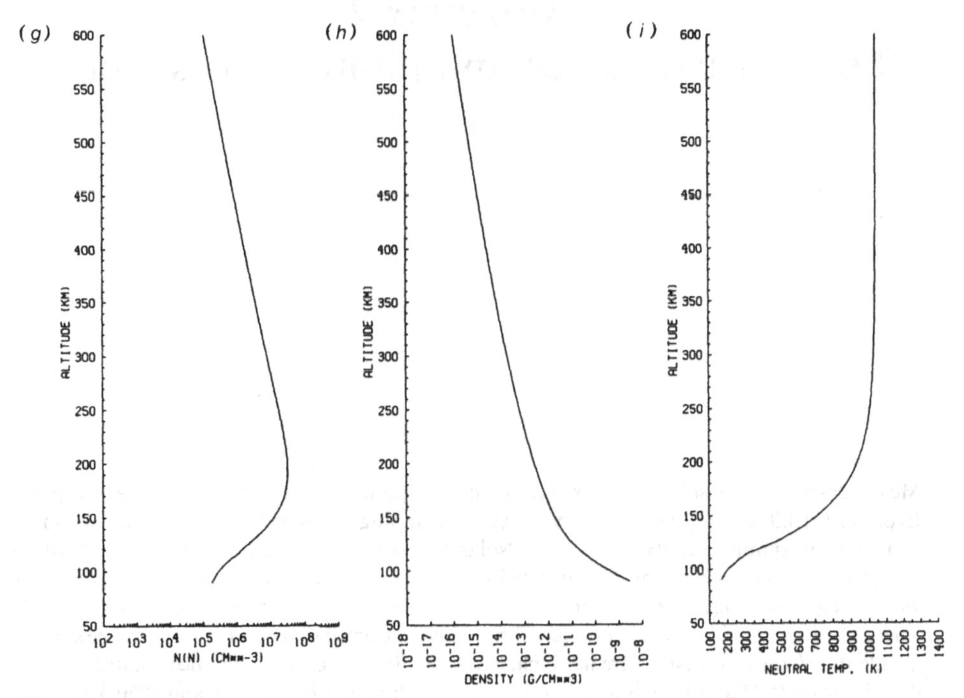

Fig. A1.3 MSIS-86 model atmosphere for 23 April, 1974 at 22 UT (lat. = 50.0°, long. = − 150.0°, zen. = 38, F107A = 86.7, F107 = 74.5, AP = 29.0): *a)* N_2, *(b)* O, *(c)* O_2, *(d)* H, *(e)* He, *(f)* Ar, *(g)* N, *(h)* density, *(i)* neutral temperature.

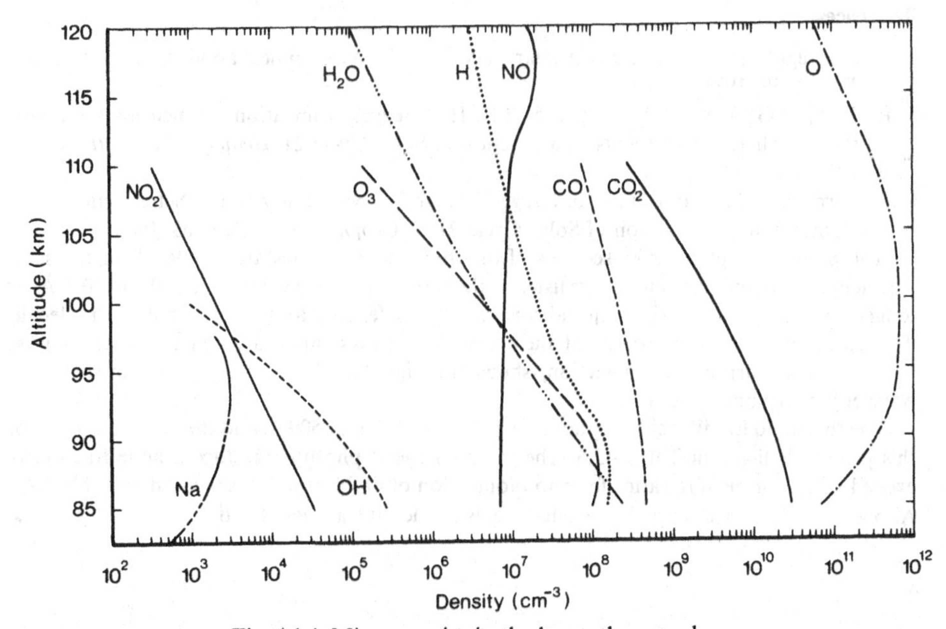

Fig. A1.4 Minor species in the lower thermosphere.

Appendix 2
The solar irradiance and photon cross sections

Measurements of the solar UV flux were made by spectrometers carried by the Atmosphere Explorer satellites over a period of several years, spanning an interval between solar minimum and solar maximum activity. A so-called 'solar UV reference spectrum' was acquired with a rocket borne spectrometer on 23 April 1974, identified in the table as day 113 of year 1974: 74/113. Table A2.1 lists the irradiance at the top of the atmosphere in photons $(\mathrm{cm}^2\,\mathrm{s}\,\Delta\lambda)^{-1}$ where $\Delta\lambda = 50$ Å. The intensities of several bright lines are listed individually, and the line sources are identified; these intensities are not included in the 50 Å bins. The irradiance obtained on five days between 1974 and 1979 is given in the wavelength region between 50 and 1050 Å which is subject to the largest variability with solar activity. Only the spectrum for day 74/113 is listed in Table A2.2 in the wavelength region 1050–1940 Å.

References:

H.E. Hinteregger, Representations of solar EUV fluxes for aeronomical applications, *Adv. Space Res.*, **1**, 39, 1981.

M.R. Torr, D.G. Torr, R.A. Ong and H.E. Hinteregger, Ionization frequencies for major thermospheric constituents as a function of Solar Cycle 21, *Geophys. Res. Lett.*, **6**, 771, 1979.

M.R. Torr, D.G. Torr and H.E. Hinteregger, Solar flux variability in the Schumann–Runge Continuum as a function of Solar Cycle 21, *J. Geophys. Res.*, **85**, 6063, 1980.

The photoabsorption cross sections of the three major species are tabulated for the same wavelength bins and individual lines listed in the wavelength interval between 50 and 1050 Å. At longer wavelengths O_2 is the major absorbing species, leading to dissociation of the molecule. The cross section is highly structured due to several band systems and an expanded set of curves of the total absorption cross section, shown in Figure 2.2.3, is given in this appendix for wavelengths beyond 1050 Å.

The threshold for dissociative photoionization of N_2 lies at 509 Å, and the fractional yield for this process is listed in Table A2.3. The more energetic photons produce a large fraction of excited states. Branching ratios for photoionization of O, N_2 and O_2 are listed in Table A2.4, A2.5 and A2.6, respectively. The excited states of the ions are identified.

Table A2.1 *Solar irradiance between 50 and 1050 Å for five days. Photoabsorption and photoionization cross sections in the same wavelength range*

		Year/Day									
		74/113	76/200	78/348	79/020	79/050					
		F10.7(10^{-22} Wm^{-2} Hz^{-1})									
		71.0	68.0	206.0	234.0	243.0	Photoabsorption (σ_a) and photoionization cross sections (10^{-18} cm^2)				
		Solar Irradiance (10^9 photons cm^{-2} s^{-1})									
$\Delta\lambda$(Å)	Ion/Atom						$\sigma_a(O) = \sigma_i(O)$	$\sigma_a(N_2)$	$\sigma_i(N_2)$	$\sigma_a(O_2)$	$\sigma_i(O_2)$
50–100		0.3984	0.4382	1.0337	1.2904	1.3710	1.06	0.60	0.60	1.18	1
100–150		0.1497	0.1687	0.3623	0.4419	0.4675	3.53	2.32	2.32	3.61	3
150–200		2.3683	1.8692	4.1772	5.3708	5.7024	5.96	5.40	5.40	7.27	7
200–250		1.5632	1.3951	4.7953	6.6473	7.1448	7.55	8.15	8.15	10.50	10
256.3	HeII, SiX	0.4600	0.5064	0.8805	1.0331	1.0832	8.43	9.65	9.65	12.80	12
284.15	FeXV	0.2100	0.0773	3.2613	5.2352	5.7229	9.26	10.60	10.60	14.80	14
250–300		1.6794	1.3556	7.5081	11.2278	12.1600	8.78	10.08	10.08	13.65	13
303.31	SiXI	0.8000	0.6000	2.9100	4.3380	4.6908	9.70	11.58	11.58	15.98	15
303.78	HeII	6.9000	7.7625	12.3424	13.8172	14.3956	9.72	11.60	11.60	16.00	16
300–350		0.9650	0.8671	4.3119	6.3164	6.8315	10.03	14.60	14.60	17.19	17
368.07	MgIX	0.6500	0.7394	1.2891	1.4661	1.5355	10.84	18.00	18.00	18.40	18
350–400		0.3140	0.2121	1.5298	2.3413	2.5423	10.70	17.51	17.51	18.17	18
400–450		0.3832	0.4073	1.0922	1.4330	1.5310	11.21	21.07	21.07	19.39	19
465.22	NeVII	0.2900	0.3299	0.6102	0.7004	0.7358	11.25	21.80	21.80	20.40	20
450–500		0.2851	0.3081	1.2120	1.6912	1.8228	11.64	21.85	21.85	21.59	21
500–550		0.4520	0.5085	1.2303	1.5496	1.6486	11.91	24.53	24.53	24.06	24
554.37	OIV	0.7200	0.7992	1.2943	1.4537	1.5163	12.13	24.69	24.69	25.59	25
584.33	HeI	1.2700	1.5875	3.4608	4.0646	4.3005	12.17	23.20	23.20	22.00	22
550–600		0.3568	0.4843	0.8732	0.9985	1.0477	11.90	22.38	22.38	25.04	25
609.76	MgX	0.5300	0.6333	1.6782	2.3443	2.4838	12.23	23.10	23.10	26.10	26
629.73	OV	1.5900	1.8484	3.2443	3.6938	3.8701	12.22	23.20	23.20	25.80	25

Table A2.1 (*Contd.*)

$\Delta\lambda$(Å)	Ion/Atom	Year/Day					Photoabsorption (σ_a) and photoionization cross sections (10^{-18} cm^2)				
		74/113	76/200	78/348	79/020	79/050					
		F10.7(10^{-22} Wm^{-2} Hz^{-1})									
		71.0	68.0	206.0	234.0	243.0					
		Solar Irradiance (10^9 photons cm^{-2} s^{-1})					$\sigma_a(O) = \sigma_i(O)$	$\sigma_a(N_2)$	$\sigma_i(N_2)$	$\sigma_a(O_2)$	$\sigma_i(O_2)$
600–650		0.3421	0.4002	0.9606	1.2843	1.3672	12.21	23.22	23.22	26.02	25
650–700		0.2302	0.2623	0.4521	0.5149	0.5388	10.04	29.75	25.06	26.27	22
703.31	OIII	0.3600	0.3915	0.6363	0.7152	0.7461	11.35	26.30	23.00	25.00	23
700–750		0.1409	0.1667	0.3439	0.4046	0.4287	8.00	30.94	23.20	29.05	23
765.15	NIV	0.1700	0.1997	0.3647	0.4178	0.4386	4.18	35.46	23.77	21.98	8
770.41	NeVIII	0.2600	0.2425	0.7760	1.1058	1.1873	4.18	26.88	18.39	25.18	9
789.36	OIV	0.7024	0.7831	1.2870	1.4501	1.5140	4.28	19.26	10.18	26.66	11
750–800		0.7581	0.8728	1.8909	2.3132	2.4541	4.23	30.71	16.75	27.09	9
800–850		1.6240	1.9311	3.9278	4.5911	4.8538	4.38	15.05	0	20.87	6
850–900		3.5370	4.4325	9.7798	11.5292	12.2187	4.18	46.63	0	9.85	4
900–950		3.0003	3.6994	7.9445	9.3132	9.8513	2.12	16.99	0	15.54	9
977.02	CIII	4.4000	4.8400	8.5523	9.7478	10.2165	0.00	0.70	0	4.00	2
950–1000		1.4746	1.7155	3.3463	3.8723	4.0779	0.00	36.16	0	16.53	12
1025.72	HI	3.5000	4.3750	9.5375	11.2000	11.8519	0.00	0.00	0	1.60	1
1031.91	OVI	2.1000	1.9425	.3929	5.7459	6.1049	0.00	0.00	0	1.00	0
1000–1050		2.4665	2.4775	4.7145	5.7798	6.0928	0.00	0.00	0	1.10	1

Table A2.2 *Solar irradiance between 1050 and 1940 Å for day 74/113*

Δλ(Å)	Atom/ion	Year/Day 74/113 Solar Irradiance (10^9 photons $cm^{-2} s^{-1}$)
1050–1100		2.9
1100–1150		0.091
1150–1200		4.4
1215.67	HI	251.0
1200–1250		8.0
1250–1300		4.1
1302.17	OI	1.10
1304.86	OI	1.13
1306.03	OI	1.23
1334.53	CII	1.8
1335.70	CII	2.5
1300–1350		4.65
1393.76	SiIV	1.3
1350–1400		6.1
1402.77	SiIV	0.91
1400–1450		9.5
1450–1500		16.2
1548.19	CIV	3.8
1500–1550		25.2
1550.77	CIV	1.9
1561.0	CI	2.5
1550–1600		35.6
1600–1650		56.0
1657.2	CI	8.5
1650–1700		121.5
1700–1750		225.0
1750–1800		357.0
1808.1	SiII	9.2
1816.93	SiII	14.2
1816.45	SiII	5.5
1800–1850		575.1
1850–1900		777.0
1900–1940		829.0

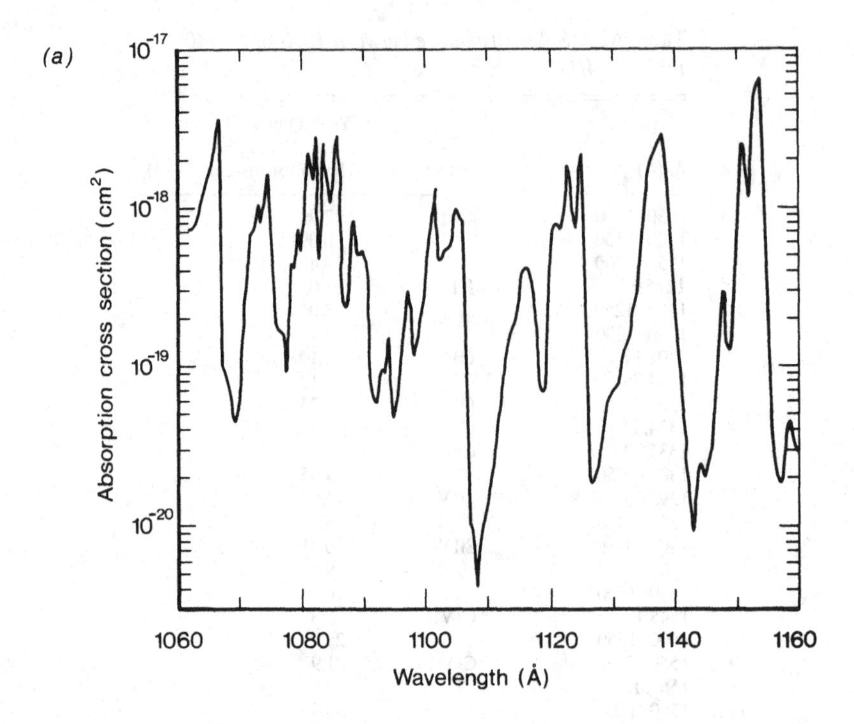
(a)
Absorption cross section (cm^2)
10^{-17}
10^{-18}
10^{-19}
10^{-20}
1060
1080
1100
1120
1140
1160
Wavelength (Å)

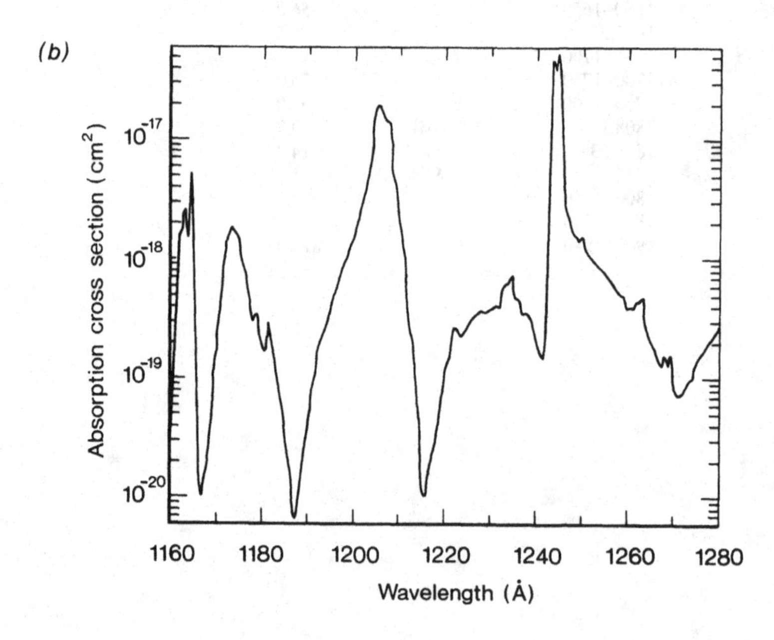
(b)
Absorption cross section (cm^2)
10^{-17}
10^{-18}
10^{-19}
10^{-20}
1160
1180
1200
1220
1240
1260
1280
Wavelength (Å)

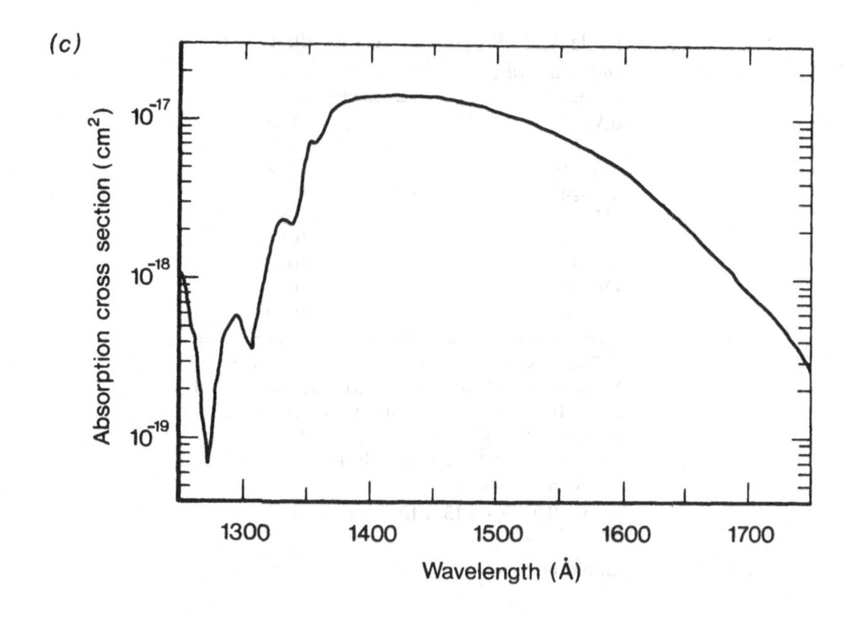

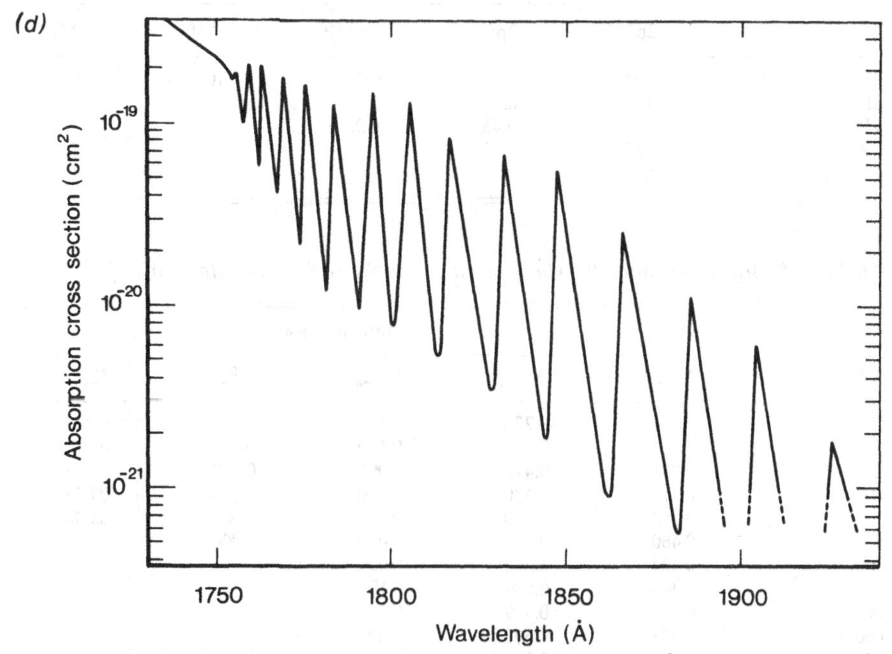

Fig. A2.1 Absorption cross section for O_2 between (*a*) 1060 and 1160 Å, (*b*) 1160 and 1280 Å, (*c*) 1250 and 1750 Å, (*d*) 1700 and 1950 Å.

Table A2.3 *Fractional yield for dissociative ionization of* N_2

λ(Å)	Yield
50–210	0.360
211–240	0.346
241–302	0.202
303–387	0.033
388–477	0.041
478–496	0.024
497–509	0.000

The cross section for dissociative ionization of N_2 at any wavelength from 50 to 509 Å is obtained by multiplying the total ionization cross section by the value of the yield found from the above table by linear interpolation, except from 387 to 477 Å, where the formula,

$$\text{yield} = 0.0329 + 8.13 \times 10^{-6} \times (\lambda - 442)^2, \quad 387 < \lambda < 477$$

should be used.

Table A2.4 *Branching ratios for photoionization of O into various states of* O^+

Δλ(Å)	$2p^3\,^4S^0$	$2p^3\,^2D^0$	$2p^3\,^2P^0$	$2p^4\,^4P$	$2p^4\,^2P$
50–314	0.25	0.36	0.23	0.10	0.06
315–435	0.26	0.41	0.26	0.07	
436–665	0.30	0.45	0.25		
666–732	0.43	0.57			
733–910	1.00				

Table A2.5 *Branching ratios for photoionization of* N_2 *into various states of* N_2^+

	Branching Ratios				
λ(Å)	$X\,^2\Sigma_g^+$	$A^2\Pi_u$	$B\,^2\Sigma_u^+$	$F\,^2\Sigma_u$	$^2\Sigma_g^+$
210	0.271	0.275	0.110	0.064	0.278
240	0.271	0.345	0.110	0.064	0.210
280	0.271	0.470	0.095	0.040	0.124
300	0.271	0.470	0.110	0.074	0.075
332	0.300	0.520	0.120	0.060	0.000
428	0.460	0.460	0.080	0.000	
500	0.404	0.506	0.090		
600	0.308	0.589	0.103		
660	0.308	0.589	0.103		
660.01	0.308	0.692	0.000		
720	0.420	0.580			
747	1.000	0.000			
796	1.000				

From 796 to 509 Å, partial photoionization cross sections can be obtained by multiplying the total ionization cross section by the interpolated branching ratio. Linear interpolation may be used to obtain branching ratios at wavelengths not listed in the table. For wavelengths shorter than 509 Å, the cross section due to dissociative ionization must first be subtracted from the total ionization cross section.

Table A2.6 *Branching ratios for photoionization of O_2 into various ion states*

	Branching ratios								
λ(Å)	$X^2\Pi_g$	$a^4\Pi_u + A^2\Pi_u$	$b^4\Sigma_g^-$	$B^2\Sigma_g^-$	$^2\Pi_u$	$c^4\Sigma_u^-$	$^2\Sigma_u^-$	$^{2,4}\Sigma_g^-$	662 Å
304	0.365	0.205	0.125	0.055	0.060	0.035	0.030	0.125	0.000
323	0.374	0.210	0.124	0.055	0.060	0.035	0.030	0.000	0.112
454	0.432	0.243	0.120	0.055	0.060	0.035	0.000		0.055
461	0.435	0.245	0.120	0.055	0.060	0.035			0.050
504	0.384	0.270	0.126	0.079	0.026	0.000			0.115
537	0.345	0.290	0.130	0.098	0.000				0.137
556	0.356	0.230	0.225	0.109					0.080
573	0.365	0.270	0.216	0.119					0.030
584	0.306	0.330	0.210	0.125					0.030
598	0.230	0.295	0.375	0.058					0.045
610	0.235	0.385	0.305	0.000					0.075
637	0.245	0.350	0.370						0.036
645	0.340	0.305	0.330						0.025
652	0.270	0.385	0.345						0.000
684	0.482	0.518	0.000						
704	0.675	0.325							
720	0.565	0.435							
737	0.565	0.435							
774	1.000	0.000							
1026	1.000								

The $X^2\Pi_g$ ground state and the $a^4\Pi_u$, $A^2\Pi_u$ and $b^4\Sigma_g^-$ excited states are bound molecular ion states and all the others are dissociating ionic states. The sum of the branching ratios of the latter six columns represents the dissociative ionization cross section, after multiplication by the total ionization cross section. The last column, labelled 662 Å, takes account of additional dissociating states that contribute to the total measured dissociative ionization cross section.

References:

K. Watanabe, *Adv. in Geophys.*, Eds, H.E. Landsberg and J. Van Mieghen, **5**, 153, 1958.

K. Kirby-Docken, E.R. Constantinides, S. Babeu, M. Oppenheimer and G.A. Victor, Photoionization of He, O, N_2 and O_2 for aeronomic calculations, *At. Data Nucl. Data Tables*, **23**, 63–82, 1979.

Appendix 3
Energy level diagrams and potential curves

Partial energy level diagrams of OI, OII, NI and NII are presented and the transitions leading to emissions that have, so far, been identified with reasonable certainty in the airglow and aurora are indicated by their wavelength (only one wavelength is shown for a multiplet). Inconclusive identifications are excluded, even though some may appear in the spectra reproduced from the research literature and shown in Chapter 7.

For the molecular species N_2, O_2, NO and their ions, fairly complete potential curves are given.

References:

A. Lofthus and P.H. Krupenie, The Spectrum of Molecular Nitrogen, *J. Phys. Chem. Ref. Data*, **6**, 113, 1977.

F.R. Gilmore, Potential energy curves for N_2, NO and O_2 and corresponding ions, *J. Quant. Spectros. Radiat. Transfer*, **5**, 369, 1965.

A.R. Striganov and N.S. Sventitskii, *Tables of Spectral Lines of Neutral and Ionized Atoms*, Plenum Press, New York, 1968.

R. Klotz and S.D. Peyerimhoff, Theoretical study of the intensity of the spin-or dipole forbidden transitions between the $c^1\Sigma_u^-$, $A'^3\Delta_u$, $A^3\Sigma_u^+$ and $X^3\Sigma_g^-$, $a^1\Delta_g$ states in O_2, *Molecular Physics*, **57**, 573, 1986.

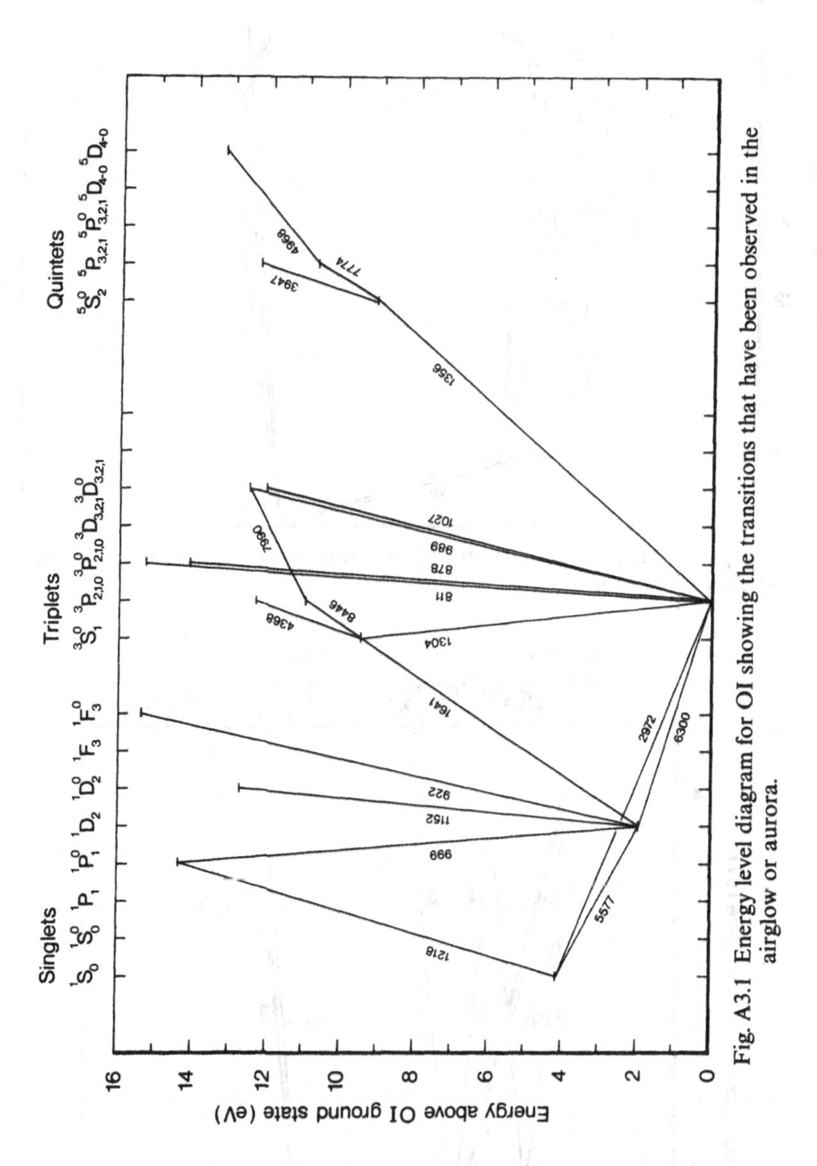

Fig. A3.1 Energy level diagram for OI showing the transitions that have been observed in the airglow or aurora.

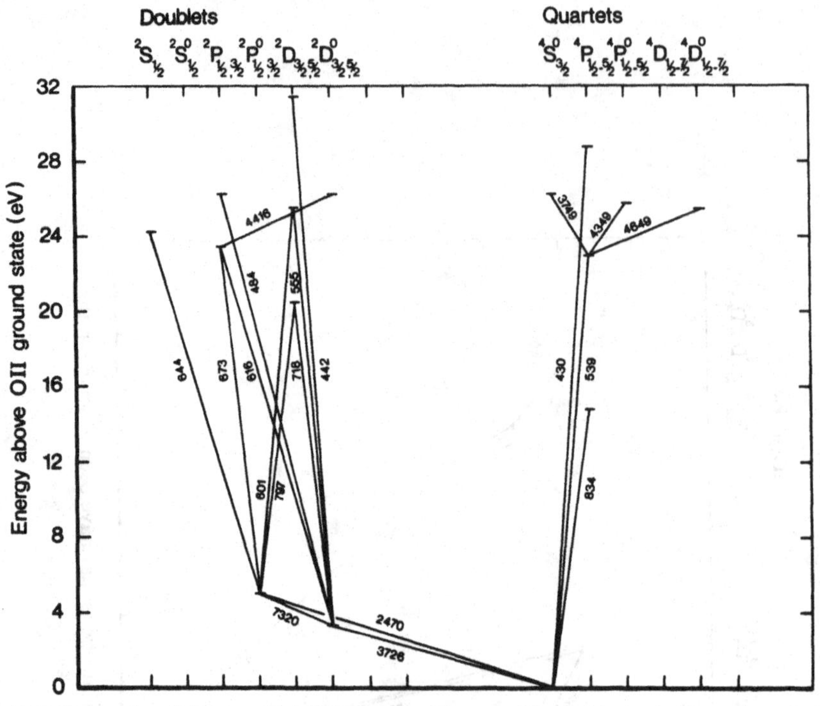

Fig. A3.2 Energy level diagram for OII showing the transitions that have been observed in the airglow or aurora.

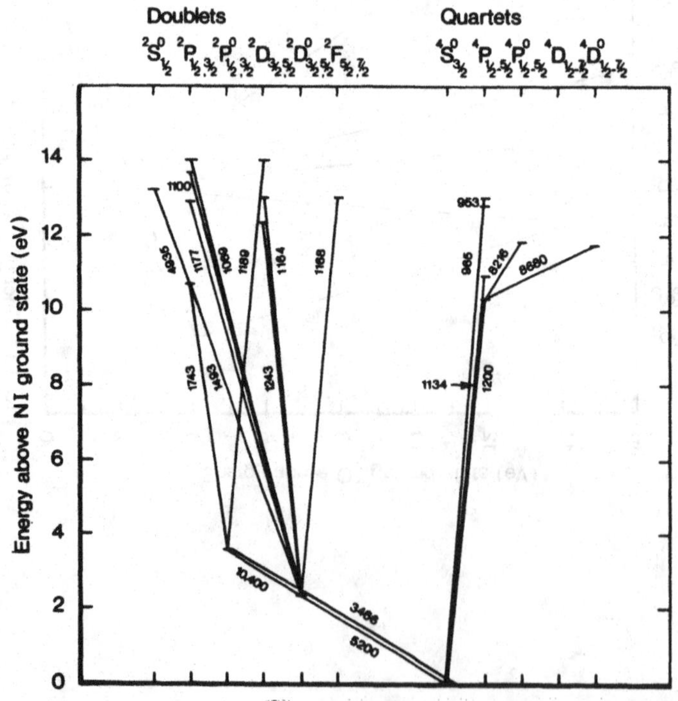

Fig. A3.3 Energy level diagram for NI showing the transitions that have been observed in the airglow or aurora.

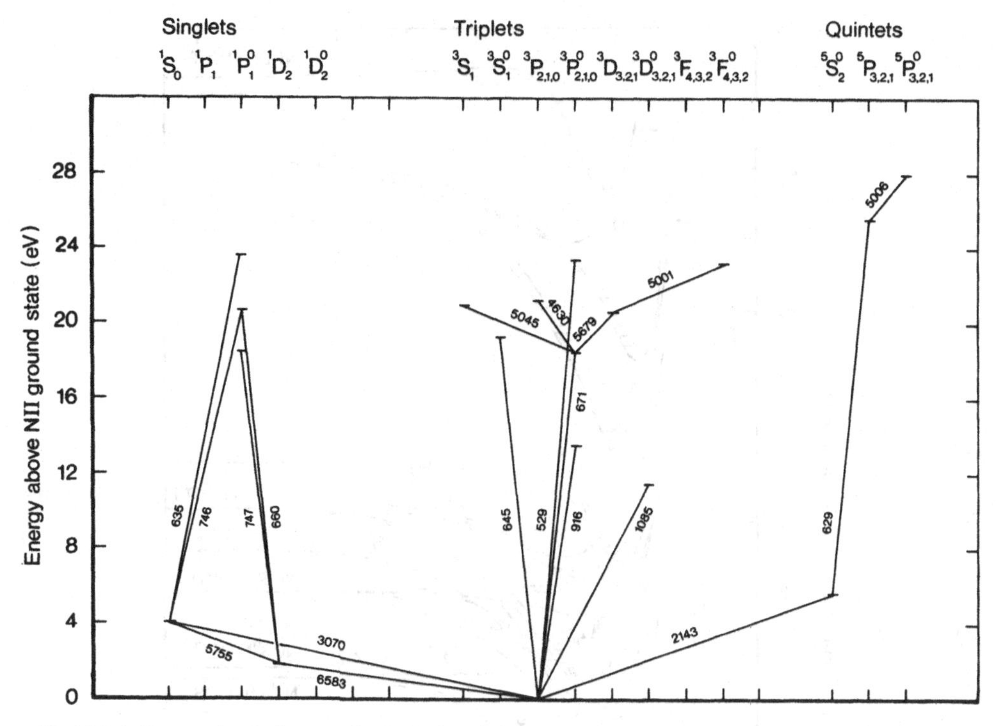

Fig. A3.4 Energy level diagram for NII showing the transitions that have been observed in the airglow or aurora.

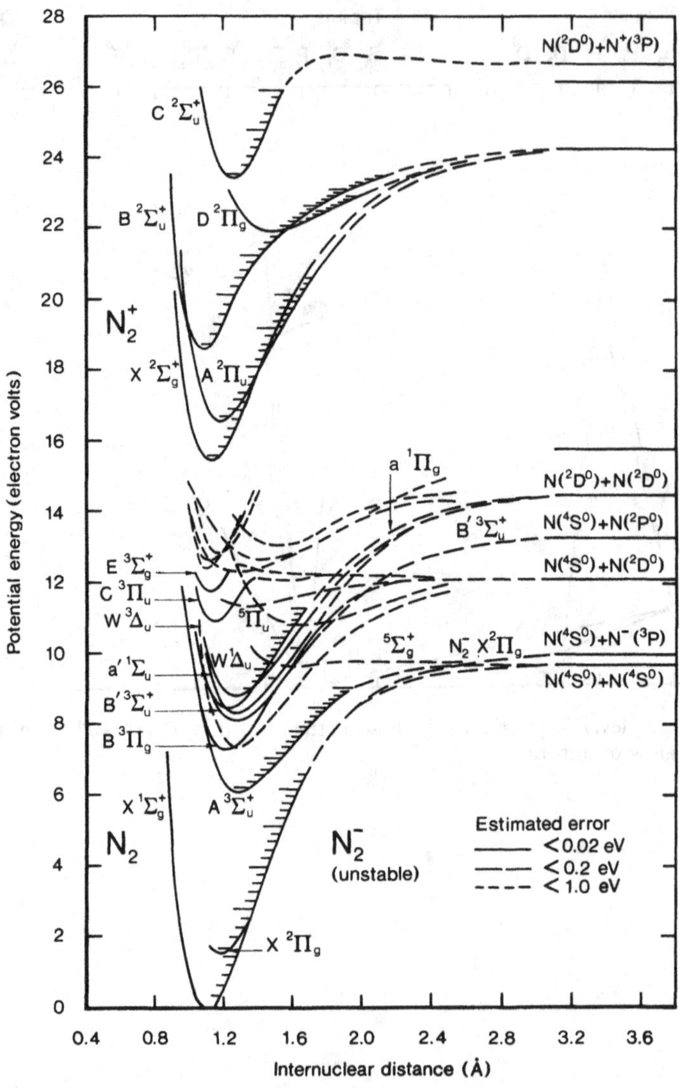

Fig. A3.5 Potential energy curves for N_2^- (unstable), N_2, and N_2^+.

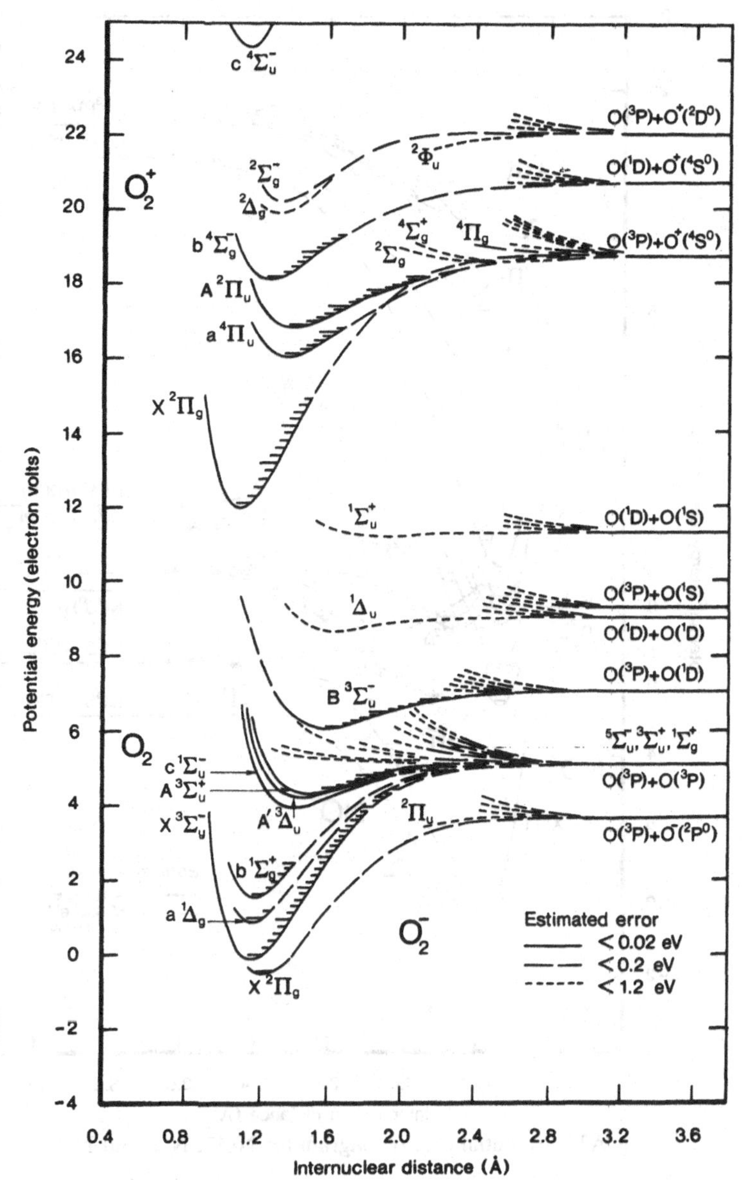

Fig. A3.6 Potential energy diagram for O_2^-, O_2, and O_2^+.

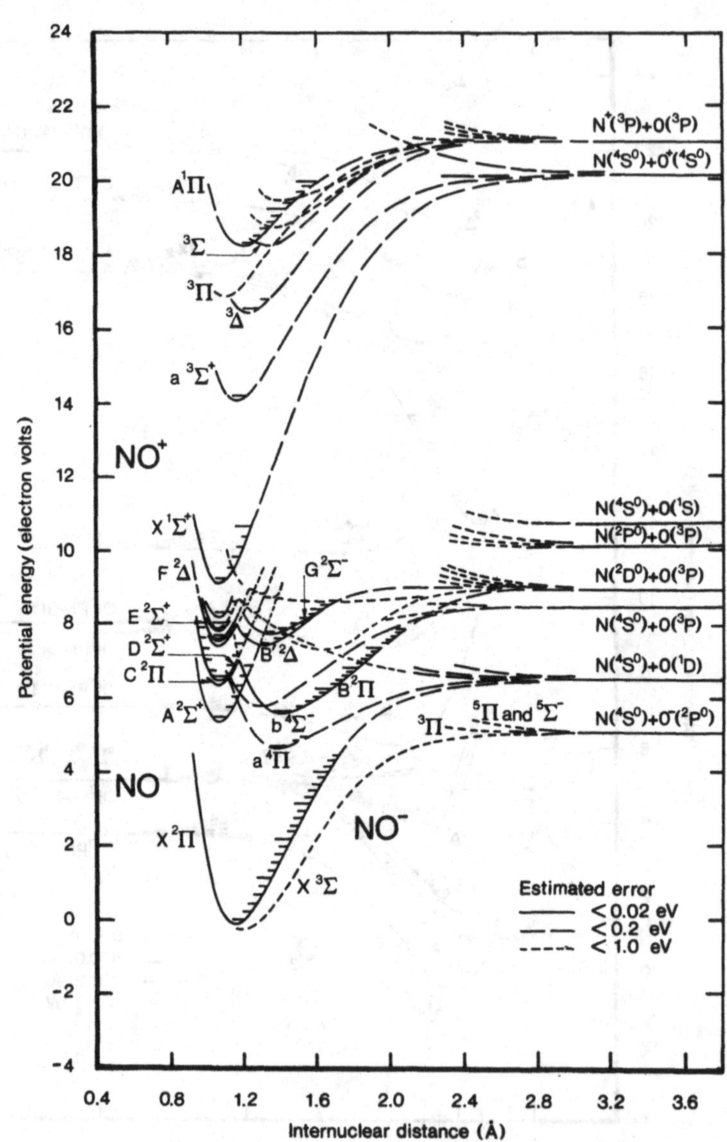

Fig. A3.7 Potential energy diagram for NO^-, NO, and NO^+.

Appendix 4
Cross sections for collisions between energetic electrons, protons and H atoms and the major atmospheric gases

Photoelectrons and auroral electrons collide with atmospheric gases in elastic and inelastic collisions. Inelastic collisions lead to excitation, dissociation and ionization. Cross sections for several important reactions between electrons and N_2, O_2 and O as a function of electron energy are shown in this appendix. These data have been compiled from numerous laboratory measurements and some theoretical calculations reported in the research literature. The compilation has been carried out by D. Lummerzheim from whose Ph.D. thesis (1987) the curves are excerpted (with permission).

The fraction of ions produced in various states of O^+, O_2^+ and N_2^+ by electron impact ionization is listed in Table A4.1, together with the corresponding ionization threshold energies. These fractions are valid at electron energies for which the respective cross sections have their maximum values.

Auroral protons with energy less than about 100 keV undergo primarily charge changing collisions with atmospheric gases. The collisions are identified in Section 3.4 as electron capture, σ_{10}, and ionization stripping, σ_{01}. Numerous laboratory measurements have been carried out for collisions between H^+ and N_2 or O_2 and between H and N_2 or O_2. Available results have been

Table A4.1 *Branching ratios for electron impact ionization of O, O_2 and N_2 into various states of the respective ions*

Ion	State	Fraction	Threshold energy (eV)
O^+	$^4S^0$	0.4	13.61
	$^2D^0$	0.4	16.92
	$^2P^0$	0.2	18.61
O_2^+	$X\,^2\Pi_g$	0.17	12.1
	$a\,^4\Pi_u$	0.37	16.1
	$A\,^2\Pi_u$	0.22	16.9
	$b\,^4\Sigma_g^-$	0.15	18.2
	$^4\Sigma$, $^2\Sigma_g$	0.09	~23.0
N_2^+	$X\,^2\Sigma_g^+$	0.50	15.58
	$A\,^2\Pi_u$	0.39	16.73
	$B\,^2\Sigma_u^+$	0.11	18.75

compiled by Van Zyl and are displayed in the accompanying figures. References to the individual results, identified in the figures, are listed by Van Zyl, Charge-state-equilibrated H^+/H flux fractions for analysis of the hydrogen aurora, Geophysical Institute, *University of Alaska Report UAGR-265*, 1978.

Collision cross sections with atomic O have not yet been thoroughly measured over a wide energy range, and it is assumed that the magnitude is approximately one half of the value for O_2 with an energy dependence following that of O_2.

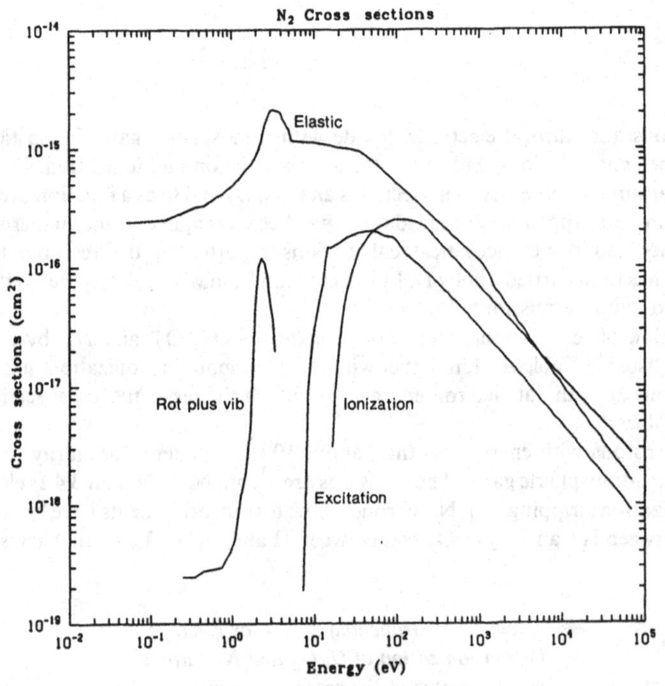

Fig. A4.1 N_2 cross sections.

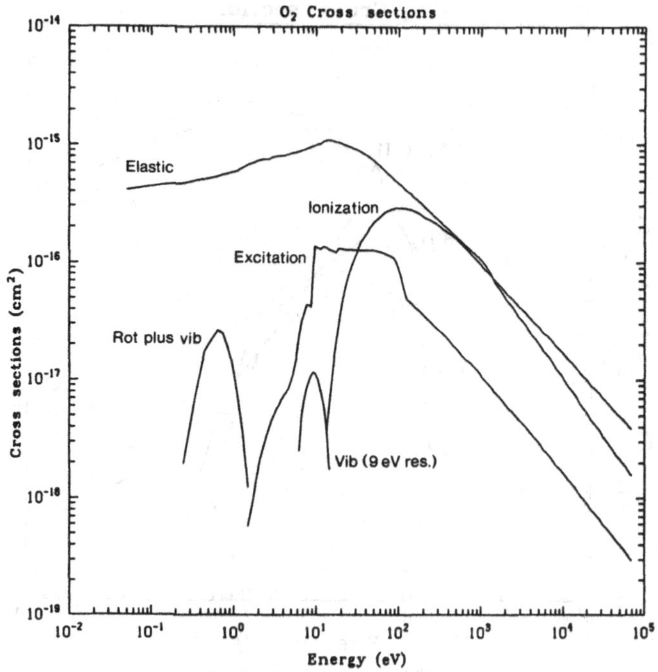

Fig. A4.2 O_2 cross sections.

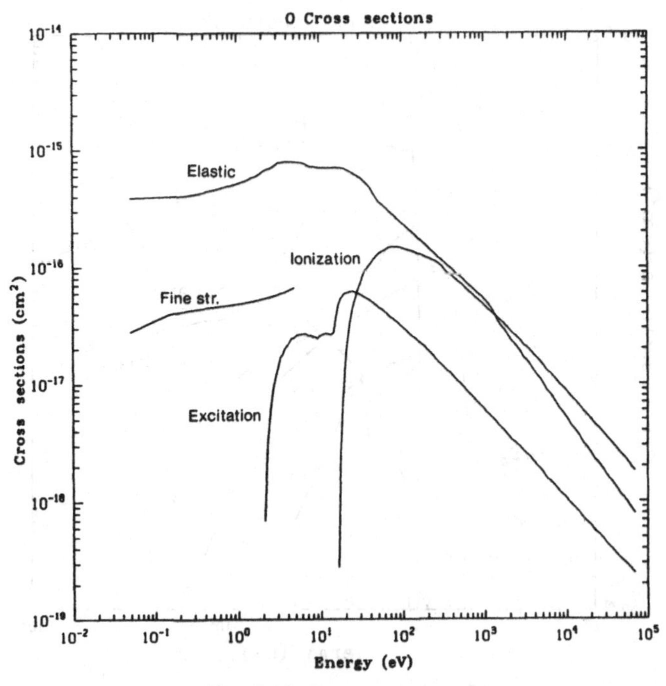

Fig. A4.3 O cross sections.

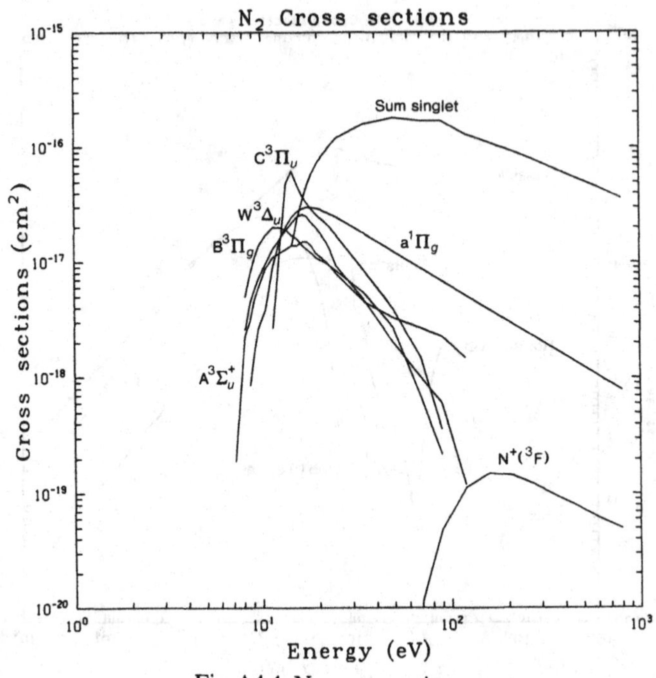

Fig. A4.4 N_2 cross sections.

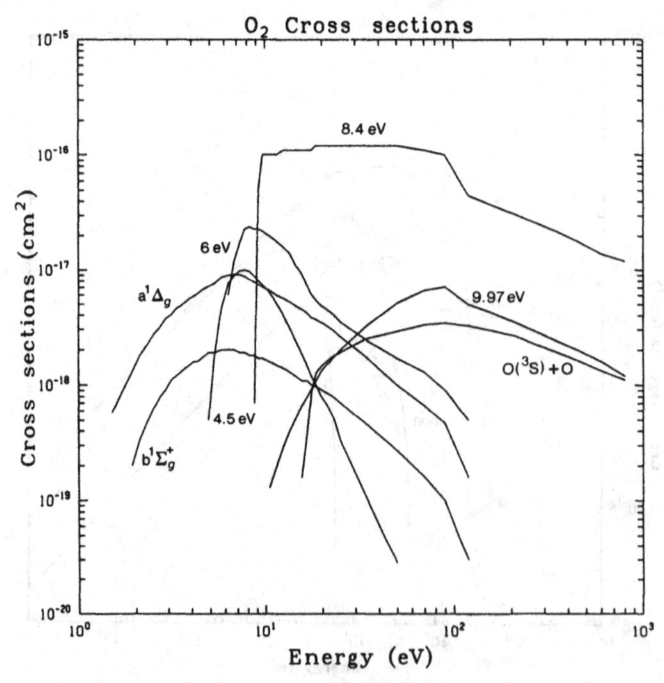

Fig. A4.5 O_2 cross sections.

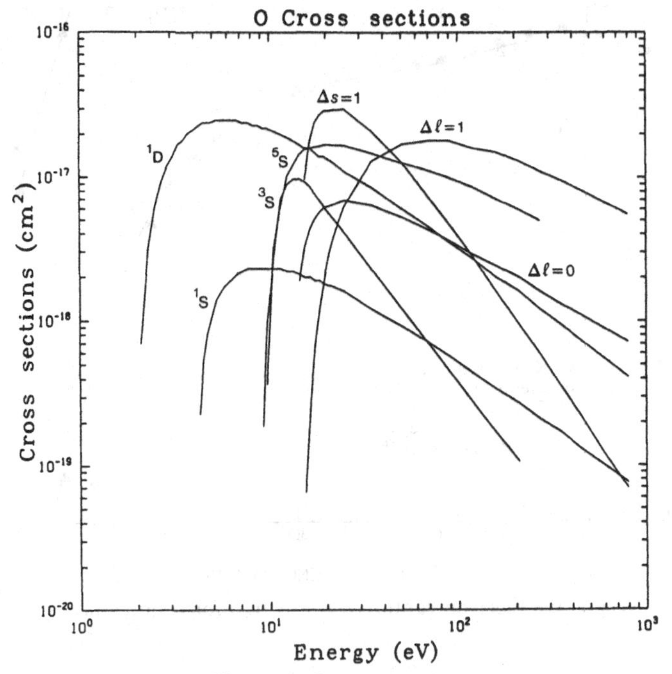

Fig. A4.6 O cross sections.

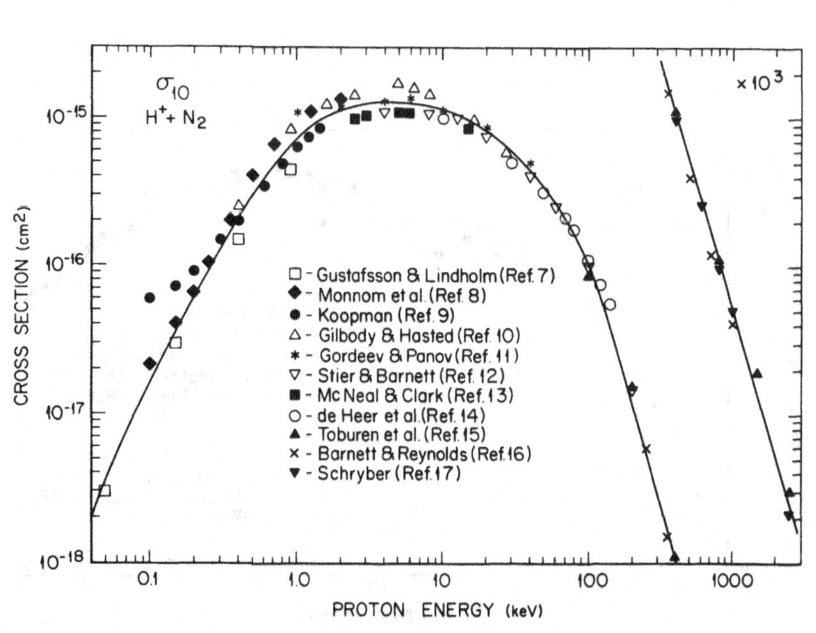

Fig. A4.7 Electron capture cross section for protons in N_2.

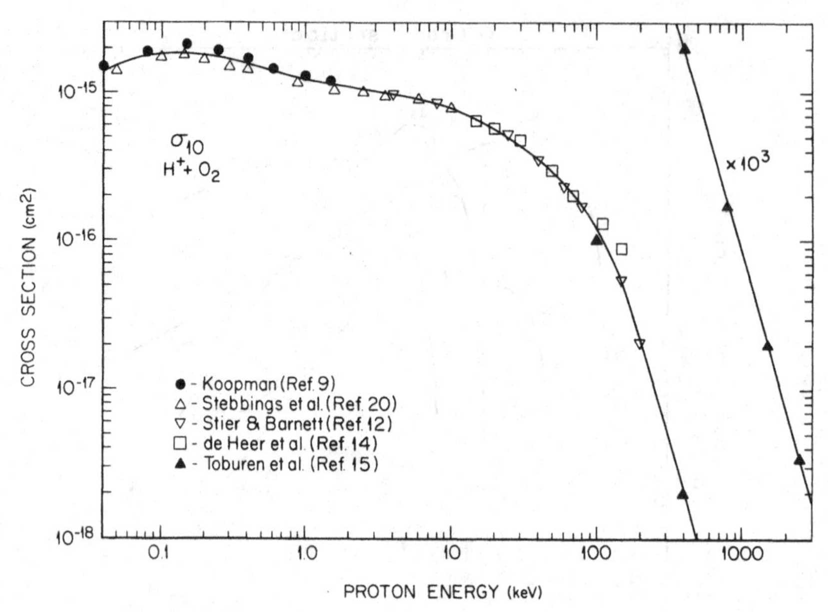

Fig. A4.8 Electron capture cross section for protons in O_2.

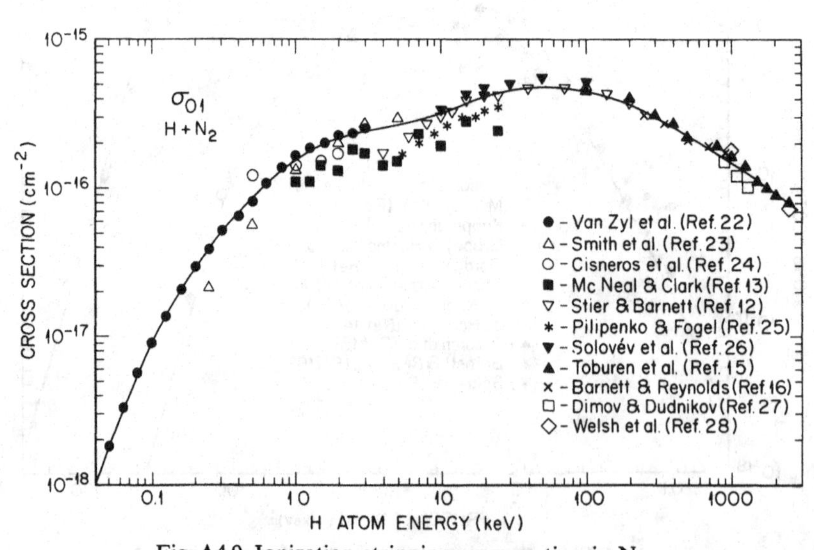

Fig. A4.9 Ionization stripping cross section in N_2.

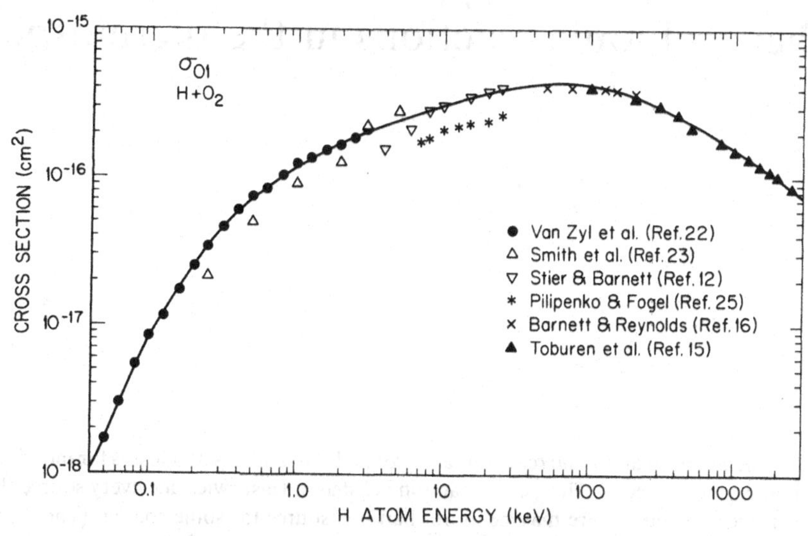

Fig. A4.10 Ionization stripping cross section in O_2.

Appendix 5
Chemical ionic reactions in the thermosphere

Exothermic chemical and ionic reactions are grouped into sources and losses for individual ions and minor neutral species. Thus, each reaction is listed at least twice: for every source there are one or more sinks, and there may be more than one source for some species. (There are a few exceptions involving minor species.) The major neutral species, N_2, O_2 and O, are end products, following the scheme of Figure 5.4.1. The chemical equations are listed with the excess kinetic energy of the reaction and the rate coefficient that is currently most widely accepted. Reaction rate coefficients are obtained, primarily, from laboratory measurements but a few are derived from atmospheric data and modelling and from theoretical work. Some are still controversial. The numerical values listed in Table A5.1 are derived from numerous references in the research literature and are applicable in the temperature range encountered in the thermosphere.

Table A5.1 *Chemical/ionic reactions in the thermosphere*

Reaction	Rate Coefficient[a]
Chemical production of N_2^+	
$O_2^+(a\,^4\Pi) + N_2 \rightarrow N_2^+ + O_2 + 0.523\,eV$	$\gamma_6 = 2.5 \times 10^{-10}$
$O^+(^2D) + N_2 \rightarrow N_2^+ + O + 1.33\,eV$,	$\gamma_3 = 8 \times 10^{-10}$
$O^+(^2P) + N_2 \rightarrow N_2^+ + O + 3.02\,eV$,	$\gamma_9 = 4.8 \times 10^{-10}$
Chemical loss of N_2^+	
$N_2^+ + O_2 \rightarrow O_2^+ + N_2 + 3.53\,eV$	$\gamma_5 = 5 \times 10^{-11}(T_R^b/300)^{-0.8}$
$N_2^+ + O \rightarrow NO^+ + N(^2D) + 0.70\,eV$	
$\rightarrow NO^+ + N(^4S) + 3.08\,eV$	$\gamma_4 = 1.4 \times 10^{-10}(T_R/300)^{-0.44}$
$N_2^+ + e \rightarrow N(^2D) + N(^2D) + 1.04\,eV$	$\alpha_2 = 1.8 \times 10^{-7}(T_e/300)^{-0.39}$
$N_2^+ + O \rightarrow O^+(^4S) + N_2 + 1.96\,eV$	$\gamma_{19} = 1.4 \times 10^{-10}(T_R/300)^{-0.44}$
$N_2^+ + NO \rightarrow NO^+ + N_2 + 6.33\,eV$	$\gamma_{18} = 3.3 \times 10^{-10}$
Chemical production of O_2^+	
$O^+(^4S) + O_2 \rightarrow O_2^+ + O + 1.55\,eV$	$\gamma_2 = 2 \times 10^{-11}(T_R/300)^{-0.4}$
$O^+(^2D) + O_2 \rightarrow O_2^+ + O + 4.865\,eV$	$\gamma_{22} = 7 \times 10^{-10}$
$N^+ + O_2 \rightarrow O_2^+ + N(^4S) + 2.486\,eV$	$\gamma_{11} = 1.1 \times 10^{-10}$
$N^+ + O_2 \rightarrow O_2^+ + N(^2D) + 0.1\,eV$	$\gamma_{33} = 2 \times 10^{-10}$
$N_2^+ + O_2 \rightarrow O_2^+ + N_2 + 3.53\,eV$	$\gamma_5 = 5 \times 10^{-11}(T_R/300)^{-0.8}$
$O_2^+(a^4\Pi) + O \rightarrow O_2^+ + O + 4.05\,eV$	$\gamma_7 = 1 \times 10^{-10}$

Table A5.1 (*Contd.*)

Reaction	Rate Coefficient[a]
Chemical loss of O_2^+	
$O_2^+ + e \rightarrow O(^1D) + O(^3P) + 5\,eV$	$\alpha_1 = 1.9 \times 10^{-7}(T_e/300)^{-0.5}$
$O_2^+ + N(^2D) \rightarrow N^+ + O_2$	$\gamma_{17} = 2.5 \times 10^{-10}$
$O_2^+ + N(^4S) \rightarrow NO^+ + O + 4.213\,eV$	$\gamma_{16} = 1.8 \times 10^{-10}$
$O_2^+ + NO \rightarrow NO^+ + O_2 + 2.813\,eV$	$\gamma_{15} = 4.4 \times 10^{-10}$
$O_2^+ + N_2 \rightarrow NO^+ + NO + 0.933\,eV$	$\gamma_8 = 5 \times 10^{-16}$
Chemical production of $O^+(^4S)$	
$O^+(^2D) + O \rightarrow O^+(^4S) + O + 3.31\,eV$	$\gamma_{13} = 1.0 \times 10^{-11}$
$O^+(^2D) + e \rightarrow O^+(^4S) + e + 3.31\,eV$	$\alpha_5 = 7.8 \times 10^{-8}(T_e/300)^{-0.5}$
$O^+(^2D) + N_2 \rightarrow O^+(^4S) + N_2 + 3.31\,eV$	$\gamma_{24} = 8 \times 10^{-10}$
$O^+(^2P) + O \rightarrow O^+(^4S) + O + 5.0\,eV$	$\gamma_{25} = 5.2 \times 10^{-11}$
$O^+(^2P) + e \rightarrow O^+(^4S) + e + 5.0\,eV$	$\alpha_7 = 4 \times 10^{-8}(T_e/300)^{-0.5}$
$O^+(^2P) \rightarrow O^+(^4S) + h\nu(2470\,Å)$	$A_3 = 0.047\,s^{-1}$
$N_2^+ + O \rightarrow O^+(^4S) + N_2 + 1.96\,eV$	$\gamma_{19} = 1.4 \times 10^{-10}(T_R/300)^{-0.44}$
$H^+ + O \rightarrow O^+(^4S) + H$	$\gamma_{20} = \gamma_{12}(8/9)[(T_i + T_n/4)/(T_n + T_i/16)]^{0.5}$
$N^+ + O_2 \rightarrow O^+(^4S) + NO + 2.31\,eV$	$\gamma_{27} = 3 \times 10^{-11}$
Chemical loss of $O^+(^4S)$	
$O^+(^4S) + O_2 \rightarrow O_2^+ + O + 1.55\,eV$	$\gamma_2 = 2 \times 10^{-11}(T_R/300)^{-0.4}$
$O^+(^4S) + N_2 \rightarrow NO^+ + N(^4S) + 1.10\,eV$	$\gamma_1 = 5 \times 10^{-13}$, $T_R < 1000\,K$ $= 4.5 \times 10^{-14}(T_R/300)^2$, $T_R > 1000\,K$
$O^+(^4S) + NO \rightarrow NO^+ + O + 4.36\,eV$	$\gamma_{21} = 8 \times 10^{-13}$
$O^+(^4S) + H \rightarrow H^+ + O$	$\gamma_{12} = 6.0 \times 10^{-10}$
$O^+(^4S) + N(^2D) \rightarrow N^+ + O + 1.45\,eV$	$\gamma_{26} = 1.3 \times 10^{-10}$
Chemical production of $O^+(^2D)$	
$O^+(^2P) + e \rightarrow O^+(^2D) + e + 1.69\,eV$	$\alpha_8 = 1.5 \times 10^{-7}(T_e/300)^{-0.5}$
$O^+(^2P) \rightarrow O^+(^2D) + h\nu(7320/7330\,Å)$	$A_4 = 0.171\,s^{-1}$
Chemical loss of $O^+(^2D)$	
$O^+(^2D) + N_2 \rightarrow N_2^+ + O + 1.33\,eV$	$\alpha_3 = 1 \times 10^{-10}$
$O^+(^2D) + O \rightarrow O^+(^4S) + O + 3.31\,eV$	$\gamma_{13} = 1.0 \times 10^{-11}$
$O^+(^2D) + e \rightarrow O^+(^4S) + e + 3.31\,eV$	$\alpha_5 = 7.8 \times 10^{-8}(T_e/300)^{-0.5}$
$O^+(^2D) + O_2 \rightarrow O_2^+ + O + 4.865\,eV$	$\gamma_{22} = 7 \times 10^{-10}$
$O^+(^2D) \rightarrow O^+(^4S) + h\nu(3726/3729\,Å)$	$A_5 = 7.7 \times 10^{-5}\,s^{-1}$
$O^+(^2D) + N_2 \rightarrow O^+(^4S) + N_2 + 3.31\,eV$	$\gamma_{24} = 8 \times 10^{-10}$
Chemical production of $O^+(^2P)$	
No sources	
Chemical loss of $O^+(^2P)$	
$O^+(^2P) + N_2 \rightarrow N_2^+ + O + 3.02\,eV$	$\gamma_9 = 4.8 \times 10^{-10}$
$O^+(^2P) + N_2 \rightarrow N^+ + NO + 0.70\,eV$	$\gamma_{30} = 1.0 \times 10^{-10}$
$O^+(^2P) + O \rightarrow O^+(^4S) + O + 5.0\,eV$	$\gamma_{25} = 5.2 \times 10^{-11}$
$O^+(^2P) + e \rightarrow O^+(^4S) + e + 5.0\,eV$	$\alpha_7 = 4.0 \times 10^{-8}(T_e/300)^{-0.5}$
$O^+(^2P) + e \rightarrow O^+(^2D) + e + 1.69\,eV$	$\alpha_8 = 1.5 \times 10^{-7}(T_e/300)^{-0.5}$
$O^+(^2P) \rightarrow O^+(^4S) + h\nu(2470\,Å)$	$A_3 = 0.047\,s^{-1}$
$O^+(^2P) \rightarrow O^+(^2D) + h\nu(7320/7330\,Å)$	$A_4 = 0.171\,s^{-1}$
Chemical production of $O_2^+(a\,^4\Pi)$	
No Sources	
Chemical loss of $O_2^+(a\,^4\Pi)$	
$O_2^+(a\,^4\Pi) + N_2 \rightarrow N_2^+ + O_2 + 0.523\,eV$	$\gamma_6 = 2.5 \times 10^{-10}$

Table A5.1 (*Contd.*)

Reaction	Rate Coefficient[a]
$O_2^+(a\,^4\Pi) + O \rightarrow O_2^+ + O + 4.05\,eV$	$\gamma_7 = 1.0 \times 10^{-10}$
$O_2^+(a\,^4\Pi) + e \rightarrow O + O + 11\,eV$	$\alpha_4 = 1.0 \times 10^{-7}$
$O_2^+(a\,^4\Pi) \rightarrow O_2^+ + h\nu(3060\,Å)$	$A_8 = 1 \times 10^{-4}\,s^{-1}$
Chemical production of N^+	
$O^+ + N(^2D) \rightarrow N^+ + O + 1.46\,eV$	$\gamma_{26} = 1.3 + 10^{-10}$
$O_2^+ + N(^2D) \rightarrow N^+ + O_2$	$\gamma_{17} = 2.5 \times 10^{-10}$
$He^+ + N_2 \rightarrow N^+ + N + He + 0.28\,eV$	$\gamma_{14} = 1.2 \times 10^{-9}$
$O^+(^2P) + N_2 \rightarrow N^+ + NO + 0.70\,eV$	$\gamma_{30} = 1.0 \times 10^{-10}$
$O^+(^2P) + N \rightarrow N^+ + O + 2.7\,eV$	$\gamma_{31} = 1.0 \times 10^{-10}$
$O^+(^2D) + N \rightarrow N^+ + O + 1\,eV$	$\gamma_{32} = 7.5 \times 10^{-11}$
Chemical loss of N^+	
$N^+ + O_2 \rightarrow NO^+ + O + 6.67\,eV$	$\gamma_{10} = 2.6 \times 10^{-10}$
$N^+ + O_2 \rightarrow N(^4S) + O_2^+ + 2.486\,eV$	$\gamma_{11} = 1.1 \times 10^{-10}$
$N^+ + O_2 \rightarrow N(^2D) + O_2^+ + 0.1\,eV$	$\gamma_{33} = 2 \times 10^{-10}$
$N^+ + O_2 \rightarrow O^+ + NO + 2.31\,eV$	$\gamma_{27} = 3 \times 10^{-11}$
$N^+ + O \rightarrow O^+ + N + 0.93\,eV$	$\gamma_{28} = 5 \times 10^{-13}$
$N^+ + H \rightarrow H^+ + N + 0.9\,eV$	$\gamma_{29} = 3.6 \times 10^{-12}$
Chemical production of H^+	
$O^+ + H \rightarrow H^+ + O$	$\gamma_{12} = 6 \times 10^{-10}$
$N^+ + H \rightarrow H^+ + N + 0.9\,eV$	$\gamma_{29} = 3.6 \times 10^{-12}$
Chemical loss of H^+	
$H^+ + O \rightarrow O^+ + H$	$\gamma_{20} = \gamma_{12}(8/9)[(T_i + T_n/4)/(T_n + T_i/16)]^{0.5}$
Chemical production of NO^+	
$N_2^+ + O \rightarrow NO^+ + N(^2D) + 0.70\,eV$	$\gamma_4 = 1.4 \times 10^{-10}(T_R/300)^{-0.44}$
$N_2^+ + O \rightarrow NO^+ + N(^4S) + 3.08\,eV$	
$N^+ + O_2 \rightarrow NO^+ + O + 6.67\,eV$	$\gamma_{10} = 2.6 \times 10^{-10}$
$O^+(^4S) + N_2 \rightarrow NO^+ + N(^4S) + 1.10\,eV$	$\gamma_1 = 5 \times 10^{-13}$, $T_R < 1000\,K$
	$= 4.5 \times 10^{-14}(T_R/300)^2$, $T_R > 1000\,K$
$N_2^+ + NO \rightarrow NO^+ + N_2 + 6.33\,eV$	$\gamma_{18} = 3.3 \times 10^{-10}$
$O^+(^4S) + NO \rightarrow NO^+ + O + 4.36\,eV$	$\gamma_{21} = 8 \times 10^{-13}$
$O_2^+ + N(^4S) \rightarrow NO^+ + O + 4.19\,eV$	$\gamma_{16} = 1.8 \times 10^{-10}$
$O_2^+ + NO \rightarrow NO^+ + O_2 + 2.813\,eV$	$\gamma_{15} = 4.4 \times 10^{-10}$
$O_2^+ + N_2 \rightarrow NO^+ + NO + 0.933\,eV$	$\gamma_8 = 5 \times 10^{-16}$
Chemical loss of NO^+	
$NO^+ + e \rightarrow O + N(^4S, ^2D) + \begin{cases} 0.38\,eV, f = 0.78 \\ 2.75\,eV, f = 0.22 \end{cases}$	$\alpha_3 = 4.2 \times 10^{-7}(T_e/300)^{-0.85} \times f$
Chemical production of $N(^4S)$	
$N_2^+ + O \rightarrow NO^+ + N(^4S) + 3.08\,eV$	$\gamma_4 = 1.4 \times 10^{-10}(T_R/300)^{-0.44}$
$N^+ + O_2 \rightarrow O_2^+ + N(^4S) + 2.46\,eV$	$\gamma_{11} = 1.1 \times 10^{-10}$
$O^+ + N_2 \rightarrow NO^+ + N(^4S) + 1.10\,eV$	$\gamma_1 = 5 \times 10^{-13}$, $T_R < 1000\,K$
	$= 4.5 \times 10^{-14}(T_R/300)^2$, $T_R > 1000\,K$
$NO^+ + e \rightarrow O + N(^4S) + 2.75\,eV$	$\alpha_3 = f \times 4.2 \times 10^{-7}(T_e/300)^{-0.85}$, $f = 0.22$
$N(^2D) + e \rightarrow N(^4S) + e + 2.38\,eV$	$\beta_6 = 5.5 \times 10^{-10}(T_e/300)^{0.5}$
$N(^2D) + O \rightarrow N(^4S) + O + 2.38\,eV$	$\beta_5 = 2 \times 10^{-12}$
$N(^2D) \rightarrow N(^4S) + h\nu(5200\,Å)$	$A_7 = 1.06 \times 10^{-5}\,s^{-1}$
$NO + h\nu(< 1910\,Å) \rightarrow N(^4S) + O$	$A_1 = 8.3 \times 10^{-6}\,s^{-1}$

Table A5.1 (*Contd.*)

Reaction	Rate Coefficient[a]
Chemical loss of $N(^4S)$	
$N(^4S) + O_2 \rightarrow NO + O + 1.385\,eV$	$\beta_1 = 4.4 \times 10^{-12} \exp(-3220/T_n)$
$N(^4S) + NO \rightarrow N_2 + O + 3.25\,eV$	$\beta_4 = 1.5 \times 10^{-12} (T_n)^{0.5}$
$N(^4S) + O_2^+ \rightarrow NO^+ + O + 4.25\,eV$	$\gamma_{16} = 1.8 \times 10^{-10}$
Chemical production of $N(^2D)$	
$N_2^+ + O \rightarrow NO^+ + N(^2D) + 0.70\,eV$	$\gamma_4 = 1.4 \times 10^{-10} (T_R/300)^{-0.44}$
$N_2^+ + e \rightarrow N(^2D) + N(^2D) + 1.04\,eV$	$\alpha_2 = 1.8 \times 10^{-7} (T_e/300)^{-0.39}$
$NO^+ + e \rightarrow O + N(^2D) + 0.38\,eV$	$\alpha_3 = f \times 4.2 \times 10^{-7} (T_e/300)^{-0.85}, f = 0.78$
$N^+ + O_2 \rightarrow O_2^+ + N(^2D) + 0.1\,eV$	$\gamma_{33} = 2 \times 10^{-10}$
$N(^2P) \rightarrow N(^2D) + h\nu(10\,400\,Å)$	$A_6 = 7.9 \times 10^{-2}\,s^{-1}$
Chemical loss of $N(^2D)$	
$N(^2D) + O_2 \rightarrow NO + O(^3P) + 3.76\,eV$	$\beta_2 = 5.3 \times 10^{-12}$
$N(^2D) + O_2 \rightarrow NO + O(^1D) + 1.74\,eV$	
$N(^2D) + O \rightarrow N(^4S) + O + 2.38\,eV$	$\beta_5 = 2 \times 10^{-12}$
$N(^2D) + e \rightarrow N(^4S) + e + 2.38\,eV$	$\beta_6 = 5.5 \times 10^{-10} (T_e/300)^{0.5}$
$N(^2D) + NO \rightarrow N_2 + O + 5.63\,eV$	$\beta_7 = 7 \times 10^{-11}$
$N(^2D) + O_2^+ \rightarrow N^+ + O_2$	$\gamma_{17} = 2.5 \times 10^{-10}$
$N(^2D) + O^+ \rightarrow N^+ + O + 1.45\,eV$	$\gamma_{26} = 1.3 \times 10^{-10}$
$N(^2D) \rightarrow N(^4S) + h\nu(5200\,Å)$	$A_7 = 1.06 \times 10^{-5}$
Chemical production of $N(^2P)$	
No sources	
Chemical loss of $N(^2P)$	
$N(^2P) + O_2 \rightarrow NO + O + 4.95\,eV$	$\beta_3 = 3.5 \times 10^{-12}$
$N(^2P) + O_2^+ \rightarrow NO^+ + O + 5.3\,eV$	$\gamma_{16} = 1.8 \times 10^{-10}$
$N(^2P) + O \rightarrow N(^4S) + O + 3.58\,eV$	$\beta_8 = 1 \times 10^{-11}$
$N(^2P) \rightarrow N(^2D) + h\nu(10400\,Å)$	$A_6 = 7.9 \times 10^{-2}\,s^{-1}$
Chemical production of NO	
$O_2^+ + N_2 \rightarrow NO^+ + NO + 0.933\,eV$	$\gamma_8 = 5 \times 10^{-16}$
$N(^2D) + O_2 \rightarrow NO + O(^3P) + 3.76\,eV$	$\beta_2 = 5.3 \times 10^{-12}$
$N(^4S) + O_2 \rightarrow NO + O + 1.385\,eV$	$\beta_1 = 4.4 \times 10^{-12} \exp(-3220/T_n)$
$N(^2P) + O_2 \rightarrow NO + O + 4.95\,eV$	$\beta_3 = 3.5 \times 10^{-12}$
$N^+ + O_2 \rightarrow O^+ + NO + 2.31\,eV$	$\gamma_{27} = 3 \times 10^{-11}$
Chemical loss of NO	
$NO + N(^4S) \rightarrow N_2 + O + 3.25\,eV$	$\beta_4 = 1.5 \times 10^{-12} (T_n)^{0.5}$
$NO + O_2^+ \rightarrow NO^+ + O_2 + 2.86\,eV$	$\gamma_{15} = 4.4 \times 10^{-10}$
$NO + N(^2D) \rightarrow N_2 + O + 5.63\,eV$	$\beta_7 = 7 \times 10^{-11}$
$NO + N_2^+ \rightarrow NO^+ + N_2 + 6.33\,eV$	$\gamma_{18} = 3.3 \times 10^{-10}$
$NO + O^+ \rightarrow NO^+ + O + 4.36\,eV$	$\gamma_{21} = 8 \times 10^{-13}$
$NO + h\nu(< 1910\,Å) \rightarrow N + O$	$A_1 = 8.3 \times 10^{-6}\,s^{-1}$
$NO + h\nu(< 1340\,Å) \rightarrow NO^+ + e$	$A_2 = 6 \times 10^{-7}\,s^{-1}$

[a] Rate coefficients have units of $cm^3\,s^{-1}$ except where otherwise shown.
[b] $T_R = (T_i + T_n)/2$.

Appendix 6
Transport coefficients, polarizability, collision frequencies, and energy transfer rates

Transport coefficients

Coefficients of binary and self diffusion, D_{12} and D_{11}, in units of $cm^2\,s^{-1}$, viscosity, η, in units of $g\,cm^{-1}\,s^{-1}$, thermal conductivity, λ, in units of $erg\,cm^{-2}\,s^{-1}\,K^{-1}$, and ion mobility, $\mathscr{K}$, in units of $cm^2\,V^{-1}\,s^{-1}$, are given at 300 K and standard sea-level atmospheric density. The thermal diffusion factor, α^T, is unitless but is also temperature dependent. Numerical values are given in Table A6.1

Table A6.1 *Transport coefficients*

Collision Partners	D_{12}	D_{11}	η	λ	$\mathscr{K}$	α^T
N_2–N_2		0.18	1.7×10^{-4}	2.4×10^3		0.37
O_2–N_2	0.18					
CO_2–N_2	0.14					
O_2–O_2		0.18	1.9×10^{-4}	2.4×10^3		0.35
CO_2–CO_2		0.10	1.4×10^{-4}	1.5×10^3		
NO–NO			1.8×10^{-4}	2.4×10^3		
He–N_2	0.61					0.36
O^+–O						0.37
N_2^+–N_2					1.8	
N^+–N_2					3.3	
O_2^+–O_2					2.25	
O^+–O_2					3.38	

Polarizability

Polarizability, α, is an atomic or molecular property that specifies the response of the electron cloud surrounding the atom to an external electric field such as arises between partners in a collision process. The interaction potential in ion–atom collisions depends on the polarizability, as discussed in Section 4.4. For example, the cross section for excitation of rotational levels includes a polarization term that depends on the molecular polarizability. Polarizability is a tensor quantity because it depends on the orientation of an atom or molecule in the electric field. The $\alpha_{\parallel}$ component corresponds to the symmetry axis of the atom parallel to the electric field and the $\alpha_{\perp}$ component corresponds to the perpendicular orientation. For molecules, $\alpha_{\parallel}$

refers to the polarizability in a direction parallel to the molecular bond and $\alpha_\perp$ perpendicular to the bond. The average value of the bond polarizability is $\bar{\alpha} = \frac{1}{3}(\alpha_\parallel + 2\alpha_\perp)$ in atomic units. The atomic unit of volume is a_0^3, a_0 being the Bohr radius. Numerical values for atmospheric species are given in Table A6.2.

Table A6.2 *Polarizability*

Species	$\alpha_\parallel$	$\alpha_\perp$	$\bar{\alpha}$
N_2	16.1	9.8	11.8
O_2	15.9	8.2	10.7
N			7.5
O			5.4

The bibliography to Chapter 4 lists several references to transport coefficients and polarizability. Additional sources are:

References:

H.J.M. Hanley and J.F. Ely, The viscosity and thermal conductivity coefficients of dilute nitrogen and oxygen, *J. Chem. Ref. Data*, Vol. 2, 735, 1973.

G.R. Freeman and D.A. Armstrong, Electron and ion mobilities, *Advances in Atomic and Molecular Physics*, Vol. 22, Academic Press, New York, 1985.

E.W. McDaniel and E.A. Mason, *The Mobility and Diffusion of Ions in Gases*, John Wiley, New York, 1973.

A.A. Radzig and B.M. Smirnov, *Reference Data on Atoms, Molecules and Ions*, Springer Verlag, Berlin, 1985.

Charge exchange collision frequencies

The ion–neutral charge exchange collision frequencies for atmospheric species are listed in Table A6.3 in units of $cm^3 s^{-1}$ as a function of $(T_i + T_n)$ in degrees Kelvin. Multiplication by the height dependent neutral densities (given in Appendix 1) yields the actual collision frequencies ν_{SCT}.

Table A6.3 *Charge exchange collision frequencies*

Temperatures $T_i + T_n$	N_2^+–N_2	$O_2^+ - O_2$	O^+–O	He^+–He	H^+–H
300				6.8×10^{-10}	2.2×10^{-9}
500	5.6×10^{-10}			8.3×10^{-10}	2.7×10^{-9}
1000	7.5×10^{-10}		5.6×10^{-10}	1.1×10^{-9}	3.6×10^{-9}
1500	8.9×10^{-10}		6.6×10^{-10}	1.3×10^{-9}	4.3×10^{-9}
2000	1.0×10^{-9}	4.9×10^{-10}	7.6×10^{-10}	1.4×10^{-9}	4.8×10^{-9}
2500	1.1×10^{-9}	5.4×10^{-10}	8.3×10^{-10}	1.6×10^{-9}	5.3×10^{-9}
3000	1.2×10^{-9}	5.9×10^{-10}	8.9×10^{-10}	1.7×10^{-9}	5.7×10^{-9}
3500	1.3×10^{-9}	6.2×10^{-10}	9.6×10^{-9}	1.8×10^{-9}	6.0×10^{-9}
4000	1.4×10^{-9}	6.6×10^{-10}	1.0×10^{-9}	1.9×10^{-9}	6.3×10^{-9}
5000	1.5×10^{-9}	7.2×10^{-10}	1.1×10^{-9}	2.1×10^{-9}	6.9×10^{-9}

Reference: P.M. Banks and G. Kockarts, *Aeronomy*, Part A, Academic Press, New York, 1973.

Energy transfer rates

The energy transfer rates in elastic collisions between electrons and neutral species $L(\mathrm{e,n})$ is

$L(\mathrm{e, N_2}) = 1.77 \times 10^{-19}\, n_e n(\mathrm{N_2})\cdot(1 - 1.21 \times 10^{-4} T_e) T_e (T_e - T_n)$

$L(\mathrm{e, O_2}) = 1.21 \times 10^{-18}\, n_e n(\mathrm{O_2})\cdot(1 + 3.6 \times 10^{-2} T_e^{1/2}) T_e^{1/2} (T_e - T_n)$

$L(\mathrm{e, O}) = 7.9 \times 10^{-19}\, n_e n(\mathrm{O})\cdot(1 + 5.7 \times 10^{-4} T_e) T_e^{1/2} (T_e - T_n)$

$L(\mathrm{e, He}) = 2.46 \times 10^{-17}\, n_e n(\mathrm{He}) T_e^{1/2} (T_e - T_n)$

$L(\mathrm{e, H}) = 9.63 \times 10^{-16}\, n_e n(\mathrm{H})\cdot(1 - 1.35 \times 10^{-4} T_e) T_e^{1/2} (T_e - T_n)$

where the units are $\mathrm{eV\,cm^{-3}\,s^{-1}}$.

The energy transfer rate between e and N_2 by vibrational excitation is represented by the empirical formula

$L^v(\mathrm{e, N_2}) = n_e n(\mathrm{N_2}) \times 2.99 \times 10^{-12} \exp[f(T_e - 2000)/2000 T_e] \exp[-g(T_e - T_n)/(T_e T_n) - 1]$

$f = 1.06 \times 10^4 + 7.51 \times 10^3 \tanh[1.10 \times 10^{-3}(T_e - 1800)]$

$g = 3300 + 1.233(T_e - 1000) - 2.056 \times 10^{-4}(T_e - 1000)(T_e - 4000)$

Reference:

R.W. Schunk and A.F. Nagy, Electron temperature in the F-region of the ionosphere: Theory and Observations, *Rev. Geophys. and Space Phys.*, **16**, 355, 1978.

Appendix 7
Physical constants and units

Numerical values of physical constants appearing in the equations included in this book are listed in Table A7.1. The cgs system of units is adopted, with exceptions noted below.

Table A7.1 *Physical constants*

Quantity	Symbol	Value
Planck constant	h	6.626×10^{-27} erg s
Speed of light	c	3×10^{10} cm s^{-1}
Boltzmann constant	k	1.38×10^{-16} erg K^{-1}
Electron rest mass	m_e	9.11×10^{-28} g
Proton rest mass	m_p	1.67×10^{-24} g
Elementary charge	e	1.60×10^{-20} emu
		4.80×10^{-10} esu
Bohr radius	a_0	5.29×10^{-9} cm

Equations arising from the quantum mechanical description of collision phenomena (used in Chapters 4 and 6) are frequently given in terms of Hartree atomic units. In this system, the electron rest mass, m_e, the elementary charge, e, and the Planck constant, $h/2\pi$, are set equal to unity. This reduces the complexity of theoretical formulae which are equally valid in atomic units and cgs–esu units upon substitution of the equivalent values listed in Table A7.2. Useful conversion factors between units of selected quantities are given in Table A7.3. Wavelengths in spectroscopy are given in angströms ($1\,Å = 10^{-8}$ cm). By convention some quantities are given in mks units, such as the electric potential in volts and the electric field in millivolts/metre (mV/m). The magnetic field (or induction) is, however, given in gauss or gammas ($1\,\gamma = 10^{-5}$ gauss = 1 nanotesla).
Commonly used geophysical parameters are listed below.

Mean radius of the Earth, $R_E = 6.371 \times 10^8$ cm
Angular velocity of rotation $\Omega = 7.29 \times 10^{-5}$ rad s^{-1}
Gravitational acceleration $g = 980.6 - 2.59 \cos 2\phi - 0.308\, z$ cm s^{-2}
where ϕ is the latitude and z the altitude above sea level in kilometres.
Earth's dipole magnetic field, horizontal $H = 0.35 \cos \phi$ gauss
vertical $Z = 0.63 \sin \phi$ gauss
where ϕ is the latitude.

Table A7.2 *Units*

Quantity	Atomic unit (au)	cgs–esu
Mass	1	m_e
Charge	1	e
Planck constant	2π	h
Length (Bohr radius)	1	$a_0 = h^2/4\pi^2 m_e e^2$
Energy	1	e^2/a_0
Dipole moment	1	$e\,a_0$
Cross section	1	a_0^2
Volume polarizability	1	a_0^3

Table A7.3 *Conversion between units*

Energy	eV	erg	cm^{-1}	au
1 eV	1	1.6022×10^{-12}	8065.48	3.6749×10^{-2}
1 erg	6.2415×10^{11}	1	5.034×10^{15}	2.2937×10^{10}
1 cm^{-1}	1.23985×10^{-4}	1.9865×10^{-16}	1	4.5563×10^{-6}
1 au	27.2116	4.3598×10^{-18}	2.1947×10^{5}	1

Pressure	dyne cm^{-2}	atm.	torr
1 dyne cm^{-2}	1	9.8693×10^{-7}	7.501×10^{-4}
1 atm.	1.013×10^{6}	1	760
1 torr	1.333×10^{3}	1.3158×10^{-3}	1

Cross section	cm^2	Å^2	a_0^2	πa_0^2
1 cm^2	1	10^{16}	3.5711×10^{16}	1.1367×10^{16}
1 Å^2	10^{-16}	1	3.5711	1.1367
1 a_0^2	2.800×10^{-17}	2.800×10^{-1}	1	3.1831×10^{-1}
1 πa_0^2	8.797×10^{-17}	8.797×10^{-1}	3.1416	1

References:

A. A. Radzig and B. M. Smirnov, *Reference Data on Atoms, Molecules and Ions*, Springer Verlag, Berlin 1985.

C. W. Allen, Astrophysical Quantities, The Athlone Press, 1970.

Index

Printed in the United States
By Bookmasters